Java 物联网、人工智能和区块链编程实战

[英] 佩里·肖(Perry Xiao) 著

王 颖 周致成 黄星河 译

清华大学出版社

北京

北京市版权局著作权合同登记号 图字：01-2020-6253

Perry Xiao
Practical Java Programming for IoT, AI, and Blockchain
EISBN：978-1-119-56001-2
Copyright © 2019 by John Wiley & Sons, Inc., Indianapolis, Indiana
All Rights Reserved. This translation published under license.

Trademarks: Wiley, Wrox, the Wrox logo, Programmer to Programmer, and related trade dress are trademarks or registered trademarks of John Wiley & Sons, Inc. and/or its afiliates, in the United States and other countries, and may not be used without written permission. Access is a registered trademark of Microsoft Corporation. All other trademarks are the property of their respective owners. John Wiley & Sons, Inc., is not associated with any product or vendor mentioned in this book.

本书中文简体字版由 Wiley Publishing, Inc. 授权清华大学出版社出版。未经出版者书面许可，不得以任何方式复制或抄袭本书内容。

Copies of this book sold without a Wiley sticker on the cover are unauthorized and illegal.

本书封面贴有 Wiley 公司防伪标签，无标签者不得销售。
版权所有，侵权必究。举报：010-62782989，beiqinquan@tup.tsinghua.edu.cn。

图书在版编目(CIP)数据

Java 物联网、人工智能和区块链编程实战 /(英)佩里·肖(Perry Xiao)著；王颖，周致成，黄星河译. —北京：清华大学出版社，2021.1
书名原文：Practical Java Programming for IoT, AI, and Blockchain
ISBN 978-7-302-56926-8

Ⅰ.①J… Ⅱ.①佩… ②王… ③周… ④黄… Ⅲ.①JAVA 语言－程序设计 ②物联网 ③人工智能 ④区块链技术 Ⅳ.①TP312.8 ②TP393.4 ③TP18 ④F713.361.3

中国版本图书馆 CIP 数据核字(2020)第 228134 号

责任编辑：王　军
封面设计：孔祥峰
版式设计：思创景点
责任校对：成凤进
责任印制：沈　露

出版发行：清华大学出版社
　　　　网　　址：http://www.tup.com.cn，http://www.wqbook.com
　　　　地　　址：北京清华大学学研大厦 A 座　　邮　编：100084
　　　　社 总 机：010-62770175　　邮　购：010-62786544
　　　　投稿与读者服务：010-62776969，c-service@tup.tsinghua.edu.cn
　　　　质 量 反 馈：010-62772015，zhiliang@tup.tsinghua.edu.cn

印 装 者：大厂回族自治县彩虹印刷有限公司
经　　销：全国新华书店
开　　本：170mm×240mm　　印　张：24.25　　字　数：557 千字
版　　次：2021 年 1 月第 1 版　　印　次：2021 年 1 月第 1 次印刷
定　　价：98.00 元

产品编号：087103-01

谨以此书献给我的家人——我的妻子 May、儿子 Zieger 和女儿 Jessica，他们使我的生活变得圆满——没有他们，生活将毫无意义。感谢我的父母和兄弟，他们与我分享了自己的生活和爱，最终成就了今天的我。我的朋友和同事们，感谢你们在我的整个职业生涯中给予我莫大的支持。

译者序

我们现在生活在数字革命时代,许多新兴的数字技术正以惊人的速度发展,比如物联网、人工智能、区块链等。这些数字技术都将越来越深入地渗透到我们生活的各个方面,并从根本上改变我们的生活方式、工作方式和社交方式。Java 作为一种现代的高级编程语言,是帮助我们学习这些数字技术以及开发数字应用程序的出色工具。

Java 是一种高级的、面向对象的通用编程语言,具有面向对象、分布式、健壮性、安全性、平台独立性与可移植性、多线程、动态性等特点。Java 最早作为浏览器插件出现于 1995 年,可以向静态单调的 Web 页面添加动态内容和交互功能,随后很快在所有具有运行 Java 小程序功能的主流 Web 浏览器中流行起来。经过数十年的发展,Java 已发展成一种完全适用于个人和企业用户的功能完备、用途广泛的强大语言。

物联网是在互联网基础上进行拓展延伸的网络,物联网使各种带有传感器的信息设备与互联网连接起来,形成的更为巨大的网络,实现了在任何时间、任何地点,人、机、物的互联互通。因此,物联网就是物物相连的互联网。

人工智能是计算机科学的一个分支,其动力是开发与人类智能相关的计算机功能。"人工智能"一词是由斯坦福大学的传奇计算机科学家 John McCarthy 提出来的。人工智能是指模仿人类的思维方式使计算机智能地思考问题,并通过研究人类大脑的思考、学习和工作方式,将研究结果作为开发智能软件和系统的基础。自诞生以来,"人工智能"的理论和技术日益成熟,应用领域也不断扩大,应用前景十分广泛。

随着计算机技术的飞速发展,信息网络已经成为社会发展的重要保证。计算机网络上运行的各式系统以及流转的敏感信息和涉密文件,必然吸引来自世界各地的各种人为攻击,主要包括信息泄露、信息窃取、数据篡改、数据删添、计算机病毒破坏等。因此,网络安全就是保护好网络系统的硬件、软件以及系统中的数据,不因偶然或恶意的原因遭受破坏、更改、泄露,让系统连续、可靠、正常地运行,网络服务不中断。

区块链起源于比特币,2008 年由中本聪提出。区块链是分布式的共享账本,也就是分布式的数据库,具有去中心化、不可篡改、全程留痕、可以追溯、集体维护、公开透明等特点。区块链具有丰富的应用场景,能够解决信息不对称问题,实现多个主体之间的信任协作与数据的一致性。目前,区块链的应用已不仅限于金融领域,在物联网、物流、数字版权等领域也都具有巨大潜力。

随着计算机与互联网的飞速发展，数据一直以惊人的速度增长。门户网站、搜索引擎、购物网站、社交软件使得数据不断膨胀。智能终端的流行让数据的流通插上翅膀，也同时收集着使用者的信息，比如个人信息、所在位置、移动轨迹、生活偏好等。5G 时代又推动着物联网的发展，而物联网设备上的传感器，无时无刻不在收集着各种大量数据。有人说数据是黄金、石油，要想把这些爆炸式增长的数据使用好，需要借助大数据技术。大数据分析可以带来很多好处：可以为政府治理社会、经济和政治问题提供决策；可以为企业分析消费者喜好和产品感知，从而更好地制订生产计划；可以分析数百万患者的既往病史，从而提供更准确的诊断；等等。使用好大数据技术，就可以更好地服务社会、造福民众。

本书第 I 部分简要介绍 Java 编程语言，使读者熟悉 Java 编程；第 II 部分提供常规编程项目的 Java 示例，比如控制台应用程序、Windows 应用程序、网络应用程序和移动应用程序；第 III 部分为本书核心，为读者介绍最新数字技术(例如物联网、人工智能、网络安全、区块链和大数据)以及各数字技术的 Java 编程应用示例。本书的目的是使用 Java 作为工具，帮助读者学习这些新的数字技术，使这些数字技术不再神秘，并让读者为未来做更充分的准备。

本书由王颖、周致成、黄星河翻译，在翻译过程中查阅了物联网、人工智能、网络安全、区块链、大数据方面的大量书籍和 Java 官方文档，力求翻译的准确性，同时润色了行文结构，尽量以通俗易懂的语言阐述相关知识点，提高可读性。但由于译者水平有限，难免有疏漏之处，望各位读者包涵并予以指正。在此要感谢清华大学出版社的编辑们，他们为本书的翻译和校对工作付出了大量的心血，本书的成功出版离不开他们的辛勤付出。

最后，衷心希望各位读者能通过本书熟悉 Java 编程，学习到物联网、人工智能、网络安全、区块链、大数据及相关技术领域的 Java 编程，并运用到实际工作中以解决问题，祝各位读者学有所获。

序言

20 世纪 90 年代，当我在中国吉林工业大学学习时，我有了自己的第一次编程经历。那时，我们使用的是大型机，因为当时还没有个人计算机(Personal Computer，PC)。大型机只是一台位于房间某处的计算机，可通过采用文本模式的不太灵活的终端来连接。当我们使用共享账户连接到大型机时，可能会发生很多奇怪的事情。你在早期创建的程序以后可能会被其他人删除或修改。打印也是一场噩梦。例如，当你打印自己的 Fortran 代码时，另一个学生也可能会发送打印请求，这时打印机将立即停止打印你的代码，并开始打印其他学生的代码。完成后，打印机将返回以继续打印其余的代码。那是什么逻辑？因此，每次当我们要打印时，都会大喊"我正在打印！"并希望其他人不要同时开始打印他们的代码。我们使用的编程语言是 Fortran。Fortran 主要用于科学计算，是一种功能非常强大的语言，但是里面的 Go To 语句会让人发疯。这使得代码很难阅读。

后来，当 IBM PC 开始可用时，我所在的大学也购买了一些 IBM PC，并建造了专用的两层楼来放置这些计算机。是的，你听说过，整个建筑物都专用于放置这些 IBM PC，并且要求设计无尘室，上面铺有红色地毯，设有接待处。你需要脱鞋，换上拖鞋，穿白色实验服才能入内。每个学生使用这些计算机的时间也都受到严格的限制。

你是否曾想过，为什么计算机硬盘的驱动器总是以驱动器 C 而不是驱动器 A 或驱动器 B 开头？原因是，一开始 IBM PC 没有硬盘，只有两个 5¼英寸的软盘驱动器。要使用计算机，就需要两张软盘：一张软盘用于 MS-DOS 操作系统，作用是启动计算机；另一张软盘用于保存数据。每张软盘可容纳 512 KB 数据，这甚至不能满足当今智能手机存储照片的要求。但在那时，这已经足够了。后来我们开始学习 BASIC 编程语言。BASIC 的使用非常简单，我们因为科学计算和绘制基于文本的图片而对 BASIC 编程充满无限的兴趣。

我从 1997 年开始学习 Java。Java 由现在隶属于 Oracle Corporation 的 Sun Microsystems 发布。由于当时的网页几乎只是静态文本和静态图片，因此我对 Java 在网页内创建动画(Java Applet)的能力十分着迷。但几年后，使我真正爱上 Java 并完全欣赏其魅力的是：我当时正在为 Java 网络编程的 MSc 模块做准备，而学习和使用 Java 非常简单，尤其是对于网络编程而言，仅仅需要几行 Java 代码，你就可以拥有一台服务器！

Java 的内存垃圾回收功能以及 Java 编程没有指针的特性也给我留下了深刻的印象。C 和

C++程序员，你们可以放心学习 Java，因为 Java 中没有指针！在使用 Java 编程语言进行有限差分分析和有限元素分析之前，我经历了一段艰难的 Java 生涯。指针是指向内存中特定位置的变量，它们对于处理矩阵至关重要。我当时由于指针处理不当，导致程序崩溃的时间占所有故障时间的 99%。因此，当我听到 Java 中没有指针时，我既惊讶又兴奋！Java 中也没有 Go To 语句，这使 Java 代码更易于理解。

Java 的异常处理也值得一提。Java 可以"引发"异常。因此，当你运行程序时，如果发生异常/错误，例如被零除，那么读取不存在的文件或连接到没有响应的远程计算机时不会挂起程序或导致整个程序崩溃，这时在计算机上，Java 只会终止程序并正常显示错误消息。

从那以后，我开发了许多用于教学和研究的 Java 程序。我非常喜欢使用 Java 语言进行编程，希望你也喜欢它。

Perry Xiao 博士

2018 年 11 月于伦敦

作者简介

Perry Xiao 博士是英国伦敦南岸大学工程学院的副教授兼课程主任。他获得了光电子学学士学位、固态物理学理学硕士学位和光物理学博士学位。他是特许工程师(Chartered Engineer，CEng)、工程技术学院(Institution of Engineering and Technology，IET)的院士(Fellow，FIET)和高等教育学院(Higher Education Academy，HEA)的高级院士(Senior Fellow，SFHEA)。Perry 从事本科和研究生课程的电子、软件、计算机网络和电信学科的教学已有近二十年的时间。他还每年指导 BEng 项目和 MSc 项目的学生。他的主要研究兴趣是为皮肤生物工程应用和工业无损检测(nondestructive testing，NDT)开发新型的红外和电子传感技术。迄今为止，Perry 已经完成了七次博士生指导，获得了两项英国专利，发表了 100 多篇科学论文，成为 9 种期刊的编辑审稿人，并获得了近 100 万英镑的研究经费。

Perry 还是英国 Biox Systems 有限公司的董事兼联合创始人，这是一家校企，设计和制造了世界上最先进的皮肤测量仪器 AquaFlux 和 Epsilon，这些仪器已在全球范围内得到广泛应用，包括领先的化妆品公司、大学、研究机构和医院等。

致谢

衷心感谢 Wiley 公司给我这次写作机会。我还要感谢 Peter Mitchell、Devon Lewis、Pete Gaughan、Athiyappan Lalith Kumar、Evelyn Wellborn 和 Compton 提供的支持。没有他们,本书就不可能顺利付梓。

　　我们生活在数字革命时代,许多新兴的数字技术正以惊人的速度发展,例如物联网(Internet of Things,IoT)、人工智能(Artificial Intelligence,AI)、网络安全、区块链等。无论我们是否喜欢,也无论我们是否准备好,这些数字技术都将越来越深入地渗透到我们生活的各个方面,这将从根本上改变我们的生活方式、工作方式和社交方式。Java 作为一种现代的高级编程语言,是帮助我们学习这些数字技术以及开发数字应用的出色工具。

　　本书的目的是使用 Java 作为工具,帮助读者学习这些新的数字技术,使这些数字技术不再神秘并让读者为未来做好更充分的准备。

本书的组织结构

　　本书分为三大部分。第 I 部分是对 Java 编程语言的基本介绍,使读者开始接触并使用 Java 进行编程;第 II 部分的各章提供了常规编程项目的 Java 示例,例如控制台应用、Windows 应用、网络应用和移动应用,所有这些都是为第III部分做准备的;第III部分是本书的核心,通过 Java 编程示例提供了有关最新数字技术(IoT、AI、网络安全、区块链和大数据)的易于阅读的指南。

第 I 部分

第 1 章:Java 简介
第 2 章:Java 编程入门

第 II 部分

第 3 章:基本的 Java 编程
第 4 章:面向 Windows 应用的 Java 编程
第 5 章:面向网络应用的 Java 编程
第 6 章:面向移动应用的 Java 编程

第Ⅲ部分

第 7 章：面向物联网应用的 Java 编程
第 8 章：面向人工智能应用的 Java 编程
第 9 章：面向网络安全应用的 Java 编程
第 10 章：面向区块链应用的 Java 编程
第 11 章：面向大数据应用的 Java 编程

附录

附录 A：Java 文档和归档工具以及在线资源
附录 B：Apache Maven 教程
附录 C：Git 和 GitHub 教程

本书所有示例的源代码可通过手机扫描封底的二维码下载。

本书读者对象

本书适合软件开发人员、设计人员和研究人员阅读。本书假设读者对计算机以及计算机的主要组件(例如 CPU、RAM、硬盘驱动器、网络接口等)有基本的了解。读者应该能够熟练地使用计算机执行基本的任务，例如打开和关闭计算机，登录和注销，运行某些程序以及复制/移动/删除文件等。本书还假设读者具有一些基本的编程经验，理想的情况是使用过 Java，但也可以是其他语言(例如 C/C ++、Fortran、MATLAB、C#、BASIC 或 Python)，并且知道基本的语法、不同类型的变量、标准输入输出、条件选择以及诸如循环和子例程的结构。最后，本书假设读者掌握计算机网络和 Internet 的一些基本概念，并且可以使用一些常用的 Internet 服务，例如万维网、电子邮件、文件的下载/上传以及在线银行/购物等。

阅读前要做的准备工作

要完成本书中的示例，你需要具备以下条件：

- 至少 124 MB 的硬盘和 128 MB 的内存，奔腾 2266 MHz 处理器以及运行 Windows 操作系统(Windows 7 或更高版本)或 Linux 操作系统(Ubuntu Linux 12.04 或更高版本、Oracle Linux 5.5 或更高版本、Red Hat Linux 5.5 或更高版本等)的标准个人计算机。当然，你也可以使用 Mac 计算机(在 Mac OS X 10.8.3 或更高版本中，具有安装权限的管理员特权，64 位浏览器)。
- Java JDK，下载网址为 http://www.oracle.com/technetwork/java/javase/downloads/index.html。
- 文本编辑器和 Java IDE(请参阅第 2 章)。
- 树莓派(可选)，下载网址为 https://www.raspberrypi.org/。

目 录

第 I 部分

第 1 章　Java 简介 ………………………… 3
　1.1　什么是 Java …………………………… 3
　1.2　Java 语言的版本 ……………………… 5
　1.3　Java 架构 ……………………………… 6
　1.4　Java 平台的版本 ……………………… 7
　1.5　Java Spring 框架 ……………………… 8
　1.6　Java 的优缺点 ………………………… 9
　　　1.6.1　优点 …………………………… 9
　　　1.6.2　缺点 …………………………… 9
　1.7　Java 认证 …………………………… 10
　1.8　小结 ………………………………… 10
　1.9　本章复习题 ………………………… 10

第 2 章　Java 编程入门 ………………… 12
　2.1　下载和安装 Java …………………… 12
　2.2　Java IDE …………………………… 15
　2.3　Java 程序 Hello World …………… 18
　2.4　Java 在线编译器 …………………… 21
　2.5　Java 在线代码转换器 ……………… 27
　2.6　Java 免费在线课程和
　　　　教程 ………………………………… 28
　2.7　Java 版本控制 ……………………… 32

　2.8　小结 ………………………………… 32
　2.9　本章复习题 ………………………… 32

第 II 部分

第 3 章　基本的 Java 编程 …………… 37
　3.1　引言 ………………………………… 38
　3.2　变量 ………………………………… 38
　　　3.2.1　常数 …………………………… 40
　　　3.2.2　String 和 StringBuffer
　　　　　　　类型 ………………………… 40
　　　3.2.3　var 变量类型 ………………… 41
　3.3　运算符 ……………………………… 42
　3.4　保留字 ……………………………… 43
　3.5　输入和输出 ………………………… 43
　3.6　循环和选择 ………………………… 46
　3.7　数组、矩阵和 ArrayList …………… 48
　3.8　读写文件 …………………………… 51
　3.9　方法 ………………………………… 53
　3.10　面向对象编程 ……………………… 55
　　　3.10.1　类和对象 …………………… 55
　　　3.10.2　实例化 ……………………… 55
　　　3.10.3　封装 ………………………… 55
　　　3.10.4　继承 ………………………… 55
　　　3.10.5　覆盖和重载 ………………… 55

	3.10.6	多态性 …………………… 56
	3.10.7	对象的可访问性 …… 56
	3.10.8	匿名内部类 …………… 56
3.11	多线程 …………………………… 59	
	3.11.1	线程的生命周期 …… 64
	3.11.2	线程的优先级 ……… 66
	3.11.3	线程调度 …………… 66
	3.11.4	线程同步 …………… 67
3.12	日期、时间、计时器和睡眠方法 ……… 68	
3.13	执行系统命令 …………… 72	
3.14	大规模的软件包和编程 ……………………… 74	
3.15	软件工程 ……………… 77	
	3.15.1	软件的开发周期 … 77
	3.15.2	缩进 …………………… 78
	3.15.3	注释 …………………… 79
	3.15.4	命名约定 …………… 80
3.16	部署 Java 应用 ………… 80	
	3.16.1	使用 Windows 批处理文件 …………………… 81
	3.16.2	使用可执行的 JAR 文件 …………………… 82
	3.16.3	使用 Microsoft Visual Studio …………… 83
	3.16.4	Java 应用的安装 …… 84
3.17	小结 ……………………… 84	
3.18	本章复习题 …………… 85	

第 4 章 面向 Windows 应用的 Java 编程 ……………………… 86

4.1	引言 ……………………… 86
4.2	Java Swing 应用 ………… 87
4.3	JavaFX 应用 …………… 91

	4.3.1	JavaFX 窗口 ………… 92
	4.3.2	在 JavaFX 中创建标签和按钮 ………………… 94
	4.3.3	JavaFX 图表 ………… 95
	4.3.4	在 JavaFX 中处理用户登录 ……………………… 97
	4.3.5	在 JavaFX 中创建图像查看器 ……………………… 99
	4.3.6	创建 JavaFX Web 查看器 …………………… 100
	4.3.7	在 JavaFX 中创建菜单 …………………………… 101
	4.3.8	创建 JavaFX 文件选择对话框 ………………… 103
	4.3.9	JavaFX 教程 ………… 105
4.4	部署 JavaFX 应用 ………… 108	
4.5	小结 ……………………… 109	
4.6	本章复习题 …………… 109	

第 5 章 面向网络应用的 Java 编程 ……………………… 110

5.1	简介 ……………………… 110

	5.1.1	局域网和广域网 …… 113
	5.1.2	思科的三层企业网络架构 …………………… 113
	5.1.3	关键网络组件 ……… 113
	5.1.4	传统网络与软件定义网络 …………………… 114
5.2	Java 网络信息编程 ……… 116	
5.3	Java 套接字编程 ………… 121	
	5.3.1	Java UDP 客户端-服务器编程 ………… 121
	5.3.2	Java TCP 客户端-服务器编程 ………… 123

	5.3.3	Java 多线程回显服务器编程·············· 126
5.4	Java HTTP 编程············ 128	
	5.4.1	Java HTTP/HTTPS 客户端············· 128
	5.4.2	Java HTTP 服务器··· 134
	5.4.3	Java 多线程 HTTP 服务器············· 136
5.5	Java 电子邮件 SMTP 编程·············· 139	
5.6	Java RMI 客户端-服务器编程·············· 143	
5.7	SDN 入门············· 146	
	5.7.1	OpenFlow 入门······ 146
	5.7.2	Floodlight 入门······ 153
	5.7.3	OpenDaylight 入门··· 153
5.8	Java 网络编程资源····· 154	
5.9	小结················· 154	
5.10	本章复习题············ 154	

第 6 章 面向移动应用的 Java 编程············ 155

6.1	引言················ 155	
6.2	Android Studio ········ 156	
6.3	Hello World 应用········ 157	
6.4	Button 和 TextView 组件的应用·············· 163	
6.5	传感器应用··········· 166	
6.6	部署 Android 应用······ 169	
6.7	Android 应用中 activity 的生命周期············ 170	
6.8	MIT App Inventor······ 171	
6.9	5G················ 179	
	6.9.1	毫米波············· 181
	6.9.2	小蜂窝············· 181

	6.9.3	大规模 MIMO ······ 182
	6.9.4	波束成形··········· 182
	6.9.5	全双工············ 182
	6.9.6	未来的 6G 和 7G····· 182
6.10	小结················ 183	
6.11	本章复习题············ 183	

第Ⅲ部分

第 7 章 面向物联网应用的 Java 编程··················187

7.1	什么是物联网·········· 187	
7.2	物联网通信协议········ 190	
	7.2.1	MQTT············· 191
	7.2.2	CoAP············· 191
	7.2.3	XMPP············· 192
	7.2.4	SOAP············· 192
	7.2.5	REST············· 192
7.3	物联网平台··········· 192	
7.4	物联网安全··········· 193	
7.5	为什么使用 Java········ 193	
7.6	使用树莓派的 Java 物联网············ 193	
	7.6.1	设置树莓派········· 196
	7.6.2	Java GPIO 示例······ 198
	7.6.3	从 Java 程序中调用 Python 程序······· 205
	7.6.4	Java PWM 示例······ 206
	7.6.5	Java PIR 和 LED 示例············· 208
	7.6.6	Java I2C 示例······· 210
	7.6.7	Java ADC 示例······ 213
	7.6.8	Java 数字传感器示例············· 217

7.6.9　Java MQTT 示例…… 221
7.6.10　Java REST 示例…… 223
7.7　Oracle Java ME 嵌入式客户端…………………… 227
7.8　适用于 Java 的物联网平台……………………… 227
　　7.8.1　Eclipse Open IoT Stack…………… 227
　　7.8.2　IBM Watson IoT…… 228
　　7.8.3　AWS IoT…………… 228
　　7.8.4　Microsoft Azure IoT…………………… 229
7.9　小结……………………… 229
7.10　本章复习题…………… 229

第 8 章　面向人工智能应用的 Java 编程…………… 231

8.1　什么是人工智能……… 231
　　8.1.1　人工智能的研究历史………………… 233
　　8.1.2　云人工智能与边缘人工智能………… 234
8.2　神经网络……………… 235
　　8.2.1　感知器………… 236
　　8.2.2　多层感知器与反向传播/前馈神经网络…………… 238
8.3　机器学习……………… 240
8.4　深度学习……………… 241
8.5　Java AI 库……………… 244
8.6　神经网络方面的 Java 示例……………………… 245
　　8.6.1　Java 感知器示例…… 245
　　8.6.2　Java 神经网络反向传播示例…………… 248

8.7　机器学习方面的 Java 示例……………………… 251
8.8　深度学习方面 Java 示例……………………… 255
8.9　适用于 Java 的 TensorFlow…………… 259
8.10　AI 资源………………… 262
8.11　小结…………………… 263
8.12　本章复习题…………… 263

第 9 章　面向网络安全应用的 Java 编程……………264

9.1　什么是网络安全……… 265
9.2　什么是加密…………… 265
　　9.2.1　私钥加密……… 266
　　9.2.2　公钥加密……… 267
9.3　哈希函数和消息摘要… 271
9.4　数字签名……………… 272
9.5　数字证书……………… 273
9.6　案例研究 1：安全电子邮件……………………… 275
9.7　案例研究 2：安全网络… 276
9.8　Java 私钥加密示例…… 276
9.9　Java 公钥加密示例…… 277
9.10　Java 数字签名/消息摘要示例………………… 279
9.11　Java 数字证书示例…… 284
9.12　其他 Java 示例………… 289
9.13　小结…………………… 289
9.14　本章复习题…………… 289

第 10 章　面向区块链应用的 Java 编程……………290

10.1　什么是区块链………… 291
10.2　如何验证区块链……… 292

10.3	如何挖掘区块 ············ 292		11.3	大数据的三个 V ········ 326
10.4	区块链的工作方式 ····· 293		11.4	大数据分析带来的
10.5	区块链的应用 ············ 294			好处 ······························ 326
	10.5.1 比特币············ 294		11.5	什么是 Hadoop ········· 326
	10.5.2 智能合约········ 298		11.6	Hadoop 的关键组件 ···· 327
	10.5.3 医疗·············· 299			11.6.1 HDFS ············ 327
	10.5.4 制造业和			11.6.2 MapReduce ···· 328
	供应链············ 299			11.6.3 Hadoop
	10.5.5 物联网·········· 300			Common ········ 328
	10.5.6 政务············· 300			11.6.4 Hadoop YARN ··· 328
10.6	关于区块链的一些			11.6.5 Hadoop 集群
	问题 ···························· 300			概述·············· 328
10.7	Java 区块链示例 ········ 300		11.7	在树莓派集群上实现
10.8	Java 区块链交易示例 ··· 305			Hadoop ······················ 329
10.9	Java BitcoinJ 示例······ 311			11.7.1 树莓派的安装和
10.10	Java Web3j 示例 ······· 312			配置·············· 330
10.11	Java EthereumJ 示例 ··· 315			11.7.2 Hadoop 的安装和
10.12	Java Ethereum 智能			配置·············· 330
	合约示例················· 316		11.8	Java Hadoop 示例 ······ 337
10.13	更进一步：选择区块链		11.9	小结 ··························· 343
	平台························· 322		11.10	本章复习题··············· 343
10.14	小结························· 323		附录A	Java 文档和归档工具以及
10.15	本章复习题··············· 323			在线资源 ······················344
第 11 章	面向大数据应用的		附录B	Apache Maven 教程·······351
	Java 编程ⷧ················ 324		附录C	Git 和 GitHub 教程 ········357
11.1	什么是大数据ⷧ··········· 324			
11.2	大数据的来源ⷧ··········· 325			

第 I 部分

第 1 章：Java 简介
第 2 章：Java 编程入门

第 1 章

Java简介

"经验只是我们给自己犯下的错误起的名字。"

——Oscar Wilde

1.1 什么是 Java
1.2 Java 语言的版本
1.3 Java 架构
1.4 Java 平台的版本
1.5 Java Spring 框架
1.6 Java 的优缺点
1.7 Java 认证
1.8 小结
1.9 本章复习题

1.1 什么是 Java

Java 是一种高级的、面向对象的通用编程语言，最初由加拿大计算机科学家 James Gosling 于 1991 年在美国加利福尼亚州的 Sun Microsystems 软件公司开发。2010 年，Sun Microsystems 软件公司被同样位于加利福尼亚州的 Oracle 公司收购。Java 是 Sun Microsystems 软件公司 Green 项目的副产品，最初被设计为一种独立于平台的语言，用于对家用电器进行编程。但是，Java 对于此类应用来说太先进了。Gosling 设计了基于 C 和 C++语言的 Java 语法，但具有较少的低级功能。Java 以备受欢迎的印尼咖啡 Java 而命名。

Java最早通过HotJava和Netscape Web浏览器作为名为Java Applets的插件出现于1995年，因为可以向静态单调的Web页面添加动态内容和交互功能，Java很快在所有支持运行Java小程序的主流Web浏览器中流行起来。如今，经过数十年的发展，Java已经发展为一种完全适用于个人和企业用户的功能完备且用途广泛的强大语言。Java与JavaScript不同，JavaScript是仅在Web浏览器中运行的一种脚本语言。

Java语言被设计为满足以下所有条件：
- 简单、面向对象且友好。
- 健壮且安全。
- 架构中立且可移植。
- 高性能。
- 解释性、线程化和动态性。

Java的主要优点在于其平台独立性。也就是说，使用Java语言编写的程序可以实现"编写一次，到处运行"。这种平台独立性是借助Java虚拟机(Java Virtual Machine，JVM)的概念实现的，如图1.1和图1.2所示。使用常规编程语言(例如C/C++)在Windows、Mac和Linux等不同的操作系统上运行程序时，需要为每种操作系统分别编译C/C++源文件。由于每个可执行文件都是在本机操作系统上运行的，因此在一个操作系统上编译的可执行文件不能在另一个操作系统上运行。Java的工作方式则有所不同。Java源代码(.java文件)被编译为Java字节码(.class文件)。字节码文件不是可执行文件，不能直接在操作系统上运行。相反，它们在JVM中运行，JVM处理操作系统之间的差异，并为不同的Java程序提供相同的运行环境。JVM是使Java平台独立的关键。JVM的缺点在于Java程序的运行速度比相应的C程序要慢得多。但对于大多数应用来说，这种差异其实并不明显。

Java是最流行的编程语言之一，特别是对于网络应用来说。据Oracle称，全世界估计有900万Java开发人员和大约30亿台运行着Java的设备。

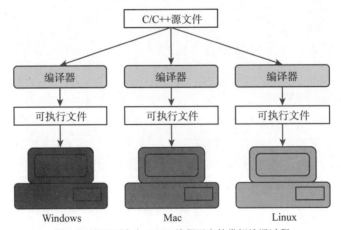

图1.1　不同平台上C/C++编程语言的常规编译过程

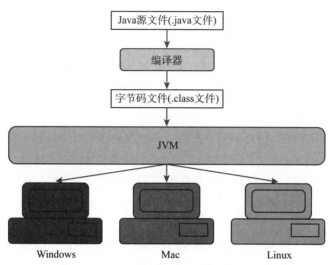

图 1.2 不同平台上的 Java 编译过程

1.2 Java 语言的版本

Java 语言有很多版本。在撰写本书时最新版本是 Java 12；等到你阅读本书时，最新版本可能已经是 Java 13。Alpha 和 Beta 是 Java 开发工具包(Java Development Kit，JDK)的最初版本，于 1995 年发布。JDK 1.0 是于 1996 年发布的第一个正式版本。Java JDK 1.2 及更高版本通常称为 Java 2。Java 2 语言、库和工具的集合又称为 Java 2 平台或 Java 2 标准版本(J2SE)。同样，后来发布的 Java 语言版本还有 Java 5、Java 6、Java 7、Java 8、Java 9、Java 10、Java 11 和 Java 12，有关详细信息，参见表 1.1。

表 1.1 Java 语言的版本历史

版本	代号	发布日期	停止支持日期
JDK Alpha和Beta		1995年	2008年
JDK 1.0	Oak	1996年1月	2008年
JDK 1.1		1997年2月	2008年
J2SE 1.2	Playground	1998年12月	2008年
J2SE 1.3	Kestre	2000年5月	2008年
J2SE 1.4	Merlin	2002年2月	2008年8月
J2SE 5.0	Tiger	2004年9月	2009年11月
Java SE 6	Mustang	2006年12月	2013年2月
Java SE 7	Dolphin	2011年7月	2015年4月
Java SE 8 (LTS)		2014年3月	2019年1月

(续表)

版本	代号	发布日期	停止支持日期
Java SE 9		2017年9月	2018年3月
Java SE 10		2018年3月	2018年9月
Java SE 11 (LTS)		2018年9月	
Java SE 12		2019年3月	

对于 Java SE 8 之后的 Java 版本，Oracle 会每三年指定一个版本作为长期支持(Long-Term-Support，LTS)版本，在这三年时间里，每六个月发布一次的非 LTS 版本也称为功能型版本。Java SE 9、Java SE 10 和 Java 12 都是非 LTS 版本，而 Java SE 8 和 Java SE 11 是 LTS 版本。当 Oracle 不再公开支持某个 Java 版本时，意味着这个 Java 版本的生命周期结束(End Of Life，EOL)。对于非 LTS 版本，EOL 是下一个新版本的发布日期，所有公开支持都将被取代。但是对于 LTS 版本，EOL 更长，即使在新版本发布之后，客户也将继续获得支持。这就是为什么使用广泛的 Java SE 8 具有比其他发行版更长的 EOL 的原因。官方计划中的下一个 LTS 版本将是 Java SE 17。本书使用的 Java 示例代码不受将来的 Java 版本的影响。

有关 Java 版本和支持指南的更多信息，请访问以下网站：

https://www.oracle.com/technetwork/java/java-se-support-roadmap.html
https://en.wikipedia.org/wiki/Java_version_history

每个 Java 版本都会分发为两个不同的软件包。

Java 运行环境(Java Runtime Environment，JRE)用于运行 Java 程序，供用户使用。JRE 由 JVM 和运行库组成。可以使用 JRE 运行 Java 程序而不需要重新编译 Java 程序。

Java 开发工具包(JDK)用于软件开发人员编译、调试和记录 Java 程序。本书将使用 JDK，因为需要编译 Java 程序。

1.3 Java 架构

图 1.3 显示了 Java 架构中 JDK、JRE 和 JVM 之间的关系。JDK 包含 JRE 和 Java 开发工具，而 JRE 包含 JVM、库类以及其他文件。JVM 内部有即时(Just-In-Time，JIT)编译器，JIT 编译器在 Java 程序执行期间(也就是运行时)将 Java 字节码编译为本地机器代码。JIT 提高了 Java 应用的性能。

图 1.4 显示了更详细的 Java 架构；这是根据原始的 Oracle Java 架构图重新创建的，通过链接 https://www.oracle.com/technetwork/java/javase/tech/index.html 可以找到。

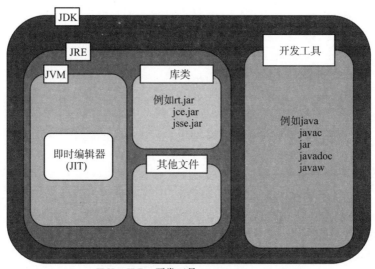

JDK = JRE + 开发工具
JRE = JVM + 库类 + 其他文件

图 1.3　Java 架构中 JDK、JRE 和 JVM 之间的关系

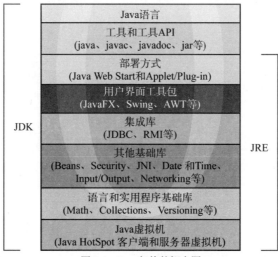

图 1.4　Java 架构的概念图

1.4　Java 平台的版本

Java 平台目前有 4 个版本：
- 适用于智能卡的 Java Card。
- 适用于移动设备的 Java ME(微型版)。
- 适用于标准个人计算机的 Java SE(标准版)。

- 适用于大型分布式企业或 Internet 环境的 Java EE(企业版)。

Java SE 是大多数人用于 Java 编程的工具。Java SE 附带完整的 Java 类库,其中包括基本类和对象/网络/安全性/数据库以及经典的 Swing 图形用户界面(Graphical User Interface,GUI)工具包。大多数 Java SE 版本还包括当前的 JavaFX 工具包,JavaFX 工具包旨在替代 Swing GUI 工具包。但是,从 Java SE 11 开始,JavaFX 工具包不再包含在 Java SDK 中,而是重新设计为单独的库。本书侧重于描述适用于标准个人计算机的 Java SE。

1.5　Java Spring 框架

Java Spring 是创建 Java 企业级应用时最受欢迎的开发框架。作为开源框架,Java Spring 最初由 Rod Johnson 编写,于 2003 年 6 月基于 Apache 2.0 许可证发布。Java Spring 框架的主要优点之一是采用了分层架构,允许开发人员选择想要使用的组件。图 1.5 显示了 Java Spring 框架的主页(https://spring.io/)。图 1.6 显示了 Java Spring 框架的指导页面(https://spring.io/guides)。

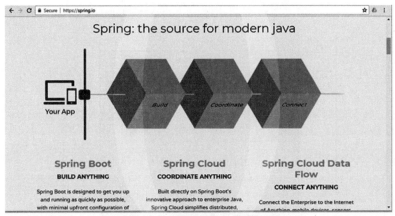

图 1.5　Java Spring 框架的主页

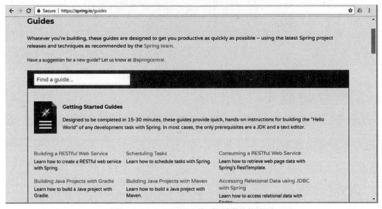

图 1.6　Java Spring 框架的指导页面

互联网上也有一些不错的 Java Spring 框架教程，比如：

https://www.tutorialspoint.com/spring/spring_overview.htm
https://howtodoinjava.com/spring-5-tutorial/
https://java2blog.com/introduction-to-spring-framework/

1.6 Java 的优缺点

Java 既有优点，也有一些缺点，这些缺点可能会影响开发语言的选择。本节将对 Java 的优缺点进行摘要分析。

1.6.1 优点

以下是 Java 的优点。

- 免费：Java 是免费使用的，但是如果用于商业应用，那么需要为安全性和某些更新付费。
- 简单：Java 相比其他编程语言更易于学习和使用。Java 还支持自动内存分配和垃圾回收。
- 平台独立：因为有了 JVM，所以 Java 程序一旦编译，就可以在任何操作系统上运行。
- 面向对象：Java 是一种完全面向对象的编程语言，能够让你创建可重用的 Java 模块(类)。第 3 章将介绍 Java 的面向对象的相关内容。
- 安全性：Java 被设计为一种十分安全可靠的编程语言。有关安全性的内容，请参见第 9 章。
- 多线程：使用 Java，可以轻松开发同时运行多个任务的多线程程序。第 3 章将介绍多线程编程。
- 联网：Java 提供了一系列功能，使开发联网应用变得更加容易。第 5 章将介绍如何开发网络应用。
- 移动开发：使用 Java 可以开发 Android 移动应用。第 6 章将介绍如何为移动设备开发应用。
- 企业级开发：使用 Java 可以开发许多企业级应用，例如 Web 服务器和其他应用服务器。

1.6.2 缺点

以下是 Java 的缺点。

- 性能：由于使用了 JVM，Java 相比其他本地编译语言(例如 C 或 C++)要慢得多。Java 还会占用更多的内存空间，并且用于调整延迟临界的选项很有限。
- GUI 开发：一般来说，使用 Java 开发 GUI 程序并不容易，尽管 JavaFX 的 GUI 工具包有了明显的改进，但是 Java Swing 工具包的观感与本地 Windows、Mac 和 Linux 应用相比有很大的不同。第 4 章将介绍如何克服困难并使用 Java Swing 和 JavaFX 开发 GUI 应用。

1.7 Java 认证

Oracle 提供了一系列的 Java 认证，通常可以分为两个级别：Associate 和 Professional，如图 1.7 所示(https://education.oracle.com/pls/web_prod-plq-dad/ou_product_category.getPageCert?p_cat_id=267)。可以先申请 Java 基金会认证的初级助理，再转到 Oracle 认证的助理，最后成为 Oracle 认证的专业人员。不同的 Java 版本需要不同的认证。例如，Java SE 7 程序员和 Java SE 8 程序员有单独的认证。Oracle 将继续推出适用于较新 Java 版本的认证。

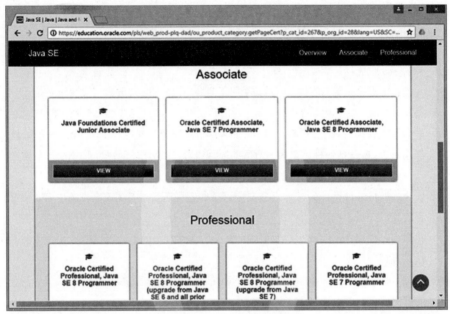

图 1.7　Oracle Java 认证级别

1.8 小结

本章简要介绍了 Java 编程语言，包括版本历史和 4 个 Java 平台版本；还介绍了很受欢迎的用于企业级应用开发的 Java Spring 框架，并总结了 Java 的优缺点；最后给出了 Java 认证的相关信息。

1.9 本章复习题

1. 什么是 Java？解释 Java 源文件和 Java 字节码文件之间的区别。
2. 什么是 HotJava？什么是 JavaScript？

3. 什么是平台独立性？
4. 哪些 Java 版本目前仍然受支持？
5. 使用图形描述 Java 架构。
6. 什么是 JDK、JRE、JVM 和 JIT？
7. Java 平台的 4 个版本是什么？
8. 什么是 Java Spring 框架？
9. Java 的优缺点是什么？
10. 目前有哪些 Java 认证？

第 2 章

Java编程入门

"工欲善其事，必先利其器。"
——出自《论语·卫灵公》

2.1 下载和安装 Java

2.2 Java IDE

2.3 Java 程序 Hello World

2.4 Java 在线编译器

2.5 Java 在线代码转换器

2.6 Java 免费在线课程和教程

2.7 Java 版本控制

2.8 小结

2.9 本章复习题

2.1 下载和安装 Java

现在开始使用 Java 进行编程。首先，你需要从 Oracle 网站下载最新的 Java 软件，如图 2.1 所示(www.oracle.com/technetwork/java/javase/downloads/index.html)，这里提供的 3 个下载选项分别是 JDK、Server JRE 和 JRE。Server JRE 和 JRE 仅用于运行 Java 程序。由于在学习本书的过程中需要编译 Java 程序，因此需要下载 JDK 软件包。你只需要按照网站上的说明在计算机上下载并安装 Java SE JDK 即可。

图 2.1　Java SE 下载网站

安装完毕后，你还需要设置一些诸如 PATH 和 JAVA_HOME 的环境变量。环境变量用于帮助程序弄明白要在哪个文件夹中安装文件、在何处存储临时文件以及在何处查找用户配置文件等。为了在 Windows 中设置环境变量，请依次选择 Control Panel | Advanced System Settings Environment Variables。在 System Variables 部分找到 PATH 环境变量，双击后进行编辑。例如，如果 Java JDK 的安装路径为 C:\Program Files\Java\jdk-10.0.2\bin，那么只需要将分号附加到末尾即可，如 C:\Program Files\Java\jdk-10.0.2\bin;。这里的分号用于分隔不同的 PATH 变量。同样在 System Variables 部分，找到 JAVA_HOME 环境变量，双击后进行编辑，将值更改为 C:\Program Files\Java\jdk-10.0.2\bin，如图 2.2 所示。

有关设置 Java 环境变量的更多信息，请参考 https://www.java.com/en/download/help/path.xml.

为了在 Linux/UNIX 和 macOS 中设置 Java 环境变量，需要编辑 shell 启动脚本。例如，如果 Java JDK 安装在 Bash shell 的/usr/local/JDK-10.0.2/目录中，那么可以编辑启动文件~/.bashrc，如下所示：

```
JAVA_HOME = /usr/local/jdk-10.0.2/
export JAVA_HOME
PATH=/usr/local/jdk-10.0.2/bin:$PATH
export PATH
```

保存并关闭后，输入以下命令以加载启动文件：

```
./.profile
```

在安装 Java 的过程中如果遇到困难，可以参考新加坡南洋理工大学提供的一份直观且简单的 Java 安装指南，如图 2.3 所示，详见 www.ntu.edu.sg/home/ehchua/programming/howto/JDK_HowTo.html.

正确安装 Java JDK 后，可以在 Windows 命令提示符下执行 java –version 命令以检查 Java 版本，如图 2.4 的上半部所示。可以在 Windows 中通过选择 Start | Run 并执行 cmd 命令来调用命令提示符，还可以通过执行 javac 命令来检查 Java 编译器，如图 2.4 的下半部所示。

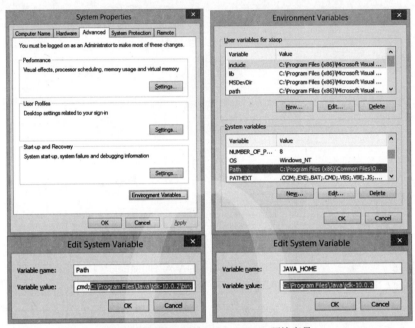

图 2.2 设置 PATH 和 JAVA_HOME 环境变量

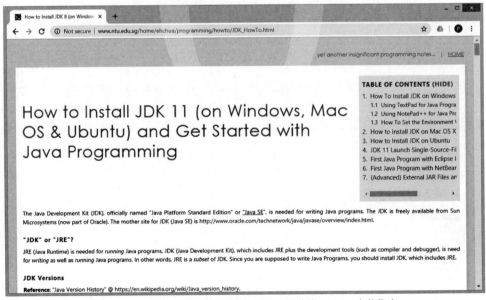

图 2.3 新加坡南洋理工大学提供的直观且简单的 Java SE 安装指南

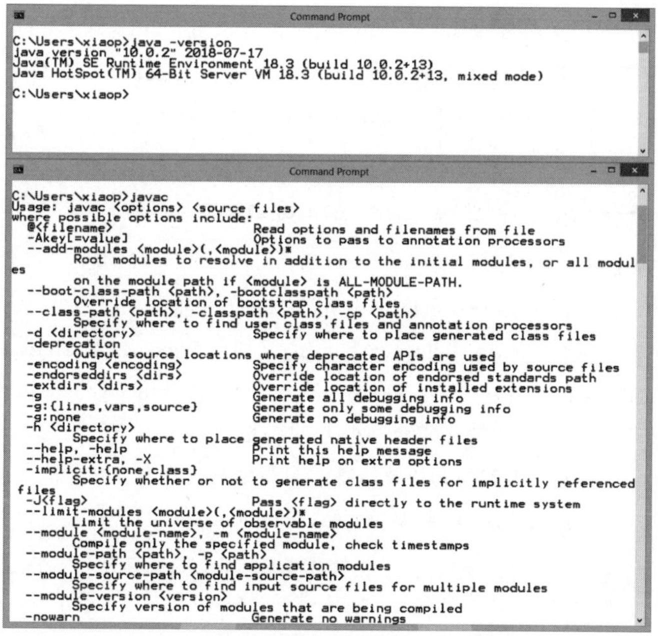

图 2.4　用于检查 Java 版本和 Java 编译器的命令

2.2　Java IDE

与使用任何编程语言一样，为了编写 Java 程序，需要准备良好的集成开发环境(IDE)，IDE 可以使编程更容易。下面介绍几个比较流行的 IDE。

第一个是 Notepad++，如图 2.5 所示。

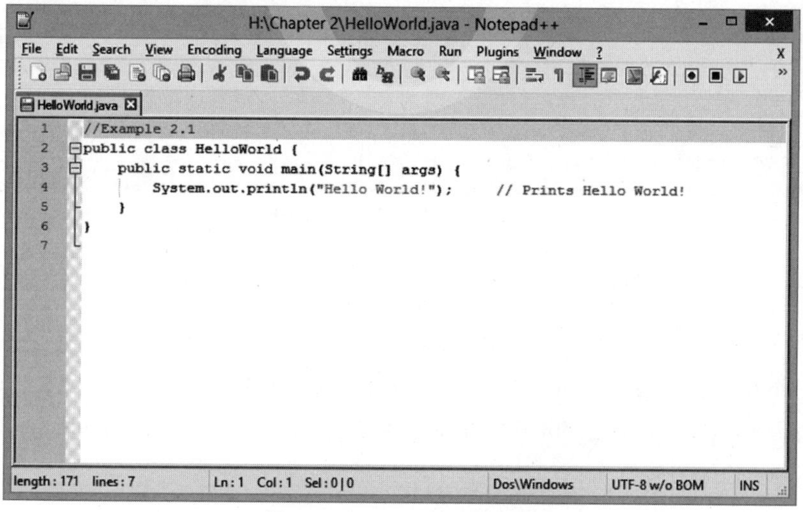

图 2.5　Notepad++文本编辑器

Notepad++是开源的文本编辑器,支持多种编程语言,包括 Java。除了支持打开多个 Java 文件、对 Java 关键字进行颜色编码和显示行号之外,Notepad++还支持代码折叠。文本编辑器都是轻量级的 IDE,不具备完整 IDE 的所有功能,但使用起来要简单得多。

Notepad++的下载链接如下:

```
https://notepad-plus-plus.org/download/v7.5.8.html
```

其他比较流行的文本编辑器包括
- Textpad(https://www.Textpad.com/):适用于 Windows 操作系统。
- Sublime Text(https://www.sublimetext.com/):适用于 Windows 和 Linux 操作系统。

第二个 IDE 是 IntelliJ IDEA,作为功能更为齐全且强大的专用 Java IDE,IntelliJ IDEA 相对也易于使用,如图 2.6 所示。与文本编辑器相比,IntelliJ IDEA 具有许多强大的功能(如智能完成、数据流分析、语言注入、检查和快速修复),此外还有许多内置的开发工具(如版本控制、反编译器、应用服务器)以及构建工具(如 Maven、Gradle、Ant 等)。

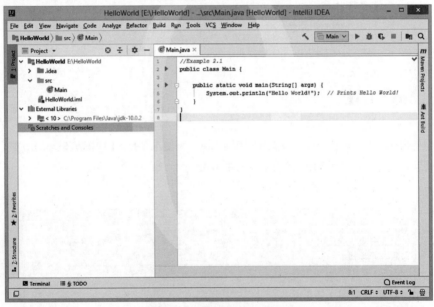

图 2.6　IntelliJ IDEA Java IDE

IntelliJ IDEA 的下载链接如下:

```
https://www.jetbrains.com/idea/
```

第三个 IDE 是 Eclipse,Eclipse 功能强大,支持 Java 和许多其他编程语言,如 C 和 C++。但对于初学者来说,Eclipse 复杂的用户界面和配置可能会让人望而生畏,如图 2.7 所示。

Eclipse 的下载链接如下:

```
www.eclipse.org/downloads/packages/release/mars/r/eclipse-ide-java-developers
```

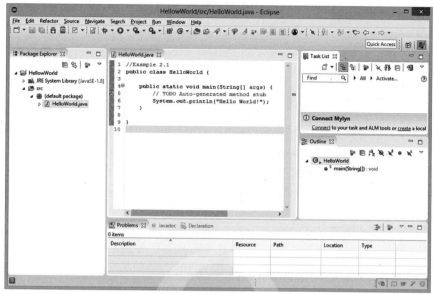

图 2.7　Eclipse Java IDE

最后一个但并非最不重要的 IDE 是 NetBeans。因为 NetBeans 可以作为包与 JDK 一起下载，所以我们不需要单独下载 JDK 软件。NetBeans 还支持拖放图形用户界面(GUI)组件，这使得开发 Windows 应用更容易，参见图 2.8。

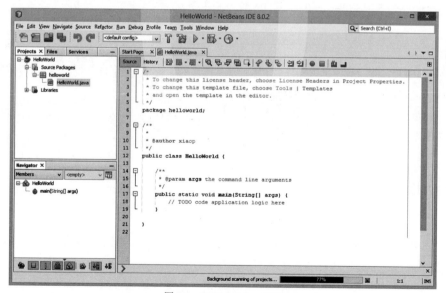

图 2.8　NetBeans Java IDE

以下是 NetBeans 的下载链接：

https://netbeans.apache.org/download/index.html

对于初学者，建议从文本编辑器(如 Notepad++)开始，随着获得越来越多的经验，然后转向 Java IDE。

2.3 Java 程序 Hello World

计算机程序在本质上是一组可以由计算机自动执行的语句(指令)。为了创建 Java 程序，首先需要创建 Java 源代码文件，Java 源代码文件是扩展名为.java 的文本文件。源代码包含一些语句，这些语句是执行某些操作的代码行。其次，需要将源代码编译成字节码(bytecode)文件，这是一种机器可理解的二进制文件，扩展名为.class，然后可以在 Java 运行环境(JRE)中执行。Java 的平台独立性是通过为每个操作系统安装 JRE 来实现的。JRE 处理各种操作系统之间的差异，并为运行字节码提供了通用的运行平台。因此，无论底层操作系统是什么，相同的字节码都可以在所有 JRE 中运行，并提供相同的结果。

例 2.1 展示的 Java 程序 HelloWorld.java 只会将 Hello World！打印到屏幕上。

例 2.1　HelloWorld.java

```
//Example 2.1
public class HelloWorld {
   public static void main(String[] args) {
      System.out.println("Hello World!");
   }
}
```

在 Java 中，每个程序至少包含一个类，并且源代码文件的文件名(包括大小写)应与类名完全相同；当程序包含多个类时，则应该与主类的名称完全相同。

与 C/C++一样，Java 使用花括号{和}对代码块进行分组，并使用//标记单行注释，而使用/*和*/标记多行注释。下面是单行注释的示例：

```
//This is a single line comment
```

下面是多行注释的示例：

```
/*
   This is a multiple line comment
   Where you can write multiple lines
*/
```

每个 Java 应用(除了 Java 小程序)都有一个且只有一个公共的静态方法 void main(String[] args)。对于 main()方法，关键字 public 意味着它可以被其他 Java 类访问，关键字 static 意味着它可以在不使用任何对象的情况下调用，关键字 void 意味着它不返回任何值。args 是一个字符串类型的数组变量，用于保存你通过命令行输入的所有参数。main()方法应该始终是静态的和公共的。在 Java 中，以分号(;)结尾的每一行称为语句；语句是一项要求计算机执行的操作。本例只有一条语句：

```
System.out.println();
```

上面这条语句用于在屏幕上打印消息。System.out.println();语句采用了一种面向对象的方式来表示调用 println()方法。System.out 是标准输出(计算机屏幕)对象，类似的还有 System.in(表示标准输入(键盘)对象)、System.err(标准错误对象)等。

每个 Java 方法或类都属于一个包。默认的包是 java.lang.*，无须显式导入，如前面的示例所示。但是，如果使用的是属于其他包的任何方法或类，则需要在程序开始时将它们导入。在下一章的示例中，你将使用 java.text.*和 java.util.*包。要运行 HelloWorld.java 程序，请首先使用喜欢的文本编辑器或 IDE 输入前面的代码，然后保存到名为 HelloWorld.java 的文本文件中。图 2.9 显示了如何使用 Notepad++创建 HelloWorld.java 程序并将文件另存为 Java 源代码文件(*.java)。请记住，文件必须与类同名。

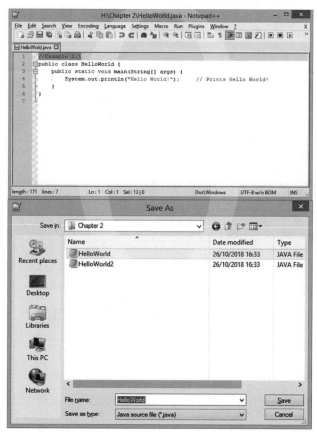

图 2.9　使用 Notepad++创建 HelloWorld.java 程序并将文件另存为 Java 源代码文件，注意文件必须与类同名

在保存 Java 源代码文件的目录中打开 Windows 命令提示符窗口，执行以下命令：

```
javac HelloWorld.java
```

结果将生成名为 HelloWorld.class 的字节码文件。然后使用以下命令运行这个字节码文件，结果如图 2.10 所示：

```
java HelloWorld
```

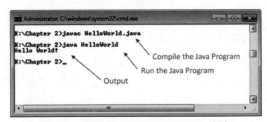

图 2.10　编译和运行 HelloWorld.java 程序

恭喜！现在，你的第一个 Java 程序正在运行。除了 Notepad ++，还可以使用 Java IDE，例如 IntelliJ IDEA。可以在 IDE 中编写、编译和运行 Java 程序。

练习 2.1　修改例 2.1，使其打印你的姓名和地址。

例 2.2 是例 2.1 中名为 HelloWorld2.Java 的 Java 程序的改进版本，它会在屏幕上打印 Hello xxx！，其中 xxx 是你在运行程序时输入的第一个参数。图 2.11 显示了如何编译和运行 HelloWorld2.java 程序。如前所述，Java 使用 String 数组变量 args 存储命令行参数，也就是你在 java HelloWorld2 之后输入的文本。在此，命令行参数是 Perry Xiao。args[0]中存储了值 Perry，而 args[1]中存储了值 Xiao。因此，System.out.println("Hello " + args[0])会把 Hello Perry！打印到屏幕上。在这里，Java 使用+符号来连接两个字符串。

例 2.2　HelloWorld2.java

```java
//Example 2.2
public class HelloWorld2 {
   public static void main(String[] args) {
       System.out.println("Hello " + args[0]);     // Prints Hello xxx!
   }
}
```

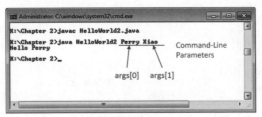

图 2.11　编译和运行 HelloWorld2.java 程序

练习 2.2　修改例 2.2，使其同时输出你的名字和姓氏。

默认情况下，javac 命令会将当前目录(也称为文件夹)用于 Java 源代码文件和已编译的类文件。为了使用其他目录或文件夹，需要指定其他选项，例如–classpath、–sourcepath、–d、–target

等。–classpath(或–cp)选项指定了在编译或运行 Java 程序时可以从哪里找到预编译的 Java 类文件(例如库)。–sourcepath 选项指定了可以从哪条路径查找 Java 源代码文件。–d 选项指定了已编译的 Java 类文件的保存位置。–target 选项指定了在编译 Java 文件时以哪个版本的 JRE 为目标。

例如，以下命令将在当前目录(.)下，使用\examples 作为子目录，使用\lib\funs.jar 文件作为类路径，编译 HelloWorld.java 程序。在 Windows 中，Java 使用分号来分隔不同的目录；而在 Linux 中，Java 使用冒号来分隔不同的目录。

```
javac -classpath .;\examples;\lib\funs.jar HelloWorld.java
```

以下命令会将 H:\examples 目录设置为源路径：

```
javac -sourcepath H:\examples H:\examples\HelloWorld.java
```

以下命令会将.\classes 子目录设置为已编译的类文件的目标路径：

```
javac -d .\classes HelloWorld.java
```

要想查看有关 javac 命令选项的更多信息，可在 Windows 命令提示符下执行 javac –help 命令。

2.4 Java 在线编译器

除前面介绍的文本编辑器和 IDE 外，还有许多在线编译器，这意味着无须下载 Java JDK 或 IDE 并将它们安装到计算机上，而只需要在计算机、平板电脑甚至手机的 Web 浏览器中编写和运行 Java 代码即可。

图 2.12 展示了 Tutorialspoint(https://www.tutorialspoint.com/compile_java_online.php)中简单易用的 Java 在线编译器。在左侧编写 Java 代码，然后在右侧查看结果，在界面的右上方，还有 Fork、Project、Edit、Setting 和 Login 菜单。可以创建并共享项目，但只支持编译一个文件，并且仅支持 Java 8。

图 2.12　Tutorialspoint 中的 Java 在线编译器

图 2.13 展示了 Codiva Java 在线编译器(https://www.codiva.io/java ##)。为了使用 Codiva，需要创建账户(参见图 2.13(a))，也可在不登录的情况下尝试 Codiva (参见图 2.13(b))。

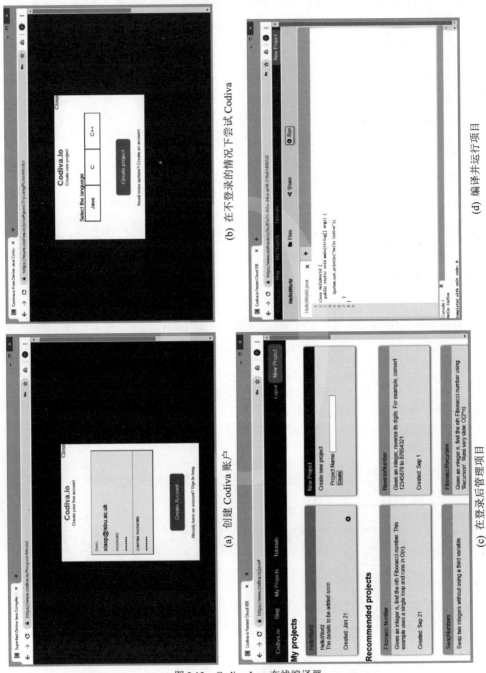

图 2.13 Codiva Java 在线编译器

登录后，除了能够创建更多的项目并查看他人的项目(参见图 2.13(c))，你还能够编译和运行自己的项目(参见图2.13(d))。Codiva 的最佳特性是自动编译，这意味着当你输入时，它会自动编译代码并显示结果。Codiva 还有自动完成功能，这使得编写代码更加容易和高效。Codiva 支持多个项目、文件、包以及 Java 8 和 Java 9，但不支持 Java 10 或 Java 11。除 Java 外，Codiva 还支持 C 和 C++语言，但不提供 UI 主题或不同的编译器设置。

图 2.14 展示了 CompileJava.net Java 在线编译器(https://www.compilejava.net/)。作为一款简单的 Java 在线编译器，CompileJava.net 仅支持一个 Java 文件，但却允许你选择不同的 IDE 后台方案，如图 2.14 所示。

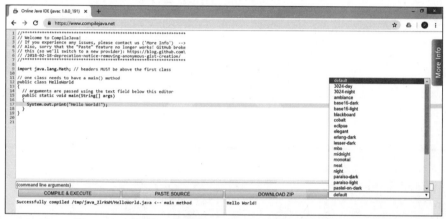

图 2.14　CompileJava.net Java 在线编译器

图 2.15 展示了流行的 Java 在线编译器 JDoodle(https://www.jdoodle.com/online-java-compilerr)。与其他 Java 在线编译器不同，JDoodle 支持 Java 8、Java 9 和 Java 10，如图 2.15 所示。可以创建账户或者使用 Google 账户登录 JDoodle。登录后，可以创建新项目并管理所有项目，也可以与他人合作。除 Java 外，JDoodle 还支持 70 多种其他编程语言。

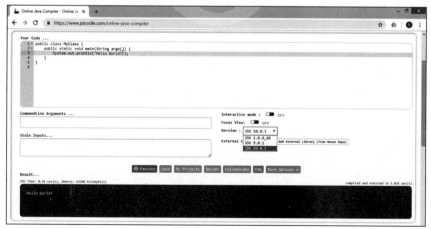

图 2.15　JDoodle Java 在线编译器

图 2.16 展示了 Browxy Java 在线编译器(www.browxy.com/)。Browxy 曾经很受欢迎，但随着其他 Java 在线编译器的追赶而开始落后。Browxy 几乎没有限制，并且支持多个 Java 文件和网络。如图 2.16 所示，可以运行 Java 网络程序以获取网络信息。你可以通过学习第 5 章进一步了解 Java 网络编程。

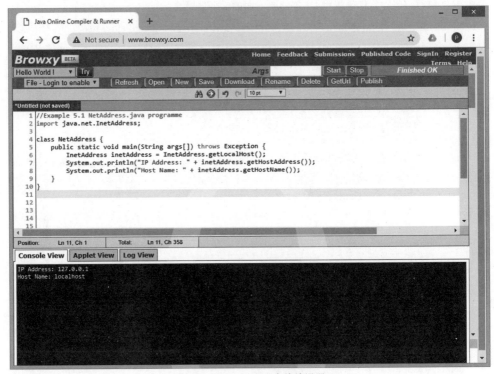

图 2.16　Browxy Java 在线编译器

Browxy 还提供了几个有趣的 Java 示例应用，例如 Hello World、Animal Game、Thread Example 等。Browxy 仅支持 Java 8。

图 2.17 展示了 OnlineGDB Java 在线编译器(https://www.online-gdb.com/online_java_compiler)。OnlineGDB 拥有独特的在线调试器功能，允许在代码中设置断点并逐步运行代码。OnlineGDB 还支持代码折叠和自动补全功能，这两项新功能使编写复杂的程序变得更加容易。可以使用 Google 或 Facebook 账户登录 OnlineGDB，登录后就可以创建新项目并管理所有项目。

图 2.18 展示了 Rextester Java 在线编译器(http://rextester.com/l/Java_online_compiler)。Rextester 代表正则表达式测试程序。Rextester 支持大约 30 种语言，包括 Java，如图 2.18 的顶部所示，并且在 C#用户中很流行。Rextester 能提供最好的实时协作支持，允许多个用户编辑同一文件，如图 2.18 的底部所示。单击页面中的 "实时协作" 按钮并对项目命名，然后 Rextester 将生成唯一的 URL，多个用户可以共享和使用同一个 Java 程序。页面的右边还有黄色的聊天框，用户可以在这里交换信息。但是，Rextester 只支持一个 Java 文件，而且只支持 Java 8。

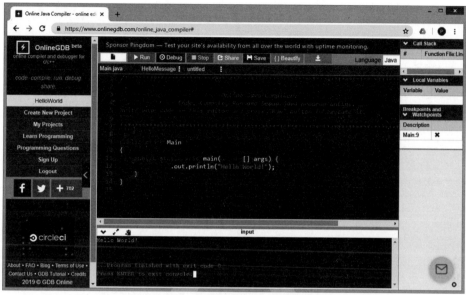

图 2.17　OnlineGDB Java 在线编译器

图 2.19 展示了 IDEOne Java 在线编译器(https://www.ideone.com/)。IDEOne 声称是第一批 Java 在线编译器之一。可以编辑代码、分叉代码(创建新代码),然后下载代码。IDEOne 支持 60 多种编程语言,包括 Java。IDEOne 还提供了一个可作为服务进行编译的 API,这意味着可以使用这个 API 创建自己的在线 IDE。

图 2.18　Rextester Java 在线编译器

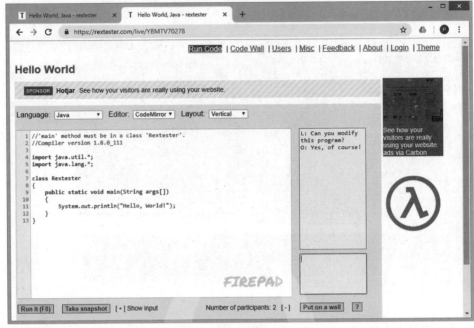

图 2.18(续)

图 2.19　IDEOne Java 在线编译器

2.5 Java 在线代码转换器

Java 在线代码转换器可以将应用的源代码从一种编程语言转换为另一种编程语言。图 2.20 展示了 Carlosag Java 在线代码转换器(https://www.carlosag.net/tools/codetranslator/)，它可以将 Java 代码转换为 C#、VB.NET 和 TypeScript 代码，反之亦然。图 2.21 展示了 mtSystems Java 在线代码转换器(https://www.mtsystems.com/)，它可以将 C 代码为 Java 代码。

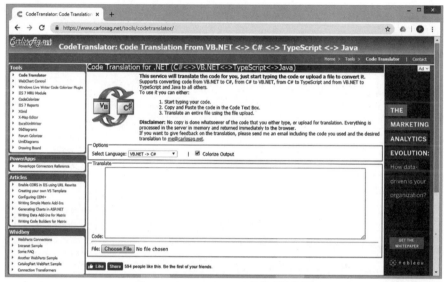

图 2.20　Carlosag Java 在线代码转换器

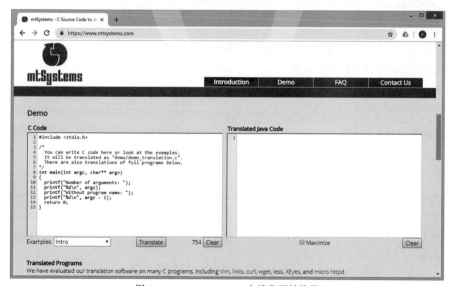

图 2.21　mtSystems Java 在线代码转换器

2.6　Java 免费在线课程和教程

网络上提供了许多免费的 Java 学习资源。图 2.22 展示了 Oracle 的 Java 教程(https://docs.oracle.com/javase/tutorial/index.html)。这些内容十分全面的教程介绍了如何以简单的方式创建应用。这些教程的主题包括 Java 基础知识以及更高级的编程技能，例如面向对象编程(OOP)、GUI、网络和 JavaBeans。

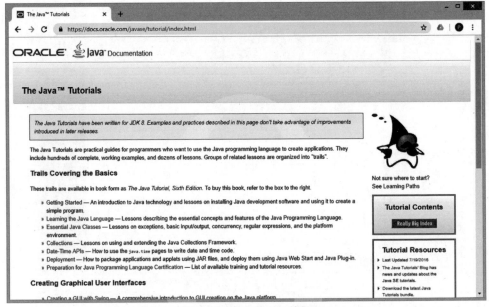

图 2.22　Oracle 的 Java 教程

图 2.23 展示了 Coursera 网站(https://www.coursera.org/specializations/java-programming)上的 Java 课程，这是最知名的在线课程网站之一。课程视频有英文、西班牙文、俄文、中文和法文等多个版本，并且课程含有字幕。

图 2.24 展示了 Codecademy 网站(https://www.codecademy.com/learn/learn-java)上的 Java 课程，这是另一个非常受欢迎的 Java 在线课程学习网站。Codecademy 网站为初学者提供了免费的 Java 编程课程，学生可将学到的 Java 知识应用到不同项目中。

图 2.25 展示了 Udacity 提供的 Java 课程(https://eu.udacity.com/course/java-programming-basics--ud282)，Udacity 是由塞巴斯蒂安·特伦(Sebastian Thrun)、大卫·史戴文斯(David Stavens)和迈克·索科尔斯基(Mike Sokolsky)创立的营利性教育组织，该组织提供了大规模的在线公开课程(MOOC)。Udacity 最初专注于提供大学风格的课程，但现在更多地专注于针对 Java 专业人士的职业课程。

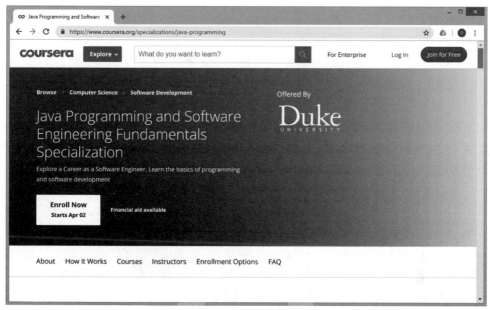

图 2.23　Coursera 网站上的 Java 课程

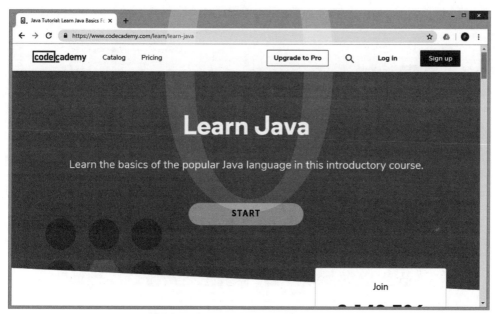

图 2.24　Codecademy 网站上的 Java 课程

图 2.26 展示了 LearnJavaOnline 网站(https://www.learnjavaonline.org/)上的 Java 教程,由于带有 Java 在线编译器,因此可以在同一网站上学习 Java 编程并运行代码。LearnJavaOnline 网站提供了基本的和高级的 Java 课程。

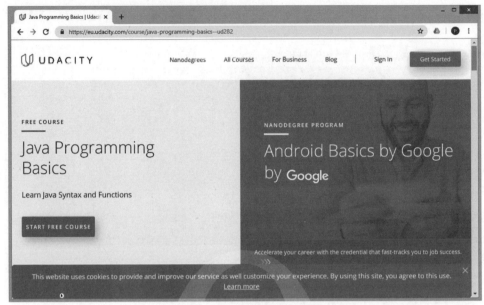

图 2.25 Udacity 提供的 Java 课程

图 2.26 LearnJavaOnline 网站上的 Java 教程

图 2.27 展示了来自 CaveofProgramming 网站(https://courses.caveofprogramming.com/p/java-for-complete-beginners)的 Java 初学者课程,该网站由拥有 14 年以上工作经验的软件开发人员 John Purcell 创建。这个网站上的课程是免费的,你可以根据自己的学习进度选择课程和制订学习计划。

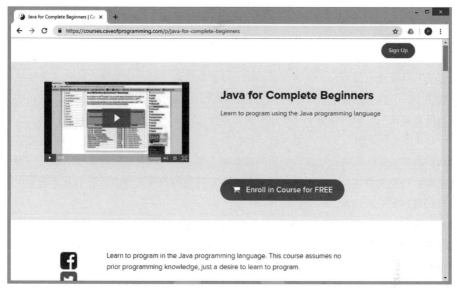

图 2.27　CaveofProgramming 网站上的 Java 课程

图 2.28 展示了 IBM 的 Java 语言基础课程(https://www.ibm.com/developerworks/java/tutorials/j-introtojava1/index.html)，其中涵盖了比较全面的 Java 内容。

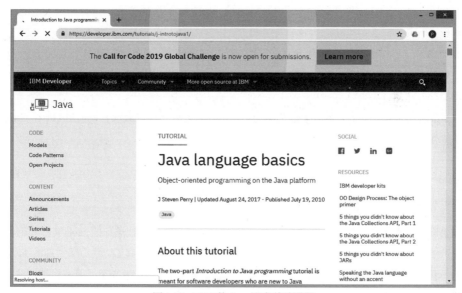

图 2.28　IBM 的 Java 语言基础课程

有关 Java 语言的详细信息，请访问 Java API 文档网站 https://docs.oracle.com/javase/10/docs/api/overview-summary.html。

2.7　Java 版本控制

对于软件开发人员而言，版本控制意味着维护软件开发的历史记录。为什么需要版本控制？请想象以下场景：你正在处理一个大的编程项目，你在添加了一个新函数之后，这个项目突然停止编译了，如何将这个项目还原到以前的工作版本？再比如以下场景：你正在与其他几个开发人员一起开发一个小组项目，他们都使用相同的源目录，你怎么才能知道谁改变了什么？或者如何才能防止刚开发的代码被其他人覆盖？答案是使用版本控制软件作为中央存储中心，开发人员可以在其中签入和签出文件，其他用户无法修改已签出并开始工作的文件，直到将它们重新签入。

具备这种功能的软件称为版本控制系统(VCS)。可以使用集中式版本控制系统(CVCS)或分布式/分散式版本控制系统(DVCS)。CVCS 使用一台中央服务器来存储所有文件，缺点是容易遭受单点故障：当服务器关闭或服务器上的文件被损坏时，所有项目都会停滞。因此，使用 DVCS 更适合，尤其是当项目开发人员位于不同城市甚至不同国家时。DVCS 不仅允许开发人员签出目录的最新版本，而且允许镜像整个存储库。如果服务器出现故障，则可以将任何开发人员的存储库复制到服务器以进行恢复。

Git 是使用最广泛的分布式版本控制系统之一，速度很快。Git 最初由著名的芬兰软件工程师 Linus Torvalds 设计和开发，他还于 1991 年在赫尔辛基大学学习时开发了 Linux 操作系统。Torvalds 基于 Linux 内核设计开发了 Git。GitHub 是使用 Git 进行版本控制的基于 Web 的服务。与其他版本控制软件相比，Git 具有免费使用、体积小、速度快以及开源的优点。另外，Git 还具有隐式备份功能和良好的安全性，易于分布式部署，并且不需要强大的硬件。

如果正在进行协作开发并且需要部署 DVCS，请参阅附录 C 以获取有关使用 Git 和 GitHub 的完整教程。

2.8　小结

本章首先介绍了如何下载和安装 Java JDK 软件，还介绍了 4 款流行的 Java 文本编辑器，并展示了如何使用它们处理简单的 Java 程序 Hello World；然后推荐了一些广受欢迎的 Java 在线编译器、Java 在线代码转换器以及一些免费的 Java 在线课程；最后介绍了版本控制软件，并讲解了如何使用 Git 和 GitHub 进行版本控制。

2.9　本章复习题

1. 从何处可以下载最新的 Java JDK 软件包？最新版本是什么？

2. 在什么情况下需要下载 Java JDK 软件包？在什么情况下需要下载 JRE 软件包？

3. 在 Java 程序 HelloWorld 中，关键字 public、static 和 void 有什么作用？

4. 如何在 Windows 中调用 Windows 命令提示符？

5. Java 程序的编译命令是什么？

6. Java 程序的运行命令是什么？

7. 什么是命令行参数？

8. 使用表格比较不同的 Java 在线编译器，并指出它们支持的功能都有哪些，例如单个文件、多个文件、创建项目、不同的 Java 版本、调试等。

9. 查找通过另一种编程语言编写的一段代码，然后使用 Java 代码转换器转换为 Java 代码。

10. 使用表格比较不同的 Java 在线课程或教程，并指出它们是否免费、是否有视频，以及是否可以自由支配学习进度或只能在固定的日期和时间学习，等等。

11. 软件版本控制是什么以及为什么需要它？

12. 什么是 Git 和 GitHub？

第 II 部分

第 3 章：基本的 Java 编程
第 4 章：面向 Windows 应用的 Java 编程
第 5 章：面向网络应用的 Java 编程
第 6 章：面向移动应用的 Java 编程

第 3 章

基本的Java编程

"如果你认为自己能做一件事，那你就能做到，反之亦然。"

——Henry Ford

3.1 引言

3.2 变量

3.3 运算符

3.4 保留字

3.5 输入和输出

3.6 循环和选择

3.7 数组、矩阵和 ArrayList

3.8 读写文件

3.9 方法

3.10 面向对象编程

3.11 多线程

3.12 日期、时间、计时器和睡眠方法

3.13 执行系统命令

3.14 大规模的软件包和编程

3.15 软件工程

3.16 部署 Java 应用

3.17 小结

3.18 本章复习题

3.1 引言

本章介绍 Java 编程的一些基本概念,这些概念是为那些在编程方面有一些经验并熟悉编程概念,但对 Java 语言还不太熟悉的开发人员而设计的。

3.2 变量

Java 支持如下 8 种基本类型的变量:byte、short、int、long、float、double、char 和 boolean。Java 中的变量名区分大小写,并且必须以字母、下画线或美元符号($)开头,而不能以数字开头。在第一个字符之后,变量名可以包含字母和数字的任意组合。空格和特殊符号,例如!、""、£、%、 &、*、#、@和~等,是不允许出现在变量名中的。在 Java 中,所有变量在使用之前都必须先声明。为了声明变量,需要指定类型,至少留一个空格,然后是变量名和分号(;)。

type variablename;

以下是名为 x 的整型变量的声明和初始化代码:

```
int x;         // Declares an integer type variable called x
x = 10;        // Initialize x
```

以下为等价形式:

```
int x = 10;    // Declaration and Initialization
```

还可以在一行中声明多个类型相同的变量:

```
int x, y, z;            // Declares three int variables, x, y, and z
x = 10; y = 5; z=1;     // Initialize x, y, and z
```

以下为等价形式:

```
int x = 10, y = 5, z=1; // Declaration and Initialization
```

例 3.1A 演示了两个整型变量的简单声明和初始化。在这里,如第 2 章所述,System.out.println() 用于在屏幕上显示文本,println("x = " + x)用于在屏幕上显示文本"x = "和变量 x 的值。+号表示将文本和变量拼接在一起。你可以在 3.5 节中找到有关输入和输出的更多信息。

例 3.1A 声明和初始化 Java 变量

===
```
//Example 3.1A Java variable example
public class VarExample {
    public static void main(String args[]) {
        int x = 10;      // declares and assigns integer variable
        int y = 5;       // declares and assigns integer variable
```

```
        int z;          // declares integer variable

        System.out.println("x = " + x );
        System.out.println("y = " + y);
        z = x + y;
        System.out.println("x + y = " + z);
    }
}
========================================================================
```

如果想在 main() 方法之外声明变量,则需要将变量声明为静态变量,因为在 main() 方法中只能使用静态变量或方法,参见例 3.1B。

例 3.1B 将 Java 变量声明为静态变量

```
========================================================================
  //Example 3.1B Java variable example
  public class VarExample2 {
    static int x = 10;   // declares and assigns integer variable
    static int y = 5;    // declares and assigns integer variable
    static int z;        // declares integer variable

    public static void main(String args[]) {
       System.out.println("x = " + x );
       System.out.println("y = " + y);
       z = x + y;
       System.out.println("x + y = " + z);
    }
}
========================================================================
```

例 3.1C 给出了一些声明和初始化不同类型变量的示例。请注意,需要将 F 放到浮点数的末尾,以表明是 float 类型数而不是 double 类型数。同样,需要将 L 放到 long 类型数的末尾以表明是 long 类型数。

例 3.1C 声明不同类型的 Java 变量

```
========================================================================
//Example 3.1C Java variable example
public class VarExample3 {
    public static void main(String args[]) {
        short x = 10, y = 5, z=1;   // declares and assigns short variables
        byte B = 17;                // declares and assigns a byte variable
        long w = 234334000000L;     // declares and assigns a long variable
        double d = 5.01;            // declares and assigns a double variable
        float f = 5.01F;            // declares and assigns a float variable
        char ch = 'a';              // declares and assigns a char variable
        String s = "Hello World!";  // declares and assigns a String variable
        boolean isDone = false;     // declares and assigns a Boolean variable

        System.out.println("ch = " + ch );
        System.out.println("isDone = " + isDone);
        System.out.println(s);
```

```
    }
}
```
==

练习 3.1 修改例 3.1A 以执行如下操作：从命令行参数 args[0]和 args[1]中获取两个整数，执行两个整数的加法、减法、乘法和除法操作并在屏幕上显示结果，最后对除法结果进行注释(提示：可使用 Integer.parseInt()将 args[0]或 args[1]字符串转换为整数)。

练习 3.2 重复练习 3.1，但执行两个 double 类型数的加法、减法、乘法和除法操作，最后对除法结果进行注释(提示：可使用 Double.parseDouble()将 args[0]或 args[1]字符串转换为 double 类型数)。

3.2.1 常数

常数是值在程序执行期间不会改变的变量。可以使用与变量相同的方式定义常量，但需要将 static 和 final 关键字放在最前面。关键字 static 表示变量在整个类中持久存在，并且在内存中仅仅保留变量的值的一个副本。final 关键字表示变量的值不能更改。由于不能更改值，因此必须在定义常量时就进行初始化。通常，常量名使用大写字母来表示，如下所示：

```
static final int TOTALNUMBER=120;
static final float PI=3.1415926F;
```

3.2.2 String 和 StringBuffer 类型

Java 还支持如下两种类型的变量：String 和 StringBuffer，可以在其中存储字符值序列。以下代码演示了如何创建 String 变量来存储学生的班级名称：

```
String ClassName;
ClassName = new String();
ClassName = "Maple Tree Class";
```

也可以把以上三个步骤合并成一个步骤：

```
String ClassName = "Maple Tree Class";
```

字符串变量是对象变量，这种变量附带了许多用于字符串操作的有用方法(有关详细信息可参阅 3.9 节)，例如 length()、charAt()、equals()、concat()、trim()、compareTo()、toUpperCase()、toLowerCase()和 substring()。

以下表达式中的 t1 和 t2 是字符串变量：

```
t1.compareTo(t2)
```

如果 t1 和 t2 的 ASCII 码相等，则返回 0；如果 t1 的 ASCII 码小于 t2 的 ASCII 码，则返回负数；如果 t1 的 ASCII 码大于 t2 的 ASCII 码，则返回正数。在 ASCII 中，每个字母、数字或特殊字符都用一个 7 位的二进制数来表示，一共有 128 个可能的字符。

以下表达式会将 t2 附加(或连接)到 t1 的末尾：

```
t1 = t1.concat(t2)
```

练习 3.3 编写一个 Java 程序,使用 String 变量存储英文句子,然后将其中的字母更改为大写并显示在屏幕上。

在 Java 中,String 类型是不可变的。String 类型的不可变性意味着 String 对象一旦创建了,就无法再更改。这在缓存、安全性、同步和性能方面带来许多好处。对于 String 类型数据的缓存,Java 使用了 String 类型池的概念,String 类型池用于存储所有 String 类型的值。由于 String 类型是不可变的,因此 Java 可以通过在 String 类型池中仅存储每个 String 类型文本的副本来优化内存的使用,进而通过节省内存来提高性能。

例如,在下面的代码中,s1 和 s2 实际上都指向 String 类型池中的同一对象 Hello:

```
String s1 = "Hello";
String s2 = "Hello";
```

但是在进行以下拼接后,s2 指向另一个对象 Hello World,而 s1 仍指向原始对象 Hello:

```
s2 = s2 + " World";
```

安全起见,Java 使用 String 变量存储用于远程数据库连接和其他网络信息的用户名及密码。String 对象的不可变性意味着一旦创建了这些对象,就无法对它们进行黑客攻击。Java 还使用 String 变量来存储哈希码,这意味着一旦在创建时将哈希码缓存起来,就无须再次计算哈希码。所有这些功能可以使 Java 更加安全。有关哈希码和 Java 安全性的更多信息,参见第 9 章。

另外,由于 String 类型是不可变的,因此可以在不同线程之间共享单个 String 实例。这样可以避免使用同步,使多线程更安全。

StringBuffer 类型提供了类似于 String 类型的功能,但前者是一种更强大的类型,可以处理动态字符串信息。当对字符串进行多处修改时,建议使用 StringBuffer 类型,这是因为 StringBuffer 类型是可变的。

例如,以下代码可以将文本追加到现有的 StringBuffer 变量中,速度相比以前的 String 拼接要快得多,例如 t1 = t1.concat(t2)或 s2 = s2 + " World",因为 String 拼接每次都会创建一个新的 String 对象。

```
StringBuffer sB = new StringBuffer("Hello");
sB.append(" World");
```

StringBuffer 类型还提供了诸如 reverse()、delete()、replace()的方法。

3.2.3 var 变量类型

var 变量类型在许多其他语言中都存在,例如 JavaScript 和 Visual Basic。var 允许在不指定变量类型的情况下声明变量,Java 将自动确定类型,从而使编程更简单,这也提高了代码的可读性。Java 开发人员长期以来对于在代码中包括样板代码以及由此导致的冗长心存抱怨。从 Java 10 开始,var 变量类型在 Java 中也可用。以下是 var 变量类型的一些示例用法:

```
//Java VAR type variable since Java 10
var str = "Java 10";                       // infers String
var list = new ArrayList<String>();        // infers ArrayList<String>
var stream = list.stream();                // infers Stream<String>s
```

3.3 运算符

与许多其他语言一样，Java 支持各种标准运算符，如算术运算符、比较运算符、逻辑运算符、位运算符和赋值运算符，如表 3.1 所示。

表 3.1 Java 运算符

运算符类型	运算符	注释
算术运算符	+	加法和字符串拼接
	−	减法
	*	乘法
	/	除法
	%	取模(或求余数)
	++	递增
	--	递减
比较运算符	<	小于
	<=	小于或等于
	>	大于
	>=	大于或等于
	==	等于*
	!=	不等于
逻辑运算符	!	逻辑非
	&&	逻辑与
	||	逻辑或
位运算符	~	按位补码
	&	按位与
	|	按位或
	^	按位异或
	<<	左移
	>>	右移
	>>>	填充 0 并右移

(续表)

运算符类型	运算符	注释
赋值运算符	=	赋值
	+=	加并赋值
	-=	减并赋值
	*=	乘并赋值
	/=	除并赋值
	%=	求模并赋值
	>>=	左移并分配
	<<=	右移并分配
	&=	按位与并赋值
	^=	按位或并赋值
	|=	按位异或并赋值

*请注意，==用于比较数值；要比较字符串，请改用 equals()方法。

3.4 保留字

下面这些 Java 保留字不能用作变量名、方法名或类名。

```
abstract       finally        short
assert         float          static
boolean        for            strictfp
break          goto           super
byte           if             switch
case           implements     synchronized
catch          import         this
char           instanceof     throw
class          int            throws
const          interface      transient
continue       long           true
default        native         try
do             new            var
double         null           void
else           package        volatile
enum           private        while
extends        protected      widefp
false          public
final          return
```

3.5 输入和输出

输入和输出在几乎所有的应用中都很重要。在 Java 中，System.out 用于在计算机屏幕上打印消息，这称为标准输出。System.in 和 System.console 用于从键盘读取文本，这称为标准输入。

例 3.2 演示了 Java 的标准输入和输出。System.console.readline()用于从键盘读取一行，并且用户输入的任何内容都将存储在名为 x 的 String 变量中。接下来，使用 System.out.print()在屏幕上显示文本。最后，System.out.println()的工作方式与 System.out.print()相同，不同之处在于：前者在显示文本后会将光标移至新行。图 3.1 展示了如何编译和运行例 3.2。

例 3.2　Java 的标准输入和输出

```
============================================================
//Example 3.2 Java standard output and input
public class InputOutputExample {
   public static void main(String[] args) {
      System.out.print("Enter something:");
      String x = System.console().readLine();
      System.out.println("You wrote: "+ x );
   }
}
============================================================
```

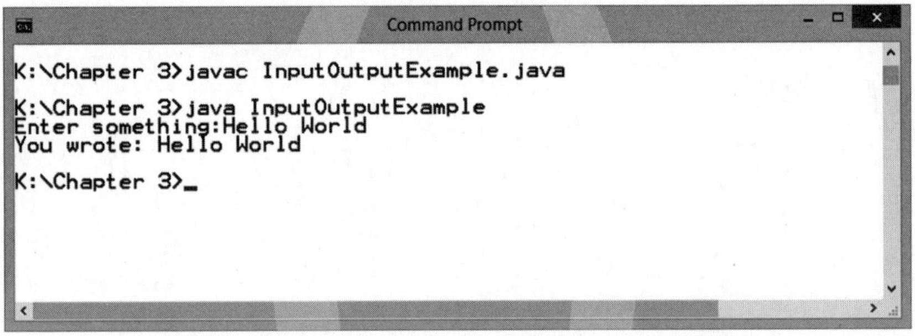

图 3.1　InputOutputExample.java 的编译、执行和输出结果

例 3.3 演示了另一种形式的 Java 输入和输出。在这段代码中，System.in 用于从键盘读取。结合 Scanner 类，既可以一次从键盘读取一行，也可以从键盘读取整数值。

下面这行代码用于导入 java.util.Scanner 库，因为这里需要使用 Scanner 类。

```
import java.util.Scanner;
```

本例首先要求用户输入姓名，然后从键盘读取姓名，最后在屏幕上显示姓名。本例还将提示并显示用户的年龄。用户输入的姓名存储在名为 name 的 String 变量中，输入的年龄存储在名为 age 的变量中。

例 3.3　另一种形式的 Java 输入和输出

```
============================================================
//Example 3.3 Java standard output and input
import java.util.Scanner;

class InputOutputExample2 {
   public static void main(String[] args) {
```

```
        System.out.println("Enter your name: ");
        Scanner scanner = new Scanner(System.in);
        String name = scanner.nextLine();
        System.out.println("Your name is " + name);

        System.out.println("Enter your age: ");
        int age = scanner.nextInt();
        System.out.println("Your age is " + age);
    }
}
========================================================================
```

在例 3.4 所示的 Java 输入和输出中，System.in 用于传入 InputStreamReader，然后传入 BufferedReader。在 Java 中，所有的输入源和输出源都被视为流，但是 Stream 类型的功能有限，并且通常需要与 BufferedReader 结合使用。使用 BufferedReader 可以每次从键盘读取一行。Integer.parseInt()用于将字符串转换为整数。throws IOException 和 try…catch 块用于处理异常。在 Java 中，必须对输入和输出等事件使用异常处理。

try…catch 块或 try…catch…finally 块是 Java 中处理异常的经典方法，如以下伪代码结构所示：try 块用于将执行某项操作的代码括起来，一个或多个 catch 块用于处理异常，finally 块用于将 try 块成功执行之后才执行的代码或异常处理程序括起来。

```
try {
    // do something
}
catch (one exception) {
    //display error message
}
catch (another exception) {
    //display error message
}
finally {
    //do something
}
```

相应的 java.io.BufferedReader、java.io.IOException、java.io.InputStreamReader 库也会被导入。你能猜出例 3.4 所示程序的作用吗？

例 3.4　Java 的标准输出和使用 BufferedReader 进行输入

```
========================================================================
//Example 3.4 Java standard output and input
import java.io.BufferedReader;
import java.io.IOException;
import java.io.InputStreamReader;

public class InputOutputExample3 {
    public static void main(String[] args) throws IOException {
        BufferedReader br = new BufferedReader(new
                InputStreamReader(System.in));
        System.out.print("Enter String");
        String s = br.readLine();
```

```
            System.out.print("Enter Integer:");
            try{
                int i = Integer.parseInt(br.readLine());
            }catch(NumberFormatException nfe){
                System.err.println("Invalid Format!");
            }
        }
    }
```

===

练习 3.4　修改例 3.2，使其询问你的姓名和地址，然后将它们显示在屏幕上。

练习 3.5　修改例 3.3，使其询问你的姓名、身高和体重，然后将它们显示在屏幕上。

练习 3.6　修改例 3.4，要求用户从键盘输入两个 double 类型的数字，然后计算它们的乘积并在屏幕上显示结果(提示：可使用 Double.parseDouble()将字符串转换为 Double 类型的数字)。

3.6　循环和选择

任何编程语言都必须具备三个关键要素：顺序、循环和选择。顺序意味着依次执行语句。循环和选择是控制结构(或流控制)。循环也称为迭代或迭代语句，是重复执行多次的语句集。Java 支持 for 循环、while 循环和 do 循环。例如，以下 for 循环将打印 Hello World! 五次：

```
for(i=0;i<5;i++){
    System.out.println("Hello World!");
}
```

类似地，可以使用 while 循环或 do 循环执行相同的操作。while 循环和 do 循环都需要在循环之前定义和初始化计数器变量(例如，int i = 0;)，并在循环过程中手动增加它(例如，i ++)。while 循环和 do 循环之间有如下细微的差别：while 循环在运行循环之前检查条件；do 循环首先运行循环，然后才检查条件。

```
int i=0;
while(i<5){
    System.out.println("Hello World!");
    i++;

}
int i=0;
do{
    System.out.println("Hello World!");
    i++;
} while(i<5)
```

例 3.5 使用 for 循环打印了所有命令行参数，它使用 args.length 来获取参数的总长度。

例 3.5　使用 for 循环

```
========================================================================
//Example 3.5 For Loops
class LoopExample {
    public static void main(String[] args) {
        int i;
        for(i=0;i<args.length;i++){
            System.out.println("Hello "+ args[i] + "!" );
        }
    }
}
========================================================================
```

练习 3.7　修改例 3.5，使其改为使用 while 循环。

练习 3.8　修改例 3.5，使其改为使用 do 循环。

选择或条件语句允许程序在不同的条件下执行不同的操作。Java 支持 if…else 条件语句(双向选择)、switch 条件语句(多项选择)和?运算符(条件赋值)。

例如，如果要在 i 等于 0 时执行某些操作，而在 i 不等于 0 时执行其他操作，则可以编写以下语句：

```
if (i==0){
...
}
else {
...
}
```

如果有两个以上的选择，并且想要根据 i 的值执行不同的操作，则可以使用 switch 条件语句。

```
switch (i){
   case 1:
   ...
   break;
   case 2:
   ...
   break;
   case 3:
   ...
   break;
   default:
   ...
   break;
}
```

?符号通常被称为三元运算符，因为它带有三个参数，并且能根据条件将两个值之一赋给变量。下面的示例在 i<0 的情况下将值 3 赋给变量 y，而在其他情况下将值 4 赋给变量 y：

```
y=(i<0) ? 3:4;
```

例 3.6 演示了 if…else 条件语句的用法。图 3.2 展示了如何编译和运行例 3.6 中的 Hello1.java 程序。

例 3.6　使用 if…else 条件语句

```
//Example 3.6 If Else
class Hello1 {
    public static void main(String[] args) {
        if(args.length!=2){
            System.out.println("Usage: java Hello1 firstname surname!");
        }
        else{
            System.out.println("Hello "+ args[0] +" "+ args[1] + "!");
        }
    }
}
```

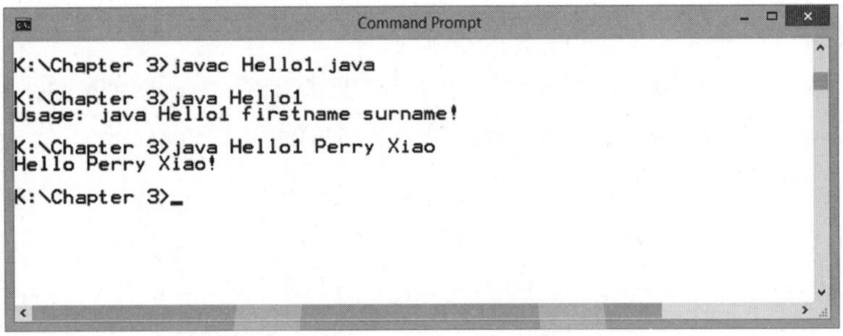

```
K:\Chapter 3>javac Hello1.java

K:\Chapter 3>java Hello1
Usage: java Hello1 firstname surname!

K:\Chapter 3>java Hello1 Perry Xiao
Hello Perry Xiao!

K:\Chapter 3>_
```

图 3.2　Hello1.java 的编译、执行和输出结果

3.7　数组、矩阵和 ArrayList

　　数组在科学和工程学编程中很重要。在 Java 中，可以创建一维数组、二维数组(矩阵)、多维数组和动态数组(列表)，这些数组可以包含一组动态元素，在程序执行期间可以不断增大和缩小。以下示例演示了如何分别声明 String 类型的一维数组、float 类型的二维数组(矩阵)和 double 类型的多维数组：

int[] x=new int[10];

float[][] y=new float [2][5];

double[][][] z=new double [5][2][3];

也可以同时声明和初始化它们：

int[] x={1,2,3,4,6,7,8,9,0};

float[][] y={{1.0F,1.0F,1.0F,1.0F,1.0F},{2.0F,3.0F,4.0F,5.0F,6.0F}};

double[][][]z={{{1,2,3},{4,5,6}},{{7,8,9},{10,11,12}},{{13,14,15},
               {16,17,18}},{{19,20,21},{22,23,24}},{{25,26,27},{28,29,30}}};

　　你可以将数组的声明分为两个步骤。例如，以下语句与前面的单行声明完全相同，但这种

方法带来的好处是允许稍后在运行程序时决定数组的大小。

```
int[] x;
x=new int[10];
```

例 3.7 演示了如何创建一个名为 x[ ]的一维整型数组和一个名为 y[ ][ ]的二维双精度型数组。图 3.3 展示了 Array1.java 的编译、执行和输出结果(注意，由于数组 y[ ][ ]是随机生成的，因此最终输出可能与图 3.3 中的不同)。在 Java 中，可以使用索引引用数组中的任何元素，例如 x[2]或 y[0][1]。数组中的所有元素必须是同一类型。

例 3.7 使用数组

```
===========================================================================
//Example 3.7 Array example
import java.util.*;

class Array1 {
    public static void main(String[] args) {
        int[] x;
        double [][] y;
        int i, j, row, col;
        if(args.length!=2){
            System.out.println("Usage: java array1 row colume");
            System.exit(0);
        }
        //initialize the arrays
        row=Integer.parseInt(args[0]);
        col=Integer.parseInt(args[1]);
        x=new int[row];
        y=new double[row][col];
        for (i=0;i<row;i++){
            x[i]=i;
        }
        for (i=0;i<row;i++){
            for (j=0;j<col;j++){
                y[i][j]=Math.random();
            }
        }
        //display the arrays
        System.out.print("x[]=");
        for (i=0;i<row;i++){
            System.out.print(x[i]+"\t");
        }
        System.out.println();
        System.out.println("y[][]=");
        for (i=0;i<row;i++){
            for (j=0;j<col;j++){
                System.out.print(y[i][j]+ "\t");
            }
            System.out.println();
        }
    }
}
===========================================================================
```

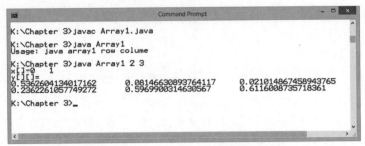

图 3.3 Array1.java 的编译、执行和输出结果

数组的缺陷在于定义后大小就是固定的。有时，我们需要数组的大小在程序执行时可以更改。可使用 Java 的 ArrayList 类创建此类数组，从而实现"可增长"的对象数组。以下示例演示了如何声明 String、Integer 和 double 类型的动态数组。同样，也可以创建任何对象的 ArrayList。

```
ArrayList<String> v=new ArrayList<String>();
ArrayList<Integer> v=new ArrayList<Integer>();
ArrayList<double> v=new ArrayList<double>();
```

尖括号<>表示这是通用的 ArrayList。Java 泛型编程是在 Java SE 5 中引入的，目的是减少错误并处理类型安全的对象。可以对非通用 ArrayList 进行编译，而不会出现任何错误，但是如果尝试运行，那么在尝试将字符串转换为整数时将出现运行时错误。在通用 ArrayList 中，<String>强制 ArrayList 仅包含字符串，因此会生成编译错误。通过这种方式，泛型编程使 Java 代码更健壮，更不易出错。

以下是非通用的 ArrayList 示例：

```
ArrayList list=new ArrayList();
list.add("hello");
Integer i = (Integer)list.get(0);
```

以下是通用的 ArrayList 示例：

```
ArrayList<String> list = new ArrayList<String>();
list.add("hello");
Integer i = (Integer)list.get(0);
```

有关 Java 泛型编程的更多信息，请访问以下网站：

```
https://docs.oracle.com/javase/tutorial/java/generics/index.html
https://en.wikipedia.org/wiki/Generics _ in _ Java
http://tutorials.jenkov.com/java-generics/index.html
```

例 3.8 展示了如何创建名为 list 的动态 ArrayList，可以使用 add()方法对它进行扩展，也可以使用 remove()方法对它进行收缩。图 3.4 给出了 Arraylist1.java 的编译、执行和输出结果。

例 3.8 使用 ArrayList

===

```
// Example 3.8 ArrayList example
```

```
import java.util.*;

class Arraylist1 {
    public static void main(String[] args) {
        ArrayList<Integer> list=new ArrayList<Integer>();
        int i, j;
        for (i=0;i<4;i++){
            list.add(i);
        }
        System.out.println("The total elements of ArrayList list
                            is:"+list.size());
        System.out.println(list);

        System.out.println("The second element is :");
            System.out.println(list.get(1));
        System.out.println("Now remove the second element:");
        list.remove(1);
        System.out.println(list);
    }
}
```
==

图 3.4　Arraylist1.java 的编译、执行和输出结果

3.8　读写文件

许多应用需要读写文件。Java 支持读写文本文件和二进制文件，但是本节仅介绍文本文件。Java 提供了多种读写文本文件的方法，最基本的方法是使用 java.io 包的 FileReader 和 FileWriter 类。

要使用 FileWriter 类写入文本文件，请执行以下步骤：

(1) 创建与要写入的文本文件关联的 FileWriter 对象。

(2) 创建与 FileWriter 对象关联的 BufferedWriter 对象。

(3) 写入缓冲区。

(4) 清空缓冲区。

(5) 关闭文本文件。

使用 BufferedWriter 类进行写入是可行的，也可以在不使用缓冲区的情况下写入文本文件。但是，使用带有 BufferedWriter 和 BufferedReader 类的缓冲区可以使读写过程更高效。BufferedWriter 类首先将几个小的写入请求打包到自己的内部缓冲区中，然后在缓冲区已满或调用 flush()方法时，将整个缓冲区一次性写入文本文件。对于文件的读写，必须通过 try…catch 块来处理异常。

例3.9演示了如何使用 FileWriter 和 BufferedWriter 类将数据写入文本文件。这里使用 args[0] 参数来获取文本文件的名称，可以在其中写入数据。如果没有提供文件名，将产生错误消息并退出程序。

例 3.9 将数据写入文本文件

```
// Example 3.9 TextWrite1.java
import java.io.*;
class TextWrite1 {
    public static void main(String[] args) {
        String filename="";
        if (args.length==1){
            filename=args[0];
        }
        else {
            System.out.println("Usage: java TextWrite1 Filename.txt ");
            System.exit(0);
        }
        try {
            FileWriter file=new FileWriter(filename);
            BufferedWriter filebuff= new BufferedWriter(file);
            for (int i=0;i<3;i++){
                filebuff.write(i+"\t"+i*i+"\n");
            }
            filebuff.flush();
            file.close();
        }
        catch(IOException e) {
            System.err.println("Error -- " + e.toString() );
        }
    }
}
```

要使用 FileReader 类读取文本文件，请执行以下步骤：

(1) 创建与要读取的文本文件关联的 FileReader 对象。

(2) 创建与 FileReader 对象关联的 BufferedReader 对象。

(3) 从缓冲区读取。

(4) 清空缓冲区。

(5) 关闭文本文件。

同样，BufferedReader 类是可选的，但这个类确实可以使文本文件的读取更高效。

例3.10演示了如何使用 FileReader 和 BufferedReader 类从文本文件读取数据。

例 3.10 从文本文件读取数据

```
// Example 3.10 TextRead1.java
```

```
import java.io.*;
class TextRead1 {
    public static void main(String[] args) {
        String filename="";
        if (args.length==1){
            filename=args[0];
        }
        else {
            System.out.println("Usage: java TextRead1 Filename.txt ");
            System.exit(0);
        }
        try {
            FileReader file=new FileReader(filename);
            BufferedReader filebuff= new BufferedReader(file);
            boolean endof=false;
            String line;
            while (!endof) {
                line=filebuff.readLine();
                if (line == null){ endof=true; break;}
                System.out.println(line);
            }
            file.close();
        }
        catch(IOException e) {
            System.err.println("Error -- " + e.toString() );
        }
    }
}
```

===

3.9 方法

方法(在其他语言中也称为函数、模块或子程序)使程序员可以将程序模块化，从而降低软件的复杂性并提高可重用性。方法通常具有一些输入参数，这些参数也称为形参，用于执行特定的任务。方法中定义的变量称为局部变量，因为它们只能在该方法中使用。形参也是局部变量。

例 3.11 演示了方法的使用。例 3.11 不是直接打印出 Hello 消息，而是调用名为 printhello() 的方法来完成工作。printhello() 方法有一个 String 类型的输入参数 msg。void 关键字意味着 printhello() 方法没有返回值。static 关键字表示 printhello() 方法存在于整个类中。这是强制性的，因为在 main() 方法中只能调用静态方法。private 关键字表示只能在当前类中调用 printhello() 方法。

例 3.11 使用方法

===
```
// Example 3.11 Method example
class Method1 {
    public static void main(String[] args) {
        printhello(args[0]);
    }
    //printhello method
```

```
            private static void printhello(String msg){
                System.out.println("Hello "+ msg + "!" );
            }
        }
```

例3.12给出了方法的另一个例子。在main()方法中，首先创建Method2类对象，然后调用pupAge()方法。

在这里，pupAge()方法是以面向对象方式调用的，因而不必是静态方法。有关面向对象编程的更多详细信息，可参见3.10节。

例3.12 以面向对象的方式使用方法

```
// Example 3.12 Method example
public class Method2 {
    public void pupAge() {
        int age = 0;
        age = age + 7;
        System.out.println("Puppy age is : " + age);
    }

    public static void main(String args[]) {
        Method2 test = new Method2 ();//create an Method2 object
        test.pupAge();          //call the pupAge() method of Method2 class
    }
}
```

如果确实需要从方法返回一些值,则需要指定将返回哪种类型的值。例如,例3.13中的add()方法接收两个双精度数字并将它们相加，然后返回它们的和。

例3.13 从方法返回值

```
// Example 3.13 Method example
class Method3 {
    public static void main(String[] args) {
        double x=3, y=5;
        double s = add (x, y);
        System.out.println("The sum of x and y is: " + s);
    }
    //add() method
    private static double add(double x, double y){
        return x+y;
    }
}
```

3.10 面向对象编程

当问题很简单时，只编写一个 Java 程序或一个类就可以解决。但是随着问题越来越复杂，最好使用"关注点分离"或"分而治之"的方法编写程序，也就是以面向对象编程(Object-Oriented Programming，OOP)的方式编写多个类。面向对象编程带来的好处是显而易见的，因为它能够将大的 Java 类划分为几个较小的、自包含的且可重用的 Java 类，所以代码更易于理解和管理且不易出错。

用最简单的术语来描述，就是 Java 中的面向对象编程意味着开发或使用具有特定属性和某些行为的自包含 Java 类。属性是参数(变量或对象)，行为由方法组成(方法在其他编程语言中也称为函数)。参数用于与外部程序交换输入/输出值，方法用于执行特定任务。下面介绍一些最重要的面向对象编程概念。

3.10.1 类和对象

类是指定相同类型实例的属性和行为的模板。对象是基于类创建的实例。现实生活中的类比可以是树(类)和橡树、棕榈树、松树(对象)，以及动物(类)和狮子、老虎、狗(对象)。在面向对象编程中，问题被描述为对象之间各种交互作用的集合。反过来，这将减少问题的复杂性并提高软件代码的可重用性。

3.10.2 实例化

通过类创建对象的过程被称为实例化。对象是类的实例。

3.10.3 封装

在面向对象编程中，所有对象都是独立的。这意味着尽管对象的方法可以由外部程序调用，但它们的详细信息是隐藏的。这种方法被称为封装，封装使面向对象编程更加容易，并且不易出错。为了在 Java 中使用封装，需要将类的变量声明为 private，并提供 public setter 和 getter 方法来修改和查看变量的值。

3.10.4 继承

在面向对象编程中，不仅可以创建类，还可以使用现有类创建子类以丰富功能，这称为继承。与其他面向对象编程语言不同，Java 仅支持单一继承，这意味着子类只能从单个父类派生，此约束旨在简化面向对象编程。如果需要多重继承，那么必须通过 Java 接口才能实现，Java 接口定义了参数和方法的集合。

3.10.5 覆盖和重载

在 Java 中，当子类继承父类的方法时，子类还可以重新定义方法的内容。这种标准的面向对象编程技术称为方法的覆盖。另一个类似的概念是方法的重载，从而可以使用完全相同的名称(但使用不同的形参)来定义两个或多个函数。

3.10.6 多态性

在面向对象编程中，当调用覆盖的方法时，将始终使用与对象的类(或子类)关联的那个方法。在下面的示例中，名为 Tree 的类有一个 getLeafShape()方法，该方法在子类 Oak 和 Pine 中被覆盖。

```
Tree tree;
Oak oak=new Oak();
Pine pine=new Pine();
tree=oak;
tree.getLeafShape();
tree=pine;
tree.getLeafShape();
```

在以上示例中，有两个相同的 tree.getLeafShape()语句。第一个调用 Oak 类的 getLeafShape()方法，第二个调用 Pine 类的 getLeafShape()方法。这种功能被称为多态性，多态性提高了软件的可重用性并增强了其信息隐藏能力。

3.10.7 对象的可访问性

在类中定义对象、变量或方法时，它们的可访问性或范围分为如下四个类别：公共(public)、受保护(protected)、包(package)和私有(private)。例如，可以从任何地方访问公共变量，但是只能在同一个类中访问私有变量。默认的访问类别为 package。表 3.2 总结了类对象的可访问性。

表 3.2 类对象的可访问性

访问类别	同一个类	子类	同一个包	其他
public	是	是	是	是
protected	是	是	否	否
package	是	是	是	否
private	是	否	否	否

3.10.8 匿名内部类

在 Java 中，匿名内部类提供了一种无需名称和实际的继承类来创建对象的方式。当使用类或接口的重载方法创建对象的实例时，匿名内部类可能会很有用。

例 3.14A 包含两个类 Math1 和 Oop1。Oop1 是主类，因为其中定义了 main()方法。Oop1 类首先使用 Math1 类创建一个名为 m 的对象(使用了 new 关键字)，然后调用 setZ()和 getZ()方法以设置 Math1 类中私有变量 Z 的值，这称为封装。

还可以将类 Math1 和 Oop1 分为两个文件，一个文件名为 Math1.java，另一个文件名为 Oop1.java，从而使代码更易于管理和理解。

例 3.14A 基本的面向对象编程

```
========================================================================
// Example 3.14A Object oriented programming example
class Math1 {
    private int z;                          //encapsulation
    public void setZ(int x){                //setter method
        z=x;
    }
    public int getZ(){                      //getter method
        return z;
    }
}

class Oop1 {
    public static void main(String[] args) {
        Math1 m = new Math1();              //instantiation (Math1:
                                            //   class m:object)
        m.setZ(5);                          //call setter method
        System.out.println(m.getZ());       //call getter method
    }
}
========================================================================
```

例 3.14B 是面向对象编程的另一个示例。例 3.14B 包含两个类 Math2 和 Oop2，其中 Oop2 是主类。Oop2 类首先使用 Math2 类创建一个名为 m 的对象(使用了 new 关键字)，然后使用点表示法(如 m.add()、m.W 等)调用公共参数和方法。这与其他编程语言(如 C ++)使用->表示法有所不同。

请注意，Math2 类有两个 add 函数，这是典型的方法重载示例。Math2()方法是一种特殊的方法，称为构造方法。每次创建对象时都会自动调用构造方法，构造方法主要用于初始化参数。在 Java 中，有很多方法可以访问类中的公共参数。

例 3.14B 面向对象编程中的方法重载

```
========================================================================
// Example 3.14B Object oriented programming example
class Math2 {
    public int W;                           //Public variable
    public Math2(){                         //constructor method
    }
    public double add(double x,double y){
        return x+y;
    }
    public int add(int x,int y){            //method overloading
        return x+y;
    }
}

class Oop2 {
    public static void main(String[] args) {
        Math2 m = new Math2();
        System.out.println(m.add(3,2));     //call add method
```

```
            System.out.println(m.add(4.0,2.1));    //call add method
            System.out.println(m.W);               //get public variable W
    }
}
============================================================================
```

例 3.14C 包含三个类：Math3 类、Math3a 类和 Oop3 类，其中 Math3a 是 Math3 的子类。从父类创建子类时，子类不仅继承父类的所有公共和受保护的参数及方法，还会引入新的参数和方法(toDegree())。在本例中，func()方法已在子类中重新定义，这是典型的方法覆盖示例。

你还可以将 Math3 类、Math3a 类和 Oop3 类分成三个文件，一个名为 Math3.java，另一个名为 Math3a.java，还有一个名为 Oop3.java。它们仍将以同样的方式运行。

例 3.14C　面向对象编程中的继承

```
============================================================================
// Example 3.14C Object oriented programming example
class Math3{                            //parent class
    public Math3(){
    }
    public double func(double x){
        return Math.cos(x);
    }
}
class Math3a extends Math3{             //subclass, inheritance
    public double func(double x){       //method overriding
        return Math.sin(x);
    }
    public double toDegree(double x){   // create a new method in
                                        // subclass
        return Math.toDegrees(x);
    }
}
class Oop3 {
    public static void main(String[] args) {
        Math3 m = new Math3();
        Math3a n = new Math3a();
        System.out.println(m.func(0.0));
        System.out.println(n.func(0.0));
        System.out.println(n.toDegree(3.14));
    }
}
============================================================================
```

例 3.15A 展示了如何在面向对象编程中使用接口。这里定义了一个名为 Animal 的接口，Animal 接口定义了一个名为 speak()的方法。Dog 类实现了 Animal 接口，并且重新定义了 speak()方法。最后，我们在 main()方法中创建一个名为 p 的对象并调用了 speak()方法。

例 3.15A　在面向对象编程中使用接口

```
============================================================================
// Example 3.15A Object oriented programming example
interface Animal{
```

```
    public void speak();
  }
  class Dog implements Animal{
    public void speak(){
      System.out.println("Woof! Woof!");
    }
    public static void main(String args[]){
      Dog p = new Dog();
      p.speak();
    }
  }
==========================================================================
```

例 3.15B 给出了例 3.15A 的匿名内部类版本。如你所见,例 3.15B 创建了一个对象,覆盖了 speak()方法,因为无须命名对象或实现 Animal 接口,从而极大简化了代码。

例 3.15B 使用匿名内部类的面向对象编程

```
==========================================================================
// Example 3.15B Object oriented programming example
interface Animal{
  public void speak();
}
class Dog2 {
    public static void main(String args[]){
        //anonymous inner class
        new Animal(){
            public void speak(){
                System.out.println("Woof! Woof!");
            }
        }.speak();
    }
}
==========================================================================
```

3.11 多线程

许多应用需要同时执行多个任务。多线程是一种用来满足上述需求的强大方法。在软件编程中,线程是在程序中执行任务的独立路径。多线程意味着在一个程序中同时执行两个或多个任务或线程。与其他传统编程语言不同,Java 本身就支持多线程。线程是 Java 语言的标准部分。

例 3.16A 展示了一个简单的名为 HelloThread0.java 的多线程程序。HelloThread0 是 Java 线程类 Thread 的子类。你需要覆盖 Thread 类的 run()方法,以便指定要在线程中执行的操作。例 3.16A 中只有一个线程,用于打印一条消息。线程对象在创建后,就可以使用 start()、yield()、join()和 stop()方法来启动和保持(给其他线程一次执行的机会),等待另一个线程完成,然后停止线程。

例 3.16A 多线程编程示例

```
//Example 3.16A Multithreading programming example
public class HelloThread0 extends Thread{
   public void run(){
      System.out.println("This is a thread...");
   }
   public static void main(String[] args) {
      HelloThread0 ht1 = new HelloThread0();
      ht1.start();
   }
}
```

例 3.16B 给出了相比前一个程序稍微复杂一点的版本。例 3.16B 中包含三个线程，每个线程可打印不同的消息，以便将它们彼此区分开来。消息则通过线程的构造方法传递到线程中。运行代码时，你会发现线程可能没有按顺序显示，原因就在于它们是独立执行的。

例 3.16B 具有三个线程的多线程编程示例

```
//Example 3.16B Multithread programming example
public class HelloThread1 extends Thread{
   String message;
   //pass message into thread
   HelloThread1 ( String message ) { this.message = message; }
   public void run(){
      System.out.println( message );
   }
   public static void main(String[] args) {
      HelloThread1 ht1 = new HelloThread1("Thread 1 ...");
      HelloThread1 ht2 = new HelloThread1("Thread 2 ...");
      HelloThread1 ht3 = new HelloThread1("Thread 3 ...");
      ht1.start();
      ht2.start();
      ht3.start();
   }
}
```

尽管例 3.16A 和例 3.16B 给出了创建多线程程序的标准方法，但是 Java 的单继承限制有时使通过实现 Runnable 接口来进行多线程编程变得更加可取。你还需要覆盖类中的 run()方法，以指定要在线程中执行的操作。例 3.16C 给出了详细信息。

例 3.16C 使用了 Runnable 接口的多线程编程示例

```
//Example 3.16C Multithread programming example
public class HelloThread2 implements Runnable {
   String message;
   HelloThread2 ( String message ) { this.message = message; }
```

```
    public void run(){
        System.out.println( message );
    }
    public static void main(String[] args) {
        HelloThread2 ht1 = new HelloThread2("Thread 1 ...");
        HelloThread2 ht2 = new HelloThread2("Thread 2 ...");
        HelloThread2 ht3 = new HelloThread2("Thread 3 ...");
        Thread t1 = new Thread(ht1);
        Thread t2 = new Thread(ht2);
        Thread t3 = new Thread(ht3);
        t1.start();
        t2.start();
        t3.start();
    }
}
==========================================================================
```

例 3.16D 给出了另一种简单且优雅地通过使用 Java 匿名内部类在 Java 中实现多线程编程的方式,你可以创建线程对象、定义 run()方法并一次性启动线程。有关 Java 匿名内部类的更多详细信息,请参阅前面有关面向对象编程的讨论。

例 3.16D 使用了匿名内部类的多线程编程示例

```
==========================================================================
//Example 3.16D Multithread programming example
public class HelloThread3{
    public static void main(String[] args) {
        new Thread(new Runnable() {
            public void run() {
                System.out.println("Thread ...");
            }
        }).start();
    }
}
==========================================================================
```

例 3.16E 和例 3.16F 展示了用于实现多线程编程的面向对象方式。其中,例 3.16E 创建了一个名为 Hello4 的单独类,该类实现了 Runnable 接口。

例 3.16E 面向对象的多线程编程示例 1

```
==========================================================================
//Example 3.16E Multithread programming example
public class HelloThread4 {
    public static void main(String[] args) {
        Hello4 ht1 = new Hello4("Thread 1 ...");
        Hello4 ht2 = new Hello4("Thread 2 ...");
        Hello4 ht3 = new Hello4("Thread 3 ...");
        Thread t1 = new Thread(ht1);
        Thread t2 = new Thread(ht2);
        Thread t3 = new Thread(ht3);
        t1.start();
        t2.start();
```

```
        t3.start();
    }
}
class Hello4 implements Runnable {
    String message;
    Hello4 ( String message ) { this.message = message; }
    public void run(){
        System.out.println( message );
    }
}
==========================================================================
```

例 3.16F 则创建了一个名为 Hello5 的单独类，该类是 Thread 类的子类。

例 3.16F 面向对象的多线程编程示例 2

```
==========================================================================
//Example 3.16F Multithread programming example
public class HelloThread5{
    public static void main(String[] args) {
        Hello5 ht1 = new Hello5("Thread 1 ...");
        Hello5 ht2 = new Hello5("Thread 2 ...");
        Hello5 ht3 = new Hello5("Thread 3 ...");
        ht1.start();
        ht2.start();
        ht3.start();
    }
}
class Hello5 extends Thread{
    String message;
    Hello5 ( String message ) { this.message = message; }
    public void run(){
        System.out.println( message );
    }
}
==========================================================================
```

例 3.16G 展示了一个更复杂的多线程程序，用于在线程中执行一些计算。例 3.16G 基于前面的面向对象的多线程编程示例，图 3.5 给出了编译、执行和输出结果。

例 3.16G 一个更复杂的多线程程序

```
==========================================================================
//Example 3.16G Multithread programming example
public class Thread1 {
    public static void main( String args[] )
    {
        CalcThread thread1, thread2;
        thread1 = new CalcThread( "Calculation1" );
        thread2 = new CalcThread( "Calculation2" );
        thread1.start();
        thread2.start();
    }
}
```

```
class CalcThread extends Thread {
    private int num,sum=0;
    public CalcThread( String name )
    {
        super( name );
        // generate a random number between 0 and 100
        num = (int) ( Math.random() * 100 );
    }

    // execute the thread
    public void run(){
        System.err.println( getName() + " started!" );
        for (int i=0;i<num;i++){
            sum+=i;
        }
        System.err.println( getName() +": sum (0 to "+num+"): " + sum );
        System.err.println( getName() + " done!" );
    }
}
```
==

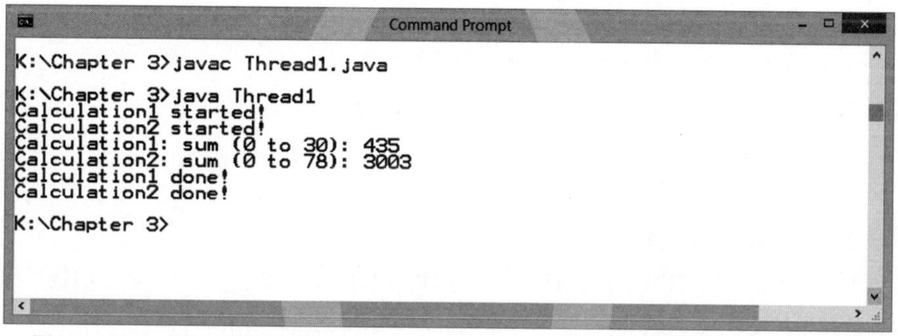

```
K:\Chapter 3>javac Thread1.java

K:\Chapter 3>java Thread1
Calculation1 started!
Calculation2 started!
Calculation1: sum (0 to 30): 435
Calculation2: sum (0 to 78): 3003
Calculation1 done!
Calculation2 done!

K:\Chapter 3>
```

图 3.5　Thread1.java 的编译、执行和输出结果。注意，你的输出结果可能与这里的不同，因为线程可以按随机顺序执行

例 3.16H 是例 3.16A 的变体，这里不是从 Thread 类继承，而是实现了 Runnable 接口。

例 3.16H　同样使用了 Runnable 接口的多线程编程示例

==
```
//Example 3.16H Multithread programming example
public class Thread2 {
    public static void main( String args[] )
    {
        Runnable calc= new CalcThread();
        Thread thread1 = new Thread( calc, "Calculation1" );
        thread1.start();
    }
}

class CalcThread implements Runnable {
    // execute the thread
    public void run(){
```

```
            int num,sum=0;
            System.err.println( "started!" );
            num = (int) ( Math.random() * 100 );
            for (int i=0;i<num;i++){
                sum+=1;
            }
            System.err.println( "sum (0 to "+num+"): " + sum );
            System.err.println( "done!" );
        }
    }
=========================================================================
```

3.11.1 线程的生命周期

线程可以处于几种已定义状态之一。当创建线程时，线程处于新生状态。当调用线程的 start() 方法时，线程进入就绪状态。当线程开始执行时，线程将进入运行状态。正在运行的线程可以进入等待状态、休眠状态或阻塞状态，并在终止时进入死亡状态。可使用 getState() 方法来确定线程的状态。

Java 线程的可能状态如下：

- 新生(NEW)
- 可运行(RUNNABLE)
- 运行(RUNNING)
- 不可运行(NONRUNNABLE)
- 终止(TERMINATED)

例 3.17A 演示了如何使用 getState() 方法获取线程的状态。这与例 3.16A 所示的线程相同，后者只是打印一条消息。在创建了线程对象之后，显示线程启动前后的状态。等待一秒后，再次显示状态。延迟是通过使用 Thread.sleep() 方法实现的；请注意，这个方法必须放在 try…catch 块中。有关 Thread.sleep() 方法的更多详细信息，可参考 3.12 节。

例 3.17A　使用了 getState() 方法的多线程编程示例

```
=========================================================================
//Example 3.17A Multithread programming example
class StateThread1 extends Thread{
    public void run(){
        System.out.println( "This is a thread..." );
    }
    public static void main(String[] args) {
        StateThread1 st1 = new StateThread1 ();
        System.out.println("thread before start(): "+ st1.getState());
//states before start()
        st1.start();
        System.out.println("thread after start(): "+ st1.getState());
//states after start()

        try {
//sleep for 1000ms
```

```
        Thread.sleep(1000);
    } catch (InterruptedException e) {
        e.printStackTrace();
    }
        System.out.println("thread after 1s: "+ st1.getState());
//states after 1s
    }
}
================================================================================
```

例 3.17A 的输出结果如下,从中可以看到线程的不同状态,在经过一秒后,线程将终止。

```
thread before start(): NEW
This is a thread...
thread after start(): RUNNABLE
thread after 1s: TERMINATED
```

例 3.17B 是例 3.17A 的改进版本。在例 3.17B 中,我们在线程中添加了 5 秒的等待时间。此外,我们通过使用放置在 try…catch 块中的 Thread.sleep(1000)方法来实现等待。在创建了线程对象之后,显示线程启动前后的状态。然后等待一秒的时间,再次显示状态。

例 3.17B 使用了 try…catch 块的多线程编程示例

```
================================================================================
//Example 3.17B Multithread programming example
public class StateThread2 extends Thread{
    public void run(){
      System.out.println( "This is a thread..." );
        try {
//sleep for 5s
            Thread.sleep(5000);
        } catch (InterruptedException e) {
            e.printStackTrace();
        }
    }
    public static void main(String[] args) {
        StateThread2 st = new StateThread2 ();
        System.out.println("thread before start(): "+ st.getState());
//states before start()
        st.start();
        System.out.println("thread after start(): "+ st.getState());
//states after start()

        try {
//sleep for 1s
        Thread.sleep(1000);
    } catch (InterruptedException e) {
        e.printStackTrace();
    }
        System.out.println("thread after 1s: "+ st.getState());
//states after 1s
    }
}
================================================================================
```

例 3.17B 的输出结果如下。如你所见，由于线程有 5 秒时间的延迟，因此 1 秒后线程处于 TIMED_WAITING 状态。

```
thread before start(): NEW
This is a thread...
thread after start(): RUNNABLE
thread after 1s: TIMED_WAITING
```

3.11.2 线程的优先级

不同的线程可以有不同的优先级。可以使用 setPriority()方法设置线程的优先级，范围为 1~10，其中 1 是 Thread.MIN _PRIORITY，10 是 Thread.MAX _ PRIORITY，5 是 Thread.NORM_PRIORITY。

例 3.17C 使用 setPriority()方法设置每个线程的优先级。如果运行代码，你将发现优先级较高的线程往往会先执行。

例 3.17C　设置线程的优先级

```
//Example 3.17C Thread Priority example
public class ThreadPriority1 extends Thread{
    String message;
    ThreadPriority1 ( String message ) { this.message = message; }
    public void run(){
        System.out.println( message );
    }
    public static void main(String[] args) {
        ThreadPriority1 ht1 = new ThreadPriority1("Thread 1 ...");
        ThreadPriority1 ht2 = new ThreadPriority1("Thread 2 ...");
        ThreadPriority1 ht3 = new ThreadPriority1("Thread 3 ...");
        ThreadPriority1 ht4 = new ThreadPriority1("Thread 4 ...");
        ThreadPriority1 ht5 = new ThreadPriority1("Thread 5 ...");
        ht1.setPriority(2);
        ht2.setPriority(4);
        ht3.setPriority(6);
        ht4.setPriority(8);
        ht5.setPriority(10);

        ht1.start();
        ht2.start();
        ht3.start();
        ht4.start();
        ht5.start();
    }
}
```

3.11.3 线程调度

许多 Java 平台使用时间分片来执行线程。通过时间分片，CPU 时间被分成一系列称为量子的小时隙。具有相同优先级的线程将以循环方式执行。具有较高优先级的线程将在具有较低优先级的线程之前执行。

3.11.4 线程同步

当多个线程在同一个对象上运行并且具有同步方法时，这些线程也需要同步，这意味着一次只允许一个线程在该对象上执行同步方法。可通过在调用同步方法时锁定对象来满足上述要求。

例 3.17D 演示了如何实现简单的线程同步。其中，Test 类定义了 printMessage()方法，该方法每次在被调用时将打印一条消息并等待 500 毫秒。main()方法中有两个线程，并且这两个线程都需要访问相同的 Test 对象。如果没有同步，你将看到这两个线程都可以访问 Test 对象。

现在取消对 synchronized void printMessage(String txt)代码行的注释，并注释掉 void printMessage(String txt)代码行。通过在 printMessage()方法的前面添加 synchronized 关键字，可以使这个方法变为同步的。这意味着当一个线程访问 Test 对象时，Test 对象将被锁定；直到第一个线程完成后，第二个线程才能访问 Test 对象。

例 3.17D 线程同步

```
================================================================
//Example 3.17D Thread Synchronization example
class Test{
  void printMessage(String txt){//method not synchronized
  //synchronized void printMessage(String txt){//method synchronized
    for(int i=1;i<=5;i++){
       System.out.println(txt);
       try{
           Thread.sleep(500);
       }catch(Exception e){System.out.println(e);}
    }
  }
}

class ThreadSyn1 extends Thread{
    String message;
      Test t;
    ThreadSyn1 ( Test t, String message ) {
       this.t=t;
       this.message = message;
       }
    public void run(){
       t.printMessage( message );
    }
    public static void main(String args[]){
       Test m = new Test();
       ThreadSyn1 t1=new ThreadSyn1(m,"A");
       ThreadSyn1 t2=new ThreadSyn1(m,"B");
       t1.start();
       t2.start();
    }
}
================================================================
```

3.12 日期、时间、计时器和睡眠方法

程序中经常需要日期、时间和精确的计时器。在 Java 中，可以使用 new Date().to string() 方法获取日期和时间字符串，也可以使用 System.currentTimeMillis()或 new Date().getTime()方法获取当前时间(以 1970 年 1 月 1 日 UTC 午夜之后的毫秒为单位)，还可以使用 System.nanoTime() 方法以纳秒为单位获取最精确的系统时间。

例3.18A 演示了如何使用 Date 类来获取日期和时间信息。图 3.6 展示了输出结果。

例 3.18A 使用 Date 类获取日期和时间

```
============================================================
//Example 3.18A Date and Time Example
import java.util.*;
public class DateExample {
    public static void main(String[] args) throws InterruptedException
    {
        //Get the current date and time
        Date date = new Date();
        System.out.println("The current Date and Time is: " + date.toString());
        //Get the current time in milliseconds
        long tm = date.getTime();
        System.out.println("The current time in milliseconds : " + tm);
    }
}
============================================================
```

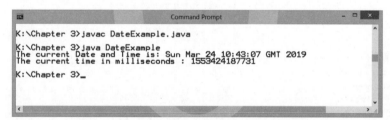

图 3.6 DateExample.java 的编译、执行和输出结果

LocalDateTime 类属于 java.time.*包，这个包提供了一些十分有用的日期和时间函数，例如 getYear()、getMonth()、getDayofMonth()和 getDayofWeek()，你还可以使用特定格式显示日期和时间。

例 3.18B 演示了如何使用 LocalDateTime 类获取日期和时间信息。图 3.7 给出了输出结果。

例 3.18B 使用 LocalDateTime 类获取日期和时间

```
============================================================
//Example 3.18B Date and Time
import java.time.LocalDateTime;          // Import the LocalDateTimeclass
import java.time.format.DateTimeFormatter;  // Import the
                                            // DateTimeFormatter class
```

```java
public class DateExample2 {
  public static void main(String[] args) {
    LocalDateTime dt = LocalDateTime.now();
    System.out.println("Year:              " + dt.getYear());
    System.out.println("Month:             " + dt.getMonth());
    System.out.println("Day:               " + dt.getDayOfMonth());
    System.out.println("Day of the week:   " + dt.getDayOfWeek());
    System.out.println("Unformatted:       " + dt);

    DateTimeFormatter f = DateTimeFormatter.ofPattern("dd-mm-yyyy hh:mm:ss");
    System.out.println("Formatted:         " + dt.format(f));
  }
}
```

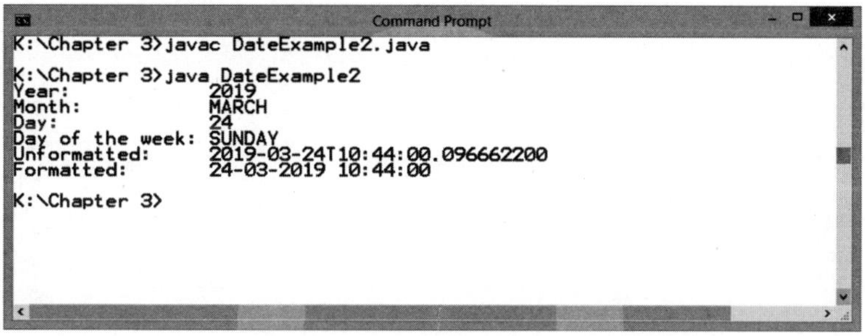

==

图 3.7 DateExample2.java 的编译、执行和输出结果

Timer 类来自 java.util 包,这个包提供了最精确的计时器函数,可以使用 schedule()方法设置计时器。想要在计时器中运行的任何类都必须是 TimerEvent 类的子类,并且必须重写 run()方法。

例 3.19A 所示的计时器程序每 1000 毫秒(1 秒)运行一次。可使用 System.currentTimeMillis() 方法来获取当前时间(以毫秒为单位)。从图 3.8 所示的输出中可以看到,计时器是准确的,误差约为 1~2 ms。

例 3.19A 一个简单的计时器程序

==

```java
//Example 3.19A Timer example
import java.util.*;
import java.awt.Toolkit;

public class Timer1 {
    long t1,t2;
    Toolkit toolkit;
    Timer timer;
    public Timer1() {
       t1=System.currentTimeMillis();
      toolkit = Toolkit.getDefaultToolkit();
       timer = new Timer();
       timer.schedule(new RemindTask(),
            0, //initial delay
```

```
                            1*1000); //subsequent rate
    }
    class RemindTask extends TimerTask {
       int i=0;
       public void run() {
         if (i <5) {
           toolkit.beep();
             t2=System.currentTimeMillis();
         t2=(t2-t1);
             System.out.println(i+ ": "+t2*0.001+" s");
           i++;
           } else {
                  System.out.println(new GregorianCalendar().getTime());
               System.exit(0);
           }
         }
       }
       public static void main(String[] args) {
         new Timer1();
       }
}
================================================================================
```

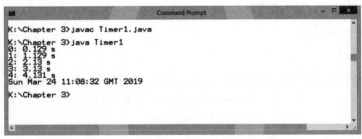

图 3.8 Timer1.java 的编译、执行和输出结果

然而，很多人更喜欢使用 ScheduledExecutorService 代替 Timer，因为前者更灵活，从而不必担心计时器阻塞程序的其余部分，并且可以更好地处理异常。例 3.19B 展示的 ScheduledExecutorService 示例将重复执行打印任务，从头开始，每 10 毫秒重复一次。

例 3.19B 使用 ScheduledExecutorService

```
================================================================================
//Example 3.19B ScheduledExecutorService example
import java.util.concurrent.*;
public class SEService {
    public static void main(String[] args) {
        ScheduledExecutorService ses
                    = Executors.newScheduledThreadPool(5);
        ses.scheduleAtFixedRate(()->{
            //do the task repeatedly, start at 0s with a gap of 100ms
            System.out.println("Time: "+ new java.util.Date());
        }, 0, 100L, TimeUnit.MILLISECONDS);
    }
}
================================================================================
```

例 3.19C 是例 3.19B 的另一个版本。在例 3.19C 中，任务被单独定义，并使用 count 变量计算任务执行了多少次。首先等待 5 秒，然后每一秒重复执行一次任务。除了重复执行任务之外，还可以只执行一次任务。为此，只需要取消注释最后两行代码并注释掉 ses.scheduleAtFixedRate()代码行即可。

例 3.19C 使用带有 count 变量的 ScheduledExecutorService

```
========================================================================
//Example 3.19C ScheduledExecutorService example
import java.util.concurrent.*;
public class SEService2 {
    private static int count = 0;
    public static void main(String[] args) {
        ScheduledExecutorService ses = Executors.newScheduledThreadPool(1);
        //define the task
        Runnable task1 = () -> {
            System.out.println("Running task1 " + count);
            count++;
        };
        //run this task repeatedly after 5 seconds
        ses.scheduleAtFixedRate(task1, 5, 1, TimeUnit.SECONDS);

        //run this task once after 5 seconds
        //ses.schedule(task1, 5, TimeUnit.SECONDS);
        //ses.shutdown();
    }
}
========================================================================
```

延迟或睡眠在许多应用中也是十分有用的功能。在 Java 中，可以使用 Thread.sleep()类来实现延迟。例 3.20 将首先打印一条消息，延迟 5000 毫秒(5 秒)，然后打印另一条消息。图 3.9 给出了输出结果。

例 3.20 实现睡眠或延迟

```
========================================================================
//Example 3.20 Sleep or delay Example
public class SleepExample {
    public static void main(String[] args)throws InterruptedException
    {
        System.out.println("Start to sleep");
        Thread.sleep(5000);
        System.out.println("Finish");
    }
}
========================================================================
```

例 3.21 是例 3.20 的变体。通过结合使用 sleep()与 System.currentTimeMillis()方法，可以显示程序延迟或休眠的毫秒数。

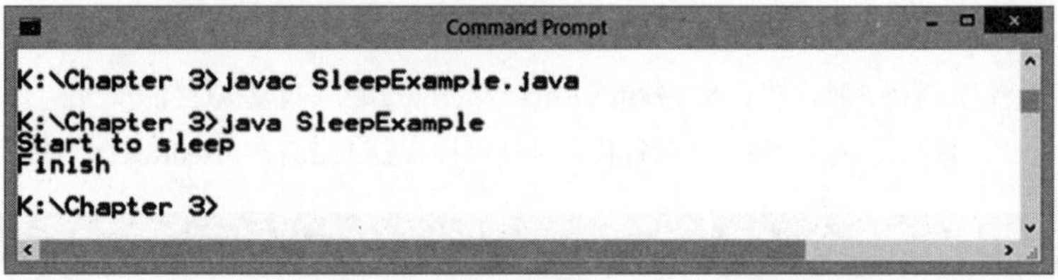

图 3.9　SleepExample.java 的编译、执行和输出结果

例 3.21　将 sleep()与 System.currentTimeMillis()方法结合使用

```
================================================================
//Example 3.21 Sleep or delay Example
public class SleepExample2 {
   public static void main(String[] args) throws InterruptedException
   {
       long start = System.currentTimeMillis();
       Thread.sleep(2000);
       System.out.println("Sleep time in ms = "+(System.currentTimeMillis()-start));
   }
}
================================================================
```

3.13　执行系统命令

在 Java 中，可以在操作系统中运行其他应用，这使你能够从 Java 程序中运行或调用系统程序和命令行程序。

例 3.22 演示了如何使用 Runtime.getRuntime().exec()来获取和执行 Windows 系统中的 Notepad.exe 程序。图 3.10 给出了输出结果。

例 3.22　使用 Runtime.getRuntime().exec()来获取和执行 Notepad.exe 程序

```
================================================================
//Example 3.22 Execute1.java program
import java.util.*;
import java.util.regex.*;

public class Execute1 {
   public static void main(String args[]) {
      try {
         String line;
         boolean more = false;
         String[] cmd = {"notepad.exe"};
         Process p = Runtime.getRuntime().exec(cmd);
      }
      catch (Exception e) {
         e.printStackTrace();
      }
   }
```

}
==

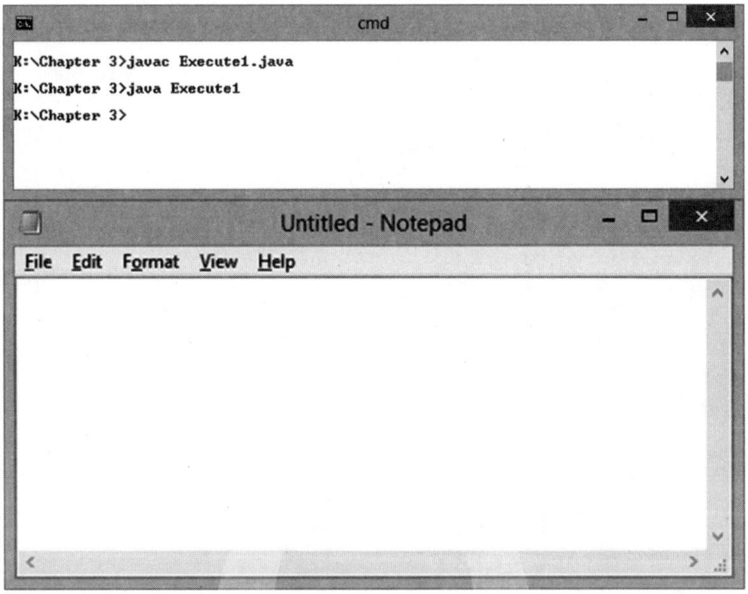

图 3.10 Execute1.java 的编译、执行和输出结果

例 3.23 演示了如何使用 Runtime.getRuntime().exec() 来执行 Windows 命令行命令 dir C:\\。cmd.exe 是 Windows 控制台(或终端)命令，/C 是 cmd.exe 的标志，表示 Windows 控制台应执行你用字符串指定的命令，然后终止。输出结果如图 3.11 所示。

例 3.23   使用 Runtime.getRuntime().exec() 执行 Windows 命令行命令

==
```java
//Example 3.23 Execute2.java program
import java.util.*;
import java.util.regex.*;
import java.io.*;
public class Execute2 {
    public static void main(String args[]) {
        try {
            String line;
            boolean more = false;
            String[] cmd = {"cmd.exe", "/c", "dir","c:\\"};
            Process p = Runtime.getRuntime().exec(cmd);
            BufferedReader input = new BufferedReader(new
                InputStreamReader(p.getInputStream()));
            while ((line = input.readLine()) != null) {
                System.out.println(line);
            }
            input.close();
        }
        catch (Exception e) {
            e.printStackTrace();
```

            }
        }
}
==

图 3.11　Execute2.java 的编译、执行和输出结果

3.14　大规模的软件包和编程

对于简单、小规模的 Java 应用，只需要编写一些 Java 程序并将它们放在 save 目录中，而不必担心目录结构。但是对于大型应用，关键的部分是创建可重用的软件组件。因此，目录结构变得非常重要。Java 软件包提供了一种将可重用的 Java 类和接口安排到不同目录结构中的机制。

例 3.24A 和例 3.24B 演示了如何在 Java 软件包中创建一个可重用的 Java 类以及如何使用它。可重用的这个 Java 类名为 SpecialFunctions.Java(详见例 3.24A)。用于测试的程序名为 SFTest1.java(详见例 3.24B)。为了在 Java 软件包中创建可重用的 Java 类，需要执行以下两个步骤。

(1) 定义一个公共类，如下所示：

```
public class SpecialFunctions
```

(2) 定义包名，如下所示：

```
package biz.biox
```

包名指定了这个可重用的 Java 类必须属于的子目录。在本例中，如果当前目录是 H:\ProjectA\，则 SpecialFunctions.java 必须位于 H:\ProjectA\biz\biox\目录中，而 SFTest1.java 必须位于 H:\ProjectA\目录中，如下所示：

```
H:\ProjectA\
|----------- SFTest1.java
|---------->\biz\
|-------------->\biox\
|-------------------- SpecialFunctions.java
```

例 3.24A SpecialFunctions.java

```
==========================================================================
// Example 3.24A SpecialFunctions.java
package biz.biox;
import java.util.*;
public class SpecialFunctions{

    public double ExpErfc(double z){
        double w;
        if (z<0.0){
            w=2*Math.exp(-z*z)-(0.3480242*erf1(-z)-0.0958798*erf1
              (-z)*erf1(-z)+0.7478556*erf1(-z)*erf1(-z)*erf1(-z));
        }
        else{
            w=(0.3480242*erf1(z)-0.0958798*erf1(z)*erf1(z)+0.7478556*erf1
              (z)*erf1(z)*erf1(z));
        }
        return w;
    }

    public double erf(double z){
        return (1.0-erfc(z));
    }

    public double erfc(double z){
        double w;
        if (z<0.0){
            w=2.0-(0.3480242*erf1(-z)-0.0958798*erf1(-z)*erf1(-z)
              +0.7478556*erf1(-z)*erf1(-z)*erf1(-z))*Math.exp(-z*z);
        }
        else{
            w=(0.3480242*erf1(z)-0.0958798*erf1(z)*erf1(z)+0.7478556*erf1(z)
              *erf1(z)*erf1(z))*Math.exp(-z*z);
        }
        return w;
    }
    private double erf1(double z){
        return 1.0/(1.0+0.47047*z);
    }
}
==========================================================================
```

例 3.24B SFTest1.java

```
==========================================================================
// Example 3.24B SFTest1.java Program
import java.text.*;
```

```
import biz.biox.SpecialFunctions;

class SFTest1{
    public static void main(String args[]){
        double x;
        SpecialFunctions sf=new SpecialFunctions();
        System.out.println("Number \t Erf \t Erfc \t ExpErfc");
        for (int i=0;i<10;i++){
            x=(i-5)/10.0;
            System.out.println(x+"\t "+sf.erf(x)+"\t "+sf.erfc(x)+"\t"
                +sf.ExpErfc(x));
        }
    }
}
```
==

要从当前目录 H:\ProjectA 编译 SpecialFunctions.java，请执行以下命令，输出如图 3.12 所示：

```
H:\ProjectA> javac .\biz\biox\SpecialFunctions.java
```

图 3.12　编译 biz.biox 包中的 SpecialFunctions.java

例 3.24B 演示了如何使用 biz.biox 包中的 SpecialFunctions.java 类。确保在 SFTest1.java 文件的开头添加以下代码行：

```
import biz.biox.SpecialFunctions;
```

执行以下命令，编译并执行 SFTest1.java 程序，输出结果如图 3.13 所示。

```
H:\ProjectA> javac SFTest1.java
H:\ProjectA> java SFTest1
```

图 3.13　SFTest1.java 的编译、执行和输出结果

3.15 软件工程

编写软件就像制造产品一样,质量应放在第一位,但如何保证软件的质量呢?答案是通过软件工程。这个概念意味着应该使用系统的工程方式开发软件,并以良好的编程风格编写代码。

使用系统的工程方式意味着不应仅在想编写程序时才开始输入代码。相反,应该始终遵循软件的开发周期。也就是说,应该先设计后编写代码。

良好的编程风格意味着软件的源代码应结构良好,并带有正确的缩进和足够的注释,还意味着应该为类、方法和变量赋予有意义的名称。以良好的编程风格编写程序将使程序易于管理、修改和调试,并且不易出错。

3.15.1 软件的开发周期

要开发高质量的软件,就必须遵循如下标准的开发周期。

步骤 1:规范。

步骤 2:设计。

步骤 3:实施(或编码)。

步骤 4:测试和调试。

步骤 5:编写文档。

步骤 6:返回到步骤 1,获取软件的下一版本。

在软件开发中,无论怎样强调规范和设计步骤的重要性都不为过。在开始编码之前,始终应该先进行正确的软件规范和软件设计。规范指定了软件将要做什么,并且软件应该简单、准确、完整且明确。设计指定了软件将如何完成要做的事情。最常用的设计工具是流程图和伪代码(或结构化语言)。

例如,例 3.24B 中的 Java 代码可以表示为以下伪代码。伪代码不需要具有预定义的语法,但应该使用纯英语和泛型编程结构。它们还应该是通用的,并且不特定于任何语言。

```
Begin
    Define double x
    Create a SpecialFunctions object sf
    Print the heading: "Number Erf Erfc ExpErfc"
    For each integer i in 1 to 10:
        x=(i-5)/10.0;
        Print x, sf.erf(x), sf.erfc(x), sf.ExpErfc(x)
    End of loop
End
```

有关流程图和伪代码的更多详细信息,请访问以下网站:

https://www.tutorialspoint.com/programming_methodologies/programming_methodologies_flowchart_elements.htm
https://www.programiz.com/article/flowchart-programming
https://www.go4expert.com/articles/pseudocode-tutorial-basics-t25593/

软件在被开发出来之后，文档往往会被忽略。但是，编写文档也是一个十分重要的步骤，因为文档有助于他人理解软件，并且让软件易于维护。幸运的是，Java 提供了名为 javadoc 的工具，从而可以直接从 Java 代码生成 Java 文档。我们使用 javadoc 生成的文档是 HTML 格式的，因此可以很容易地放在网站上。接下来我们介绍良好编码风格的一些因素，也就是文档形式，如缩进、注释和命名约定，它们有助于提高代码的可读性和维护性。

3.15.2 缩进

尽管缩进对程序的编译和执行没有影响，但是缩进有助于理解代码。例 3.25A 演示了带有正确缩进的 Java 程序(与例 3.6 中的 Hello1.java 相同)，例 3.25B 演示了缩进错误或没有缩进的情况。可以想象那些具有数百行代码和嵌套循环的程序，如果没有适当的缩进，将很难理解程序的结构。因此，强烈建议在编写程序时，无论大小，都应始终使用缩进来显示代码的逻辑结构。

例 3.25A 具有正确缩进的 Java 代码

```
// Example 3.25a Java code with correct indentation
class Hello1a {
    //indentation inside the class
    public static void main(String[] args) {
        //indentation inside the main()
        if(args.length!=2){
            //indentation inside if else
            System.out.println("Usage: java Hello1a firstname surname!");
        }
        else{
            //indentation inside if else
            System.out.println("Hello "+ args[0] +" "+ args[1] + "!");
        }//end of if else
    }//end of main
}//end of class
```

例 3.25B 缩进错误或没有缩进的 Java 代码

```
// Example 3.25B Java code with poor or no indentation
class Hello1b {
public static void main(String[] args) {
if(args.length!=2){
System.out.println("Usage: java Hello3 firstname surname!");
}
else{
System.out.println("Hello "+ args[0] +" "+ args[1] + "!");
}
}
}
```

3.15.3 注释

像缩进一样，注释在源代码中也很重要，即使它们不影响程序的编译和执行。足够的注释可以使程序更易于理解，尤其是对于其他人而言，这在团队项目中至关重要。像 C/C++一样，Java 支持单行注释和多行注释。例 3.26 再次给出了 Hello1.java 程序，其中带有注释。

例 3.26 带有注释的 Java 代码

```
========================================================================
//Example 3.26 Java code with comments

/**
 * Program: Comment1.java
 * Author: Perry Xiao
 *
 */
class Comment1 {
    /*
    This is the main method of the program
    */
    public static void main(String[] args) {
        //if argument length is not two, display usage information
        if(args.length!=2){
            System.out.println("Usage: java Hello3 firstname surname!");
        }
        //if argument length is two, display hello message
        else{
            System.out.println("Hello "+ args[0] +" "+ args[1] + "!");
        }
    }
}
========================================================================
```

请注意，对于多行注释，以下两种方式之间没有区别：

```
/*
 This is a multiple line
 comment
 */
```

和

```
/**
 * This is a multiple line
 * comment
 */
```

但是，由于 javadoc 工具可以识别以/**开头的注释，因此我们建议对类和方法使用/**和*/注释，因为它们将被 javadoc 工具自动生成程序文档。可以在附录 A 中找到有关 Java 文档和 javadoc 工具的更多信息。

3.15.4 命名约定

为类、方法和变量提供合理的名称是很重要的，特别是当它们有成百上千个的时候。许多编程语言中最流行的命名约定是匈牙利命名法，可简单地描述为任何变量名都必须具有前缀和后缀。前缀应标识变量的类型，后缀应提供变量的含义。通常前缀是小写形式，后缀是一串单词，单词之间没有空格，每个单词的第一个字母大写。例如，可以将存储环境温度的 double 型变量命名为以下任意一种：

```
doubleAmbientTemperature
dblAmbientTemperature
dAmbientT
```

以下是变量命名的更多示例：

```
trPlotTitle        //A String variable that stores the plot title
bolDeviceOn        //A boolean variable that stores a true or false value for
                   //representing the on/off status of a device
intNumberOfPoints  //An integer variable that stores the total number of
                   //points
```

对于类和方法，最好遵循 Java 命名约定。Java 类名应该是不带空格的单词的集合，每个单词的首字母大写；Java 方法名也应该是不带空格的单词的集合，除了第一个单词的首字母以外，其他单词的首字母大写。下面是一些例子：

```
getCurrentDirectory()    //A method that returns the current directory
setCurrentDirectory()    //A method that sets the current directory
toString()               //A method that returns a string representing the object
FileReader()             //A class that reads character files
FileWriter()             //A class that writes character files
```

你不需要遵循匈牙利表示法，因为 Java 已经被强类型化。但是，无论选择使用什么命名约定，都应该在整个程序中保持一致。

有关 Java 命名约定的更多详细信息，请访问以下网站：

```
https://www.oracle.com/technetwork/java/codeconventions-135099.html
http://gee.cs.oswego.edu/dl/html/javaCodingStd.html
https://google.github.io/styleguide/javaguide.html
```

3.16 部署 Java 应用

对于 Linux/UNIX 用户，到目前为止，通过文本模式的终端或控制台编译和运行 Java 程序似乎没有问题；但是对于习惯了图形界面和鼠标的 Windows 用户，打开文本模式的 Windows 命令提示符，通过输入 java HelloWorld 之类的命令来运行 Java 程序并不是很有吸引力。因此，本节将介绍创建可执行 Java 程序的几种不同方法，只需要双击文件即可运行。然后，你将看到如何部署 Java 应用。

3.16.1 使用 Windows 批处理文件

编译和运行 Java 代码的最简单方法是使用 Windows 批处理文件。可以使用 Notepad++或任何其他文本编辑器来创建批处理文件。图 3.14 展示了如何创建名为 oop1run.bat 的 Windows 批处理文件以编译并运行以前的 Oop1.java 程序。其中包含三行命令：第一行用于编译代码；第二行用于执行代码；第三行是 stop 命令，用于保持屏幕以便观察结果。批处理文件创建后，只需要双击即可编译并运行程序，如图 3.15 所示。这种方法简单、快速且有效。

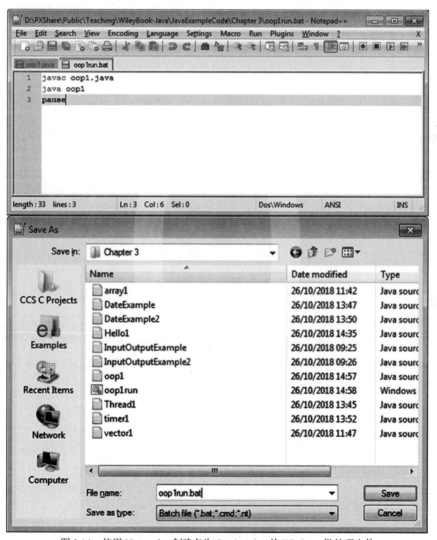

图 3.14　使用 Notepad++创建名为 Oop1run.bat 的 Windows 批处理文件

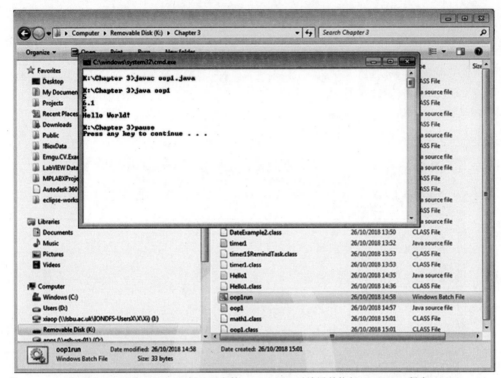

图 3.15　双击 Windows 批处理文件 oop1run.bat 以编译并执行 Oop1.java 程序

3.16.2　使用可执行的 JAR 文件

Java 技术支持可执行的 JAR 文件这一概念，这意味着可以创建 JAR 文件(Java 归档文件，相当于压缩文件)，JAR 文件可以在双击时运行。

要想创建可执行的 JAR 文件，必须首先创建 mainClass 清单文件，其中包含两行内容，用于指定 mainClass 在 JAR 文件中的位置。在本例中，mainClass 文件名为 Oop1.mf，其中的内容如下：

```
Main-Class: Oop1
Class-Path: Oop1.jar
```

现在，可以通过在 Windows 命令提示符窗口中输入以下命令来创建可执行的 JAR 文件：

```
jar cmf Oop1.mf Oop1.jar *.class
```

以上命令将创建一个名为 Oop1.JAR 的 JAR 文件。

要想执行这个 JAR 文件，只需要输入以下命令：

```
java -jar Oop1.jar
```

也可以通过双击 Oop1.jar 文件来执行。图 3.16 给出了 Oop1.mf 文件和相应的 Oop1.jar 文件的内容。

第 3 章 基本的 Java 编程 83

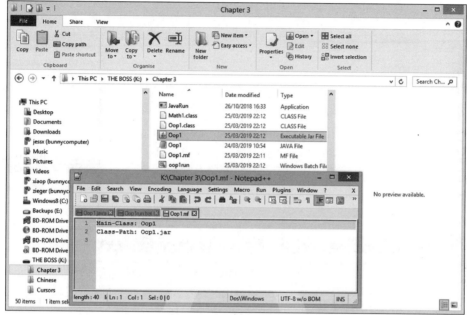

图 3.16　Oop1.mf 文件和相应的 Oop1.jar 文件的内容

3.16.3　使用 Microsoft Visual Studio

你可以使用 Microsoft Visual Studio 编写一个简单的 Windows 命令提示符程序，以使用系统调用来编译和运行 Java 程序。Microsoft Visual Studio 是业界流行的软件开发工具，可用于轻松地创建安装包。如图 3.17 所示，可创建 C#程序，通过使用 System.Diagnostic.Process.Start()执行 javac Oop1.java 命令来编译 Java 程序，并通过执行 java Oop1 命令来运行 Java 程序。

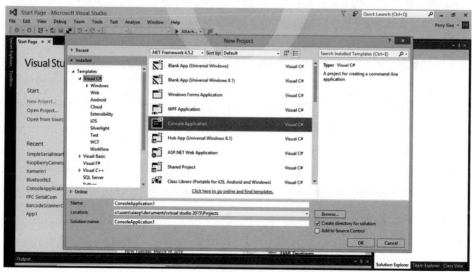

图 3.17　C#程序可通过执行 Java 命令来编译和运行 Java 程序

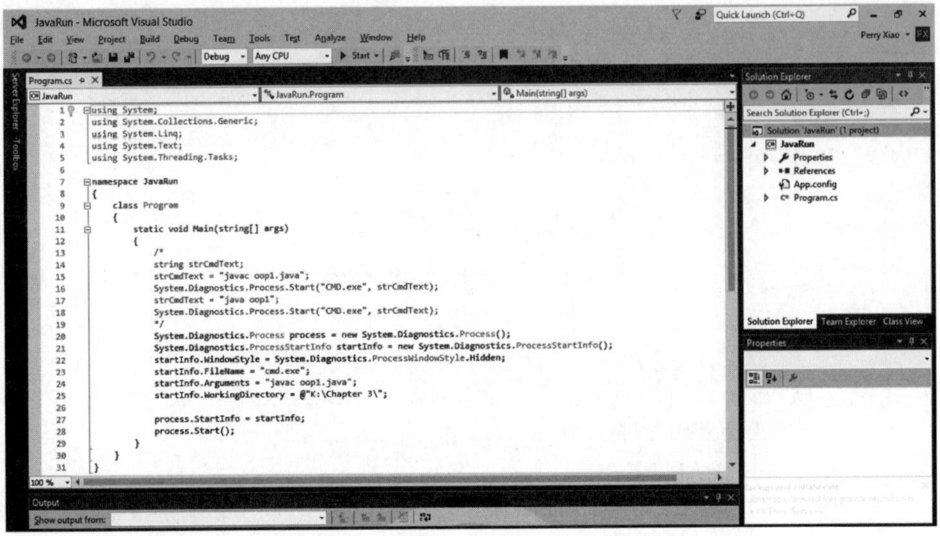

图 3.17(续)

3.16.4　Java 应用的安装

当开发商用 Java 应用或者将程序提供给他人使用时，需要使用专业软件来创建安装包以部署 Java 应用。

Install Creator 2 是 ClickTeam 提供的一种简单的、用户友好且功能强大的专业软件部署和安装工具。Install Creator 2 提供了一系列标准安装选项，例如指定要安装的版本、语言、软件注册代码和安装目录，并且可以生成丰富的安装过程。使用 Install Creator 2，还可以指定一系列与安装相关的操作，例如配置主要的可执行程序，将 DLL、EXE 和 ActiveX 文件复制到系统目录，将 DLL 和其他元素注册到 Windows 注册表，安装设备驱动程序以及在 Windows INI 和注册表文件中创建新条目。有关更多详细信息，请访问网址 https://www.clickteam.com/install-creator-2。

Advanced Installer 是另一个 Windows Installer 创作工具。Advanced Installer 提供了友好且易于使用的图形用户界面，用于基于 Windows Installer 技术创建和维护安装程序包(EXE、MSI 等)。有关更多详细信息，请访问网址 https://www.advancedinstaller.com/。

Install4J 是 EJ 技术的应用。Install4J 是功能强大的跨平台 Java 安装程序构建器，可为 Java 应用生成本机安装程序和应用启动器。有关更多详细信息，请访问网址 https://www.ej-technologies.com/products/install4j/overview.html。

Oracle 还提供了有关如何部署 Java 应用的教程。有关详细信息，请访问网址 https://docs.oracle.com/javase/tutorial/deployment/selfContainedApps/index.html。

3.17　小结

本章对 Java 编程做了基本介绍，内容包括变量、运算符、保留字、输入和输出、循环和选择、数组、矩阵和向量、文件读写、面向对象编程、多线程、方法、日期和时间、系统命令、软件工程和部署。

本章为你学习后续章节奠定了基础。变量、运算符和控制结构(例如循环和选择)是 Java 编程语言的基本构建块。你将在整本书中看到它们的用法。数组和列表对于存储复杂的用户数据很有用。文件读写工具允许程序从文件读取信息并将信息保存到文件中。面向对象编程是 Java 语言的主要功能之一，本书的所有示例程序都以某种方式使用了面向对象编程。多线程编程对于许多网络应用也至关重要，因为需要使用不同的线程来处理不同的连接。使用系统的工程方式开发 Java 软件是很好的实践。如果需要将 Java 程序分发给其他人，最好将程序打包到安装包中。

3.18 本章复习题

1. 标准输入和输出是什么？哪些 Java 类用于处理标准输入，哪些 Java 类用于处理标准输出？
2. Java 支持哪 8 种基本变量类型？
3. 什么是常数？什么是 var 类型变量？
4. String 和 StringBuffer 有什么区别？
5. 什么是 ASCII？
6. Java 支持哪些运算符？
7. Java 中有几个保留字？
8. 什么是循环和选择？
9. Java 数组和 Java 向量有什么区别？
10. 如何用 Java 读写文本文件？
11. 面向对象编程的优点是什么？
12. Java 线程的生命周期是什么？
13. 什么是软件的开发周期？
14. 如何创建可执行的 Java 程序？

第 4 章

面向Windows应用的Java编程

"成功的路上没有捷径。"

——Bewverly Sills

4.1 引言

4.2 Java Swing 应用

4.3 JavaFX 应用

4.4 部署 JavaFX 应用

4.5 小结

4.6 本章复习题

4.1 引言

到目前为止，你已经学习了基本的面向控制台应用的 Java 编程，控制台应用就是没有图形的基于文本的应用。但是在许多情况下，你可能更喜欢使用拥有图形用户界面(Graphical User Interface，GUI)的应用。在 Java 编程中，可以使用 Java 抽象窗口工具包(Abstract Window Toolkit，AWT)、Java Swing 和 JavaFX 部件工具箱来开发 GUI 程序。GUI 编程一直是 Java 的致命弱点，因为没有内置的图表功能，所以使用 Java 开发 GUI 程序很麻烦。多亏有了最新版本的 JavaFX，这一切得以改变。与 Java Swing 相比，JavaFX 改进了事件处理方式，具有特殊效果，允许使用层叠样式表(Cascading Style Sheets，CSS)创建外观，具有更一致的控件，更易于制作动画，并支持现代触摸屏设备。JavaFX 是基于 Java 的下一代客户端应用平台，可以用于构建基于主机操作系统、移动操作系统和嵌入式系统的应用。本章将首先介绍 Java Swing 应用，然后介绍 JavaFX 应用。

4.2 Java Swing 应用

Java Swing 是 GUI 部件工具箱，作用是提供相比早期 Java AWT 部件工具箱更复杂的 GUI 组件集。例 4.1 展示了一个标准的 Java Swing GUI 应用，名为 GUIApplication.java。为了创建 Java Swing GUI 程序，除了需要扩展 javax.swing.*库中的 JFrame 类之外，还需要实现 createAndShowGUI() 方法，该方法用于创建并设置窗口。为了显示窗口，需要将 createAndShowGUI() 方法作为 main() 方法中的线程来调用。图 4.1 显示了 SwingApplication1.java 程序的输出窗口，在本例中，这个窗口是空白的。

例 4.1　一个标准的 Java Swing GUI 应用

```
========================================================================
//Example 4.1 Java Swing example
import javax.swing.*;

public class SwingApplication1 extends JFrame{
    public SwingApplication1 () {
        super("Hello World Swing Application");
    }

    /**
     * Create the GUI and show it. For thread safety,
     * this method should be invoked from the
     * event-dispatching thread.
     */
    private static void createAndShowGUI() {
        //Make sure we have nice window decorations.
        JFrame.setDefaultLookAndFeelDecorated(true);

        //Create and set up the window.
        JFrame frame = new SwingApplication1();
        frame.setDefaultCloseOperation(JFrame.EXIT_ON_CLOSE);
        //Set the window size and location
        frame.setSize(300, 300);
        frame.setLocationRelativeTo(null);        //Display the window.
        frame.pack();
        frame.setVisible(true);
    }
    public static void main(String[] args) {
        //Schedule a job for the event-dispatching thread:
        //creating and showing this application's GUI.
        javax.swing.SwingUtilities.invokeLater(new Runnable() {
            public void run() {
                createAndShowGUI();
            }
        });
    }
}
========================================================================
```

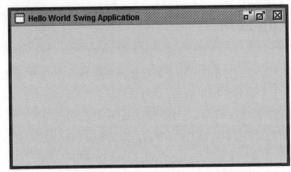

图 4.1　SwingApplication1.java 程序的输出窗口

练习 4.1　修改例 4.1 中的 Java 程序，使用绿色作为输出窗口的背景色。

接下来，可以在 SwingApplication1.java 程序中添加一些组件，例如标签和文本字段，如例 4.2 所示。窗口使用的布局为 FlowLayout，这意味着可以从左到右以及从上到下将组件添加到窗口中。你还可以在文本字段中添加一些动作，以便用户每次在文本字段中输入内容并按 Enter 键时，它们都会被复制到标签中。为此，代码必须从 JFrame 类扩展，并且还需要实现 ActionListener 以及导入 javax.swing.*、javax.awt.*和 javax.awt.event.*库。ActionListener 是事件处理程序的一种，将在出现某种动作(例如，按下按钮、选中复选框等)时触发。图 4.2 展示了 SwingApplication2.java 程序的输出窗口。

例 4.2　带有标签和文本字段的 Java Swing GUI 应用

```
================================================================
//Example 4.2 Java Swing example with Label and text field
import java.awt.*;
import java.awt.event.*;
import javax.swing.*;

public class SwingApplication2 extends JFrame implements ActionListener
{

    private JLabel label;
    private JTextField tf;

    public SwingApplication2 () {
       super("Java Swing with Label");
       setLayout(new FlowLayout());

       label=new JLabel("This is a swing label!");
       add(label);

       tf =new JTextField (20);
       tf.addActionListener(this);
       add(tf);
    }

    /**
     * Create the GUI and show it. For thread safety,
```

```
 * this method should be invoked from the
 * event-dispatching thread.
 */
private static void createAndShowGUI() {
    //Make sure we have nice window decorations.
    //JFrame.setDefaultLookAndFeelDecorated(true);

    //Create and set up the window.
    JFrame frame = new SwingApplication2 ();
    frame.setDefaultCloseOperation(JFrame.EXIT_ON_CLOSE);

    //Display the window.
    frame.pack();
    frame.setVisible(true);
}

public static void main(String[] args) {
    //Schedule a job for the event-dispatching thread:
    //creating and showing this application's GUI.
    javax.swing.SwingUtilities.invokeLater(new Runnable() {
        public void run() {
            createAndShowGUI();
        }
    });
}
public void actionPerformed (ActionEvent e){
    label.setText(tf.getText());
}
}
==========================================================================
```

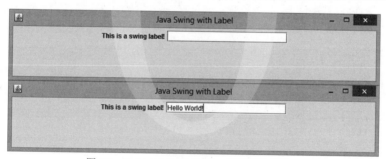

图 4.2 SwingApplication2.java 程序的输出窗口

练习 4.2　修改例 4.2 中的 Java 程序，将文本字段中的文本以大写形式复制到标签中。

例 4.3 展示了另一个带有文本字段、按钮和标签的 Java Swing GUI 应用。由于为按钮添加了一个动作，因此每次单击按钮时，都会计算你在文本字段中输入的数字的平方并显示在标签上。图 4.3 显示了输出窗口。

例 4.3　具有标签和文本字段的 Java Swing GUI 应用

```
==========================================================================
//Example 4.3 Java Swing example with label and text field
import java.awt.*;
```

```java
import java.awt.event.*;
import javax.swing.*;

public class SwingApplication2 extends JFrame implements ActionListener
{
    private JLabel label;
    private JTextField tf;
    private JButton button;
    public SwingApplication2 () {
        super("Java Swing with Label");
        setLayout(new FlowLayout());

        tf =new JTextField (20);
        add(tf);

        button=new JButton ("Equals");
        button.addActionListener(this);
        add(button);

        label=new JLabel("This is a swing label!");
        add(label);
    }

    /**
     * Create the GUI and show it. For thread safety,
     * this method should be invoked from the
     * event-dispatching thread.
     */
    private static void createAndShowGUI() {
        //Make sure we have nice window decorations.
        //JFrame.setDefaultLookAndFeelDecorated(true);

        //Create and set up the window.
        JFrame frame = new SwingApplication2 ();
        frame.setDefaultCloseOperation(JFrame.EXIT_ON_CLOSE);

        //Display the window.
        frame.pack();
        frame.setVisible(true);
    }

    public static void main(String[] args) {
        //Schedule a job for the event-dispatching thread:
        //creating and showing this application's GUI.
        javax.swing.SwingUtilities.invokeLater(new Runnable() {
            public void run() {
                createAndShowGUI();
            }
        });
    }
    public void actionPerformed (ActionEvent e){
        float s;
        if (e.getSource()==button){
            s=Float.parseFloat(tf.getText());
            s=s*s;
```

```
            label.setText("The square of "+ tf.getText() + " is " + s);
        }
    }
}
```
==

图 4.3　SwingApplication3.java 程序的输出

当程序中有很多 GUI 组件时，需要使用 Java Layout Manager，它包含在标准的 Java JDK 中。Java Layout Manager 支持许多布局，例如 FlowLayout、BorderLayout、GridLayout、BoxLayout、CardLayout、GridBagLayout、GroupLayout 和 Spring-Layout，默认使用的布局为 FlowLayout。

练习 4.3　修改例 4.3 中的 Java 程序，使用 GridLayout 布局，使得文本字段在顶部，按钮在中间，标签在底部。

有关 Java Swing 的更多信息和教程

有关 Java Swing 的更多信息和教程，请参见以下链接：

```
https://docs.oracle.com/javase/tutorial/uiswing/
https://docs.oracle.com/javase/tutorial/uiswing/layout/visual.html
https://docs.oracle.com/javase/8/docs/technotes/guides/swing/index.html
https://www.tutorialspoint.com/swing/
https://www.javatpoint.com/java-swing
https://netbeans.org/kb/docs/java/quickstart-gui.html
```

4.3　JavaFX 应用

JavaFX 是 Java 库，被设计用于开发复杂的 Java GUI 应用。JavaFX 既可用于开发标准的桌面应用，也可用于开发 RIA(Rich Internet Application)，RIA 能够提供与桌面应用类似的功能和体验。JavaFX 内置的应用可以在多种平台上运行，包括网络、手机和台式机。RIA 不需要任何其他软件即可运行。其他两项主要的 RIA 技术是 Adobe Flash 和 Microsoft Silverlight。

JavaFX 最初由 Sun Microsystems 开发，JavaFX 1.0 于 2008 年 12 月 4 日发布。自 2014 年 3 月发布 JavaFX 8.0 以来，JavaFX 已成为 Java 不可或缺的一部分。Oracle 将继续支持 JavaFX 直到 2022 年 3 月。JavaFX 的最新版本是 JavaFX 11.0，该版本于 2018 年 9 月 18 日发布。JavaFX 现在已作为独立的库发布，不再是标准 Java SDK 的一部分。未来，Gluon 公司将支持把 JavaFX 作为 JDK 之外的可下载模块。

有关最新的下载和安装 JavaFX 的说明信息，请访问网址 https://openjfx.io/openjfx-docs/#introduction。

创建 JavaFX 应用的最简单方法是使用 Java IDE，例如 IntelliJ IDEA，IntelliJ IDEA 将在默认情况下启用 JavaFX 插件。图 4.4 展示了如何使用 IntelliJ IDEA 创建 JavaFX 项目。

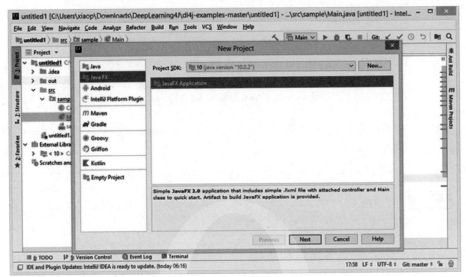

图 4.4　使用 IntelliJ IDEA 创建 JavaFX 项目

典型的 JavaFX 应用包含三个主要组件：stage、scene 和 nodes。Stage 表示窗口，其中包含 JavaFX 应用的所有 GUI 组件。stage 有两个参数：width 和 height，可分为内容区域和装饰区域（标题栏和边框）。

stage 有五种类型：装饰、未装饰、透明、统一和多功能。

scene 是包含所有组件的容器，nodes 是可以添加到 scene 中的 GUI 组件。nodes 可以是 2D 或 3D 对象（圆形、矩形、多边形等）、GUI 控件（标签、按钮、复选框、选择框、文本区域等）、容器（布局窗格，例如边框窗格、网格窗格、流窗格等）或媒体元素（例如音频、视频或图像）。

4.3.1　JavaFX 窗口

例 4.4 展示了一个简单的 JavaFX 程序，它仅仅创建一个空的窗口。为了创建 JavaFX 程序，你需要扩展 Application 类（该类是 javafx.application 包的一部分）并导入一系列其他 JavaFX 库；你还需要实现 start() 方法以设置所有 GUI 组件，并使用 launch() 方法来运行程序。对于例 4.4，你只需要使用 IntelliJ IDEA 创建一个名为 JavaFXApplication1 的 JavaFX 项目，然后将代码复制到 main.java 中即可（本章的所有 JavaFX 示例都将遵循相同的过程）。图 4.5 展示了 IntelliJ IDEA 中的示例代码以及运行后的输出窗口。

例 4.4　一个简单的 JavaFX 程序

```
//Example 4.4 JavaFX example
package sample;
```

```java
import javafx.application.Application;
import javafx.fxml.FXMLLoader;
import javafx.scene.*;
import javafx.stage.Stage;

import javafx.scene.control.*;
import javafx.scene.layout.*;
import javafx.event.*;

public class Main extends Application {

    @Override
    public void start(Stage primaryStage) throws Exception{
        primaryStage.setTitle("Hello World!");

        FlowPane root = new FlowPane();

        primaryStage.setScene(new Scene(root, 300, 250));
        primaryStage.show();
    }


    public static void main(String[] args) {
        launch(args);
    }
}
```
==

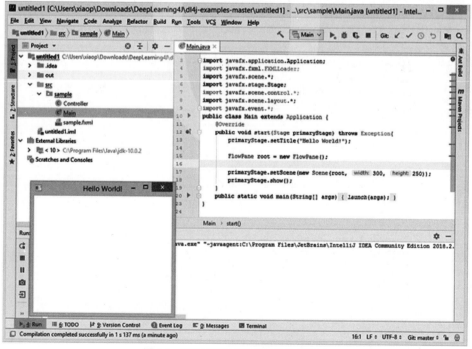

图 4.5　示例代码以及运行后的输出窗口

> **练习 4.4**　修改例 4.4 中的 Java 程序，使用红色作为窗口的背景色。

4.3.2　在 JavaFX 中创建标签和按钮

例 4.5 是例 4.4 的改进版本，这个版本带有标签和按钮。例 4.5 使用了 FlowPane 布局，从而可以像在流布局中那样在其中添加 GUI 对象。这里还为按钮添加了一个动作，每次单击按钮时，就会在标签上显示 "Hello World!"。为了运行例 4.5，你只需要使用 IntelliJ IDEA 创建一个名为 JavaFXApplication2 的 JavaFX 项目，然后将代码复制到 main.java 中即可。图 4.6 给出了输出结果。

例 4.5　带有标签和按钮的 JavaFX 程序

```
//Example 4.5 JavaFX example with label and button
package sample;
import javafx.application.Application;
import javafx.fxml.FXMLLoader;
import javafx.scene.*;
import javafx.stage.Stage;

import javafx.scene.control.*;
import javafx.scene.layout.*;
import javafx.event.*;


public class Main extends Application {

    @Override
    public void start(Stage primaryStage) throws Exception{
        primaryStage.setTitle("Hello World!");
        Label label = new Label("My Label");
        Button btn = new Button();
        btn.setText("Say 'Hello World'");
        btn.setOnAction(new EventHandler<ActionEvent>() {

            @Override
            public void handle(ActionEvent event) {
                //System.out.println("Hello World!");
                label.setText("Hello World!");
            }
        });

        FlowPane root = new FlowPane();
        root.getChildren().add(label);
        root.getChildren().add(btn);
        primaryStage.setScene(new Scene(root, 300, 250));
        primaryStage.show();
    }


    public static void main(String[] args) {
        launch(args);
```

```
        }
}
======================================================================
```

与 Java Swing 相似，JavaFX 也支持许多不同的布局，称为布局窗格，例如 FlowPane、BorderPane、GridPane、StackPane、HBox、VBox、TextFlow、AnchorPane、TitlePane 等。

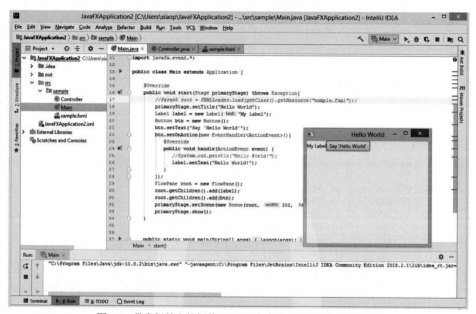

图 4.6 带有标签和按钮的 JavaFX 程序以及运行后的输出窗口

练习 4.5 在例 4.5 所示的 Java 程序中添加另一个按钮，当单击这个按钮时，清除标签上显示的文本。

4.3.3 JavaFX 图表

很长时间以来，Java 都没有提供内置的图表功能。借助 JavaFX，用户现在可以使用内置的 JavaFX 图表功能轻松地创建条形图、饼图、折线图、散点图等。例 4.6 展示了如何使用 LineChart、NumberAxis 和 XYChart 类在 JavaFX 中创建折线图。要想创建图表，首先需要为 x 轴和 y 轴创建 NumberAxis 对象，之后需要创建一个添加了 NumberAxis 对象的 LineChart 对象，然后再创建一个 XYChart.Series 对象，并将数据作为 XYChart.Data 对象添加到其中，最后将 XYChart.Series 对象添加到 LineChart 对象即可。

为了运行这个示例，你需要使用 IntelliJ IDEA 创建一个名为 JavaFXApplication3 的 JavaFX 项目，然后将代码复制到 main.java 程序中。图 4.7 给出了输出结果。

例 4.6 使用 JavaFX 创建折线图

```
======================================================================
//Example 4.6 JavaFX Line chart example
```

```java
package sample;

import javafx.application.Application;
import javafx.scene.Group;
import javafx.scene.Scene;
import javafx.stage.Stage;
import javafx.scene.chart.LineChart;
import javafx.scene.chart.NumberAxis;
import javafx.scene.chart.XYChart;

public class Main extends Application {
    @Override
    public void start(Stage stage) {
        //Defining the x axis
        NumberAxis xAxis = new NumberAxis(0, 50, 10);
        xAxis.setLabel("Time [s]");

        //Defining the y axis
        NumberAxis yAxis = new NumberAxis (0, 20, 2);
        yAxis.setLabel("Voltage [V]");

        //Creating the line chart
        LineChart linechart = new LineChart(xAxis, yAxis);

        //Prepare XYChart.Series objects by setting data
        XYChart.Series series = new XYChart.Series();
        series.setName("Measurement Results");

        series.getData().add(new XYChart.Data(0, 15));
        series.getData().add(new XYChart.Data(10, 3));
        series.getData().add(new XYChart.Data(20, 6));
        series.getData().add(new XYChart.Data(30, 12));
        series.getData().add(new XYChart.Data(40, 2));
        series.getData().add(new XYChart.Data(50, 10));

        //Setting the data to Line chart
        linechart.getData().add(series);

        //Creating a Group object
        Group root = new Group(linechart);

        //Creating a scene object
        Scene scene = new Scene(root, 600, 400);

        //Setting title to the Stage
        stage.setTitle("Line Chart");

        //Adding scene to the stage
        stage.setScene(scene);

        //Displaying the contents of the stage
        stage.show();
    }
    public static void main(String args[]){
        launch(args);
    }
}
```
==

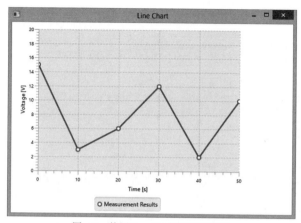

图 4.7 使用 JavaFX 创建的折线图

练习 4.6　修改例 4.6 中的 Java 程序，改为在条形图中显示数据。

4.3.4 在 JavaFX 中处理用户登录

处理用户登录是所有 GUI 应用的基本功能。例 4.7 展示了使用 GUI 组件(例如标签、文本字段和按钮)的 JavaFX 登录示例。该例使用的是 GridPane 布局，允许将 GUI 对象添加到网格中。为了运行这个示例，你需要使用 IntelliJ IDEA 创建一个名为 JavaFXApplication4 的 JavaFX 项目，然后将代码复制到 main.java 程序中。图 4.8 给出了输出结果。

例 4.7　使用 JavaFX GUI 组件进行登录

```
========================================================================
//Example 4.7 JavaFX GUI component Login example
package sample;

import javafx.application.Application;
import static javafx.application.Application.launch;
import javafx.geometry.Insets;
import javafx.geometry.Pos;

import javafx.scene.Scene;
import javafx.scene.control.Button;
import javafx.scene.control.PasswordField;
import javafx.scene.layout.GridPane;
import javafx.scene.text.Text;
import javafx.scene.control.TextField;
import javafx.stage.Stage;

public class Main extends Application {
    @Override
    public void start(Stage stage) {
      //creating label email
      Text text1 = new Text("Username");

      //creating label password
```

```
        Text text2 = new Text("Password");
        
        //Creating Text Filed for email
        TextField textField1 = new TextField();
        
        //Creating Text Filed for password
        PasswordField textField2 = new PasswordField();
        
        //Creating Buttons
        Button button1 = new Button("Login");
        
        //Creating a Grid Pane
        GridPane gridPane = new GridPane();
        
        //Setting size for the pane
        gridPane.setMinSize(400, 200);
        
        //Setting the Grid alignment
        gridPane.setAlignment(Pos.CENTER);
        
        //Arranging all the nodes in the grid
        gridPane.add(text1, 0, 0);
        gridPane.add(textField1, 1, 0);
        gridPane.add(text2, 0, 1);
        gridPane.add(textField2, 1, 1);
        gridPane.add(button1, 0, 2);
        
        //Creating a scene object
        Scene scene = new Scene(gridPane);
        
        //Setting title to the Stage
        stage.setTitle("Bank Login");
        
        //Adding scene to the stage
        stage.setScene(scene);
        
        //Displaying the contents of the stage
        stage.show();
    }
    public static void main(String args[]){
        launch(args);
    }
}
```

图 4.8 使用 JavaFX GUI 组件进行登录

在 JavaFX 中，可以使用 Alert 类，通过弹出消息框来显示错误消息。例如：

```
new Alert(Alert.AlertType.ERROR, " An error occurred!").showAndWait();
```

再例如：

```
Alert alert = new Alert(Alert.AlertType.ERROR);
alert.setHeaderText("An error occurred!");
alert.showAndWait();
```

Alert 类还可以用来显示带有 OK 和 Cancel 按钮的标准消息框。例如：

```
Alert alert = new Alert(Alert.AlertType.WARNING,
          "Are you sure to continue? ",
          ButtonType.OK,
          ButtonType.CANCEL);
alert.setTitle("Warning");
Optional<ButtonType> result = alert.showAndWait();

if (result.get() == ButtonType.OK) {
   //do something
}
else{
   //do something else
}
```

练习 4.7 为例 4.7 所示 Java 程序中的 Login 按钮添加一个动作，以便单击 Login 按钮时，检查用户名和密码并显示警告消息。

4.3.5 在 JavaFX 中创建图像查看器

一幅图胜似千言万语，尤其是在网上，图片查看器已成为当今计算环境中最常用且方便的工具之一。例 4.8 展示了一个使用 Image 和 ImageView 类的 JavaFX 图像查看器。为了查看图像，首先需要创建一个 Image 对象，从而使用 FileInputStream 类从文件中加载图像。你还需要抛出 FileNotFoundException 异常。下一步是创建一个 ImageView 对象，并将刚才创建的那个 Image 对象添加到其中。为了运行这个示例，你需要使用 IntelliJ IDEA 创建一个名为 JavaFXApplication5 的 JavaFX 项目，然后将代码复制到 main.java 程序中。图 4.9 给出了输出结果。

例 4.8 在 JavaFX 中创建图像查看器

```
========================================================================
//Example 4.8 JavaFX image viewer example
package sample;

import java.io.FileInputStream;
import java.io.FileNotFoundException;
import javafx.application.Application;
import javafx.scene.Group;
import javafx.scene.Scene;
import javafx.scene.image.Image;
```

```java
import javafx.scene.image.ImageView;
import javafx.stage.Stage;

public class Main extends Application {
   @Override
   public void start(Stage stage) throws FileNotFoundException {
   stage.setTitle("Image Viewer");

   Image image = new Image(new FileInputStream("xxxx"));
   ImageView imageView = new ImageView(image);

   //Creating a Group object
   Group root = new Group(imageView);

   Scene scene = new Scene(root, 900, 600);
   stage.setScene(scene);
   stage.show();
   }
   public static void main(String args[]) {
      launch(args);
   }
}
```

==

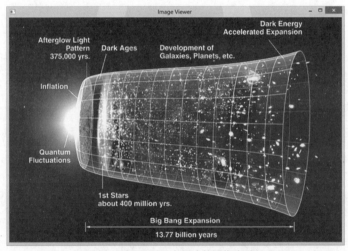

图 4.9　在 JavaFX 中创建的图像查看器

4.3.6　创建 JavaFX Web 查看器

可以使用 JavaFX 连接到网络并查看内容。例 4.9 展示了一个使用 WebView 类创建的 JavaFX Web 查看器。为了运行这个示例，你需要使用 IntelliJ IDEA 创建一个名为 JavaFXApplication6 的 JavaFX 项目，然后将代码复制到 main.java 程序中。图 4.10 给出了输出结果。

例 4.9　创建 JavaFX Web 查看器

==

```
//Example 4.9 JavaFX Web viewer example
```

```java
package sample;

import javafx.application.Application;
import javafx.stage.Stage;
import javafx.scene.Scene;
import javafx.scene.web.WebView;
import javafx.scene.Group;

public class Main extends Application {

    public static void main(String[] args) {
        launch(args);
    }

    public void start(Stage stage) {
        stage.setTitle("Web Viewer");

        WebView webView = new WebView();

        webView.getEngine().load(" http://google.com ");

        Group root = new Group(webView);
        Scene scene = new Scene(root, 800, 500);

        stage.setScene(scene);
        stage.show();
    }
}
```
==

图 4.10　创建的 JavaFX Web 查看器

4.3.7　在 JavaFX 中创建菜单

菜单是 GUI 应用的另一种常见元素。例 4.10 展示了如何使用 Menu、MenuBar、MenuItem

和 SeparatorMenuItem 类创建简单的 JavaFX 菜单。创建 JavaFX 菜单的步骤如下：

(1) 创建一个 MenuBar 对象。

(2) 创建一个 Menu 对象以及一些 MenuItem 对象；在本例中，需要创建的 MenuItem 对象有 Open、Separator 和 Exit。将如下动作添加到 Exit 对象，以便选择这个菜单项时，程序能够终止：Platform.exit()。

(3) 将 MenuItem 对象添加到 Menu 对象。

(4) 将 Menu 对象添加到 MenuBar 对象。

(5) 创建 BorderPane 布局，然后将 MenuBar 对象添加到布局的顶部。

(6) 将 BorderPane 布局添加到场景中。

在这里，对于 Menu 对象，setOnAction()用于向 Menu 对象添加操作，而 Platform.exit()用于终止程序。例如，exitMenuItem.setOnAction(actionEvent -> Platform.exit())会将程序终止操作添加到 exitMenuItem 对象中。为了运行这个示例，你需要使用 IntelliJ IDEA 创建一个名为 JavaFXApplication7 的 JavaFX 项目，然后将代码复制到 main.java 程序中。图 4.11 给出了输出结果。

例 4.10　在 JavaFX 中创建菜单

```
============================================================================
//Example 4.10 JavaFX Menu example
package sample;

import javafx.application.Application;
import javafx.application.Platform;
import javafx.stage.Stage;
import javafx.scene.Scene;
import javafx.scene.layout.BorderPane;
import javafx.scene.control.Menu;
import javafx.scene.control.MenuBar;
import javafx.scene.control.MenuItem;
import javafx.scene.control.SeparatorMenuItem;
import javafx.scene.Group;

public class Main extends Application {
    public static void main(String[] args) {
        launch(args);
    }

    @Override
    public void start(Stage primaryStage) {
        primaryStage.setTitle("JavaFX Menu");
        MenuBar menuBar = new MenuBar();
        menuBar.prefWidthProperty().bind(primaryStage.widthProperty());

        // File menu - new, save, exit
        Menu fileMenu = new Menu("File");
        MenuItem newMenuItem = new MenuItem("Open");
        MenuItem exitMenuItem = new MenuItem("Exit");
        exitMenuItem.setOnAction(actionEvent -> Platform.exit());
```

```
        fileMenu.getItems().addAll(newMenuItem, new SeparatorMenuItem(),
            exitMenuItem);
        menuBar.getMenus().add(fileMenu);

        BorderPane bp = new BorderPane();
        bp.setTop(menuBar);
        Scene scene = new Scene(bp, 400, 300);

        primaryStage.setScene(scene);
        primaryStage.show();
    }
}
========================================================================
```

图 4.11　在 JavaFX 中创建的菜单

4.3.8　创建 JavaFX 文件选择对话框

许多应用都需要读写文件。在 JavaFX 中，可以使用 FileChooser 类打开文件选择对话框，用户可以在其中选择要读写的文件。例 4.11 展示了一个使用 FileChooser 类的 JavaFX 菜单应用。这个 JavaFX 菜单应用提供了一个按钮，允许用户选择要查看的图像文件，而不是加载相同的图像文件。同样，这里必须使用 try…catch 块抛出或处理 FileNotFoundException 异常。为了运行这个示例，你需要使用 IntelliJ IDEA 创建一个名为 JavaFXApplication8 的 JavaFX 项目，然后将代码复制到 main.java 程序中。图 4.12 给出了输出结果。

例 4.11　使用 FileChooser 类创建 JavaFX 文件选择对话框

```
========================================================================
//Example 4.11 JavaFX File Chooser example
package sample;

import javafx.application.Application;
import javafx.application.Platform;
import javafx.stage.Stage;
import javafx.scene.Scene;
import javafx.scene.control.Button;
import javafx.scene.layout.BorderPane;
import javafx.stage.FileChooser;
```

```
import javafx.scene.image.Image;
import javafx.scene.image.ImageView;
import javafx.scene.Group;
import java.io.File;
import java.io.FileInputStream;
import java.io.FileNotFoundException;

public class Main extends Application {
    public static void main(String[] args) {
        launch(args);
    }

    @Override
    public void start(Stage primaryStage)throws FileNotFoundException {
        primaryStage.setTitle("JavaFX File Chooser");
        ImageView imageView = new ImageView();
        FileChooser fileChooser = new FileChooser();
        Button button = new Button("Select a Image File");
        button.setOnAction(e -> {
            File file = fileChooser.showOpenDialog(primaryStage);
            if (file != null) {
                System.out.println("Selected file: " + file);
                try {
                    Image image = new Image(new FileInputStream(file));
                    imageView.setImage(image);
                }
                catch (FileNotFoundException ex){
                    //
                }
            }
        });

        BorderPane bp = new BorderPane();
        bp.setTop(button);
        bp.setCenter(imageView);
        Scene scene = new Scene(bp, 700, 500);

        primaryStage.setScene(scene);
        primaryStage.show();
    }
}
```
==

练习 4.8 创建一个包含文本字段、按钮和 Web 查看器的 Java 程序。首先使用 GridPane 布局在窗口的顶部添加一个文本字段和一个按钮，然后在中心位置添加一个 Web 查看器，最后为按钮添加一个动作，以便在单击按钮时显示你在文本字段中输入的 URL 指向的网页。

练习 4.9 修改例 4.8 中的 Java 图像查看器，使其具有的 File 菜单包含 Open、Save 和 Exit 菜单项。添加动作，以便：选择 Open 菜单项时，打开文件选择对话框，从中可以选择要查看的图像文件；选择 Save 菜单项时，将当前图像保存到文件中；选择 Exit 菜单项时，关闭程序。

图 4.12　创建的 JavaFX 文件选择对话框

4.3.9　JavaFX 教程

互联网上有很多优秀的 JavaFX 教程。图 4.13 展示了 Oracle 提供的官方 JavaFX 教程 (https://docs.oracle.com/javafx/2/get_started/jfxpub-get_started.htm)，该教程非常全面，包括所有 JavaFX 文档。

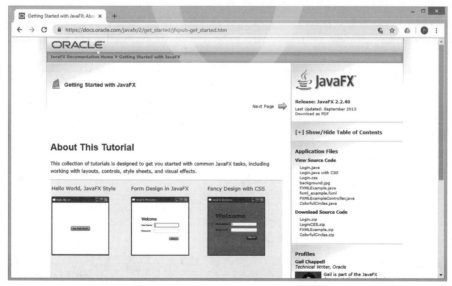

图 4.13　Oracle 提供的官方 JavaFX 教程

图 4.14 展示了 Tutorialspoint 网站(https://www.tutorialspoint.com/javafx/)上的 JavaFX 教程。该教程非常全面且讲解清晰，其中涉及许多带有示例的 JavaFX 主题。

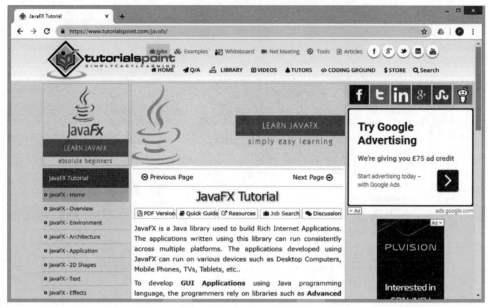

图 4.14　Tutorialspoint 网站上的的 JavaFX 教程

图 4.15 展示了另一个 JavaFX 教程，该教程来自 Jenkov Apps(http://tutorials.jenkov.com/javafx/scatterchart.html)，其中包含不同的 JavaFX 主题和代码示例。

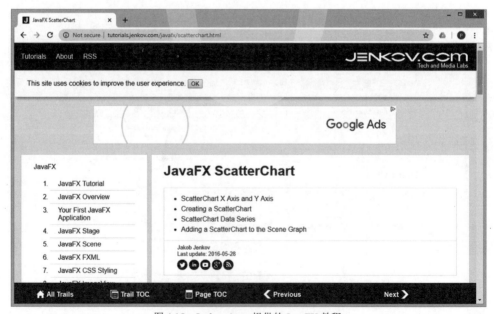

图 4.15　Jenkov Apps 提供的 JavaFX 教程

图 4.16 展示了来自 Eclipse 的 JavaFX 教程(http://wiki.Eclipse.org/Efxclipse/Tutorials)，该教程主要讨论如何使用 Eclipse IDE 开发 JavaFX 应用。

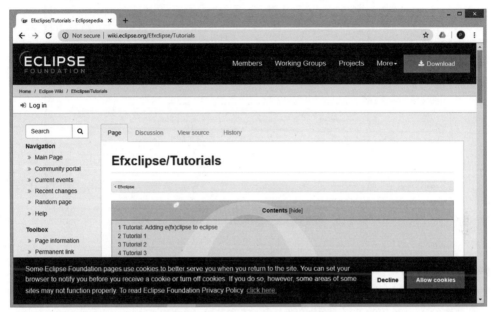

图 4.16　来自 Eclipse 的 JavaFX 教程

图 4.17 展示了 Oracle 提供的 JavaFX 官方文档网站(https://docs.oracle.com/javafx/2/)。图 4.18 展示了来自 Oracle 的对应 JavaFX API 的文档网站(https://docs.oracle.com/javafx/2/api/index.html)。

图 4.17　Oracle 提供的 JavaFX 官方文档网站

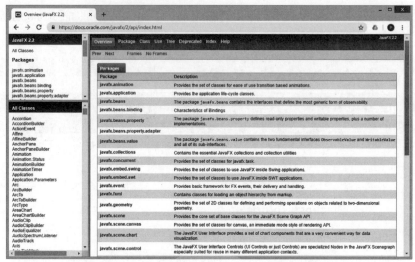

图 4.18　来自 Oracle 的对应 JavaFX API 的文档网站

4.4　部署 JavaFX 应用

部署 JavaFX 应用的最简单方法是将其部署为独立的、可执行的 JAR 文件，如第 3 章的 3.16 节所述。此外，JavaFX 应用可以在浏览器中运行、在网络上运行以及作为独立的应用运行。诸如 Eclipse 和 NetBeans 的 Java IDE 均提供了 JavaFX 部署功能。

图 4.19 展示了 Oracle 提供的 JavaFX 部署指南(https://docs.oracle.com/javafx/2/get_started/basic_deployment.htm)。图 4.20 展示了来自 code.makery(https://code.makery.ch/library/javafx-tutorial/part7/)的 JavaFX 部署指南。

图 4.19　Oracle 提供的 JavaFX 部署指南

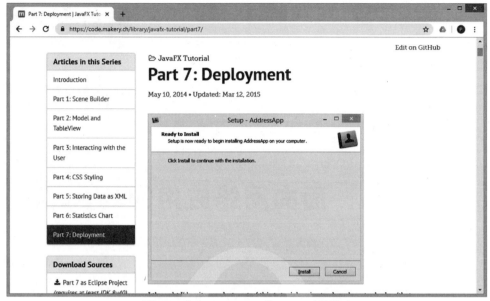

图 4.20　来自 code.makery 的 JavaFX 部署指南

4.5　小结

本章介绍了 Java GUI 编程。我们首先介绍了如何使用 Java Swing 工具包进行 Java GUI 开发，然后介绍了如何使用最新的 JavaFX 工具包进行 Java GUI 开发。本章还提供了相关的学习网站，其中包括一些有趣的 JavaFX 教程以及 JavaFX 部署指南。

4.6　本章复习题

1. Java 控制台应用和 Java GUI 应用之间有什么区别？
2. 什么是 Java AWT？什么是 Java Swing？
3. Java Swing 支持哪些布局？默认布局是什么？
4. 什么是 JavaFX？
5. 在 JavaFX 中，stage、scene 和 nodes 之间的关系是什么？
6. JavaFX 支持哪些类型的图表？
7. JavaFX 提供了哪些布局窗格？
8. Menu、MenuBar 和 MenuItem 之间是什么关系？

第 5 章

面向网络应用的Java编程

"再也没有比看到别人做成了你说不可能办到的事情更尴尬的了。"

——Sam Ewing

5.1 简介
5.2 Java 网络信息编程
5.3 Java 套接字编程
5.4 Java HTTP 编程
5.5 Java 电子邮件 SMTP 编程
5.6 Java RMI 客户端-服务器编程
5.7 SDN 入门
5.8 Java 网络编程资源
5.9 小结
5.10 本章复习题

5.1 简介

计算机网络已成为我们生活中不可或缺的一部分。借助 Internet(快速增长的全球化计算机网络)，我们只需要单击几下鼠标，即可在线进行购物、办理银行业务、发送电子邮件、进行消息传递以及预订航班或酒店。Internet 彻底改变了我们的生活。根据 Internet Live Stats 网站(http://www.internetlivestats.com/internet-users/)所做的统计，现在有超过 40 亿的 Internet 用户。Internet 现象背后的魔力是一种称为 TCP/IP 的技术，TCP/IP 技术是在 20 世纪 60 年代通过美国高级研究计划局(Advanced Research Projects Agency，ARPA)的研究项目 ARPANET 开发出来的。ARPA 是隶属美国国防部的研究机构，因此 TCP/IP 也被称为国防部(Department of Defense，DoD)模型。

TCP/IP 模型包含一整套协议，但主要围绕两个核心协议而构建：传输控制协议(Transmission Control Protocol，TCP)和 Internet 协议(Internet Protocol，IP)。IP 定义了一种设备寻址方案，并根据 IP 地址在设备之间传递数据；TCP 则提供了对传递的控制。TCP/IP 模型分为四层：应用层、传输层、网络层和链路层(参见图 5.1)。

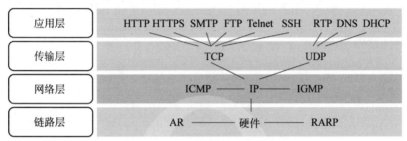

图 5.1 TCP/IP 模型及其包含的协议

应用层提供用户或应用与底层通信协议之间的接口。应用层中的主要协议是用于 Web 服务的超文本传输协议/安全超文本传输协议(Hypertext Transfer Protocol / Hypertext Transfer Protocol Secure，HTTP/HTTPS)、用于文件上载/下载的文件传输协议(File Transfer Protocol，FTP)、用于电子邮件的简单邮件传输协议(Simple Mail Transfer Protocol，SMTP)、用于从电子邮件服务器检索电子邮件的邮局协议(Post Office Protocol，POP3)、用于远程登录的安全外壳(Secure Shell，SSH)、用于将域转换为 IP 地址的域名系统(Domain Name System，DNS)以及用于自动主机配置的动态主机配置协议(Dynamic Host Configuration Protocol，DHCP)。

传输层负责管理数据传输，并提供诸如面向连接的通信、可靠性、流控制和多路复用的服务。传输层中的主要协议是用于可靠传输的 TCP 协议和用于实时传输的用户数据报协议(User Datagram Protocol，UDP)。

网络层基于 IP 地址将数据(称为数据包)从源设备传送到目标设备。网络层中的主要协议是用于传送数据的 IP 协议。网络层中还包括用于错误报告的 Internet 消息控制协议(Internet Control Message Protocol，ICMP)和用于多播的 Internet 组管理协议(Internet Group Management Protocol，IGMP)。

链路层负责处理网络设备的物理组件，例如以太网、令牌环、光纤和 WiFi 连接，有时又称为网络接口层。链路层也可以细分为数据链路层和物理层。链路层中的主要协议是用于将 IP 地址解析为硬件地址或媒体访问控制(Medium Access Control，MAC)地址的地址解析协议(Address Resolution Protocol，ARP)、反向地址解析协议(Reverse Address Resolution Protocol，RARP)以及用于收集有关相邻设备信息的邻居发现协议(Neighbor Discovery Protocol，NDP)。

网络中的所有计算机、手机、平板电脑或其他设备都必须安装 TCP/IP 套件软件才能运行，如图 5.2 所示。每台设备还需要拥有属于自己的 IP 地址和端口号才能与网络中的其他设备通信。

IP 地址用于唯一标识设备,有点像电话号码。IP 地址有两种类型:IPv4 地址和 IPv6 地址。IPv4 地址是 32 位的二进制地址,通常以点分十进制格式表示,例如 10.0.0.1、136.148.1.27 或 192.168.0.1。IPv6 地址是新的 128 位的二进制地址,通常以冒号分隔的十六进制格式表示,例如 FE80:CD00:0000:0CDE:1257:0000:211E:729C、2001:0DB8:85A3:0000:0000:8A2E:0370:7334 或 FF01:0:0:0:0:0:0:1。目前,40 亿个 IPv4 地址已经用尽,因此 IPv6 地址将是计算机网络发展的未来方向。我们总共有 3.4×10^{38} 个 IPv6 地址,这足以为地球上的每一粒沙子分配一个 IP 地址。端口号用于唯一标识你在设备上运行的程序,例如 Web 浏览器、电子邮件代理和聊天程序。端口号是 16 位的二进制数,范围为 0~65 535。端口 0~1023 大家已十分熟知,它们也称为系统端口,例如:

- 20: FTP(数据传输)
- 21: FTP(命令控制)
- 22: SSH
- 25: SMTP
- 53: DNS
- 67: DHCP 服务器
- 68: DHCP 客户端
- 80: HTTP
- 110: POP3
- 443: HTTPS

端口 1024~49 151 被称为注册端口。端口 49 152~65 535 被称为临时端口。在编写 Java 程序时,不应使用系统端口;相反,应该使用注册端口或临时端口。

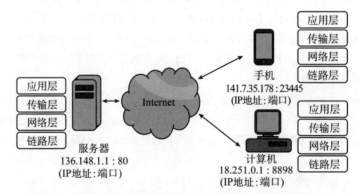

图 5.2 通过 Internet 进行交换的 TCP/IP 模型、IP 地址和端口号

有关 TCP/IP 的更多详细信息,请访问以下资源:

https://en.wikipedia.org/wiki/Internet_protocol_suite
http://www.pearsonitcertification.com/articles/article.aspx?p=1804869
https://docs.oracle.com/cd/E19683-01/806-4075/ipov-10/index.html
https://www.cisco.com/c/en/us/support/docs/ip/routing-information-protocol-rip/13769-5.html

5.1.1 局域网和广域网

我们可以将计算机网络从物理上分为局域网(Local Area Network，LAN)和广域网(Wide Area Network，WAN)。LAN 是一栋建筑物或同一园区内一组互联的计算机。WAN 是在更大范围内分布的 LAN 的集合。城市内的 WAN 也称为城域网(Metropolitan Area Network，MAN)。LAN 通常属于单个组织，并且具有较高的连接进度，例如快速以太网(100 Mbps)或千兆以太网(1000 Mbps)。MAN 或 WAN 通常属于不同的组织，并且具有相对较低的连接速度。Internet 是典型的全球公共 WAN。

5.1.2 思科的三层企业网络架构

具有庞大计算机网络的大型组织通常采用思科的三层企业网络架构，如图 5.3 所示，从上到下分别是核心层、分布层和访问层。

访问层为最终用户设备(例如计算机、打印机或服务器)提供网络访问。访问层中典型的网络设备是交换机，它们以星型网络拓扑或树型网络拓扑(由分层结构连接的几个星型网络拓扑组成)连接所有设备。

分布层负责桥接访问层和核心层之间的间隙。典型的网络设备是路由器，它们将访问层中的所有交换机连接到核心层。分布层还通过使用网状网络拓扑为网络连接提供冗余。

核心层将所有分发路由器连接在一起，并提供 Internet 访问。核心层也称为主干层。典型的网络设备是网关路由器，也称为默认网关。大型组织可能有多个网关路由器用来为 Internet 连接提供更多带宽和冗余。

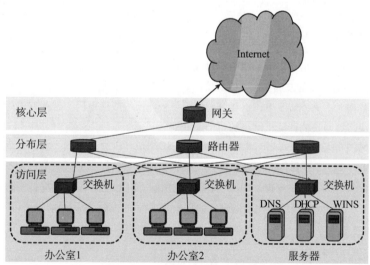

图 5.3 思科的三层企业网络架构

5.1.3 关键网络组件

前面的图 5.3 显示了以下关键网络组件：交换机、路由器、网关以及各种服务器(例如 DNS 和 DHCP 服务器)。对于 Windows 计算机网络，还需要 Windows 网络名称服务(Windows Internet

Name Service，WINS)服务器。
- 交换机是一种用于连接用户的客户端计算机的设备，通常具有大量相同类型的接口，称为端口。交换机可以具有 8 个端口、16 个端口、64 个端口甚至 128 个端口。
- 路由器是连接不同网络(而不是用户计算机)的设备，通常具有一些不同类型的接口，例如以太网接口、快速以太网接口、千兆以太网接口、串行接口、光纤分布式数据接口(Fiber Distributed Data Interface，FDDI)、异步传输模式(Asynchronous Transfer Mode，ATM)接口和综合服务数字网(Integrated Services Digital Network，ISDN)接口。之所以称为路由器，是因为它们还执行路由功能以寻找到达目的地的最佳路由。
- 网关或默认网关是本地计算机网络连接到 Internet 的地方。网关通常也是实现网络代理服务、网络策略和网络防火墙的地方。
- DNS 是另一个关键组件。计算机基于 IP 地址工作，但是我们人类并不擅长记住数字，我们更喜欢名称。DNS 服务器负责将人类可理解的域名转换为机器可理解的 IP 地址，反之亦然。
- DHCP 服务器会自动为计算机提供网络配置信息。以下是配置计算机以访问 Internet 所需的四种必备信息。
  - IP 地址：用于标识设备。
  - 子网掩码：用于标识设备的子网。
  - DNS 服务器：用于标识 DNS 服务器在哪里。
  - 默认网关：用于标识网关路由器在哪里。

5.1.4 传统网络与软件定义网络

传统的计算机网络由交换机和路由器组成。每台交换机或路由器都有两层：控制层和数据层(或转发层)，如图 5.4 的左侧所示。控制层确定向何处发送通信数据，数据层则执行这些决策并转发通信数据。在这种情况下，需要分别配置每台交换机或路由器。这种方式耗时且容易出错，难以解决日益增长的计算机网络的可伸缩性问题。

软件定义网络(Software-Defined Networking，SDN)是一项新兴技术，现在已成为计算机网络/IT 行业内十分热门的技术。借助 SDN，交换机和路由器的控制层可从单台设备脱离出来，并使用专用服务器进行集中管理，如图 5.4 的右侧所示。SDN 位于现有计算机网络顶部的软件层。服务器通过直接对交换机和路由器进行编程来集中控制流量。这样做能带来以下几个好处。

- 中央管理：可以利用控制器对网络进行配置，监视和故障排除；也可以从控制器查看完整的网络拓扑。
- 网络设备更轻量级：由于所有控制都是集中执行的，因此可以使用精简的交换机和路由器来降低网络成本。
- 网络虚拟化：网络虚拟化给 IT 行业带来一场有关虚拟化存储和计算实体的革命，在有效利用资源方面发挥了关键作用。

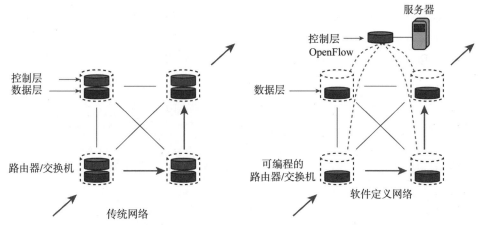

图 5.4　传统网络和软件定义网络的比较

最常用的 SDN 通信协议是 OpenFlow，OpenFlow 是由开放网络基金会管理的开放标准协议(你将在本章的后面学习如何获取和使用 OpenFlow)。OpenFlow 需要同时在交换机/路由器和控制器中实现。OpenFlow 可在交换机/路由器与控制器之间传送消息。为了控制网络交换机/路由器，控制器将使用 OpenFlow 将转发规则和安全规则等推送到交换机/路由器中，以便它们在网络流量到达时做出决策。交换机/路由器需要在 OpenFlow 表中维护此类规则。规则也称为流，并存储在流表(flow table)中。为了监视交换机/路由器，控制器使用 OpenFlow 的各种请求和响应消息来获取统计信息和事件消息，以向控制器发送有关交换机/路由器上发生的更改或故障的更新信息。

Floodlight 是很受欢迎的 SDN 控制器项目(在本章的后面，你将学习如何获取和使用 Floodlight)。Floodlight 是基于 Java 的 Apache 许可的 OpenFlow 控制器，可用于帮助构建软件定义网络。Floodlight 由开源的开发人员社区开发，并且易于构建和运行。Floodlight 支持大多数虚拟和物理的 OpenFlow 交换机。Floodlight 可以处理混合的 OpenFlow 和非 OpenFlow 网络。Floodlight 的目的是实现高性能的控制器项目，因此从一开始就被设计为多线程的，并且支持 OpenStack(链接)云编排平台。

OpenDaylight 是另一个基于 Java 的开源 SDN 控制器项目(在本章的后面，你将学习如何获取和使用 OpenDaylight)。OpenDaylight 是基础项目，可在此基础上构建许多其他控制器。OpenDaylight 由 Linux 基金会负责。OpenDaylight 的目标是促进软件定义网络和网络功能虚拟化(Network Functions Virtualization，NFV)。NFV 是当前的另一个热门话题。SDN 与 NFV 密切相关，但有所不同。SDN 致力于将网络的控制层与转发层分离，并提供对网络的集中控制。NFV 专注于改进和虚拟化网络服务本身，例如 DNS、缓存、负载均衡、防火墙、入侵检测等。SDN 和 NFV 都致力于推进一种基于软件的联网方法，使网络的可扩展性更强，从而更灵活和更加通用。

总之，SDN 提供了流量的可编程性、敏捷性，能够实现策略驱动的网络监管和实施网络自动化。SDN 还允许通过创建框架来支持更多的数据密集型应用，例如大数据和虚拟化。SDN

既可以减少网络设备的资本支出(CAPEX)，又可以减少网络的运营和维护支出(OPEX)，因此越来越多的公司开始采用 SDN。

有关 SDN 的更多详细信息，请访问以下资源：

```
https://en.wikipedia.org/wiki/Software-defined_networking
https://en.wikipedia.org/wiki/OpenFlow
https://en.wikipedia.org/wiki/List_of_SDN_controller_software
https://www.cisco.com/c/en_uk/solutions/software-defined-networking/overview.html
https://www.juniper.net/uk/en/products-services/sdn/
https://github.com/mininet/openflow-tutorial/wiki
http://www.projectfloodlight.org/floodlight/
https://www.opendaylight.org/
http://sdnhub.org/tutorials/
```

5.2 Java 网络信息编程

Java 具有丰富的联网功能，适合开发运行在联网和分布式计算机环境中的应用。本节将重点介绍如何使用 Java 开发基于网络的应用。

使用 Java 可以轻松编写程序以获取网络信息。例 5.1 展示了如何使用 InetAddress 类获取计算机的 IP 地址和主机名。图 5.5 给出了编译、执行和输出结果。

例 5.1　NetAddress1.java

```java
//Example 5.1 NetAddress1.java program
import java.net.InetAddress;

class NetAddress1 {
    public static void main(String args[]) throws Exception {
        InetAddress inetAddress = InetAddress.getLocalHost();
        System.out.println("IP Address: " + inetAddress.getHostAddress());
        System.out.println("Host Name: " + inetAddress.getHostName());
    }
}
```

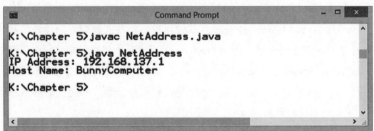

图 5.5　NetAddress.java 的编译、执行和输出结果

例 5.2 展示了如何使用 InetAddress 和 NetworkInterface 类获取计算机的 IP 地址和 MAC 地址。图 5.6 给出了编译、执行和输出结果。子网 24 表示 255.255.255.0。

例 5.2　NetInfo.java

```java
//Example 5.2 NetInfo.java program
import java.net.*;

class NetInfo {
    public static void main(String args[]) throws Exception {
        //Get IP address
        InetAddress ip;
        ip = InetAddress.getLocalHost();
        System.out.println("IP address : " + ip.getHostAddress());
        NetworkInterface network = NetworkInterface.getByInetAddress(ip);

        //Get subnet mask
        InetAddress localHost = Inet4Address.getLocalHost();
        NetworkInterface networkInterface = NetworkInterface.
           getByInetAddress(localHost);
        System.out.println("Subnet Mask : "+networkInterface.
           getInterfaceAddresses().get(0).getNetworkPrefixLength());

        //Get MAC address
        byte[] mac = network.getHardwareAddress();
        StringBuilder sb = new StringBuilder();
        for (int i = 0; i < mac.length; i++) {
                sb.append(String.format("%02X%s", mac[i], (i < mac.
                        length - 1) ? "-" : ""));
        }
        System.out.println("MAC address : " + sb.toString());
    }
}
```

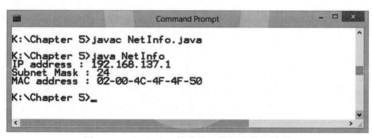

图 5.6　NetInfo.java 的编译、执行和输出结果

练习 5.1　基于例 5.1 和例 5.2 中的 Java 程序，对 InetAddress 类进行一些研究，然后编写一个 Java 示例程序来使用 InetAddress 类定义的其他方法，例如 getAllByNam()、getLoopbackAddress()、hashCode()和 isLinkLocalAddress()。

练习 5.2　基于例 5.1 和例 5.2 中的 Java 程序，对 NetworkInterface 类进行一些研究，然后编写一个 Java 示例程序以使用 NetworkInterface 类定义的其他方法，例如 getDisplayName()、getSubInterfaces()、isPointToPoint()和 supportsMulticast()。

例 5.3 展示了使用 InetAddress 和 NetworkInterface 类的网络信息程序的较长版本。图 5.7 给出了编译、执行和输出结果。

例 5.3 NetAddress2.java

```java
//Example 5.3 NetAddress2.java program
import java.net.*;
import java.util.*;
public class NetAddress2
{
    public static void main(String[] args)
    {
        try
        {
            System.out.println("getLocalHost: " + InetAddress.
                getLocalHost().toString());

            System.out.println("All addresses for local host:");
            InetAddress[] addr = InetAddress.getAllByName(InetAddress.
                getLocalHost().getHostName());
            for(InetAddress a : addr)
            {
                System.out.println(a.toString());
            }
        }
        catch(UnknownHostException _e)
        {
            _e.printStackTrace();
        }
        try
        {
            Enumeration<NetworkInterface> nicEnum = NetworkInterface.
                getNetworkInterfaces();
            while(nicEnum.hasMoreElements())
            {
                NetworkInterface ni=nicEnum.nextElement();
                System.out.println("Name: " + ni.getDisplayName());
                System.out.println("Name: " + ni.getName());
                Enumeration<InetAddress>addrEnum = ni.getInetAddresses();
                while(addrEnum.hasMoreElements())
                {
                    InetAddress ia= addrEnum.nextElement();
                    System.out.println(ia.getHostAddress());
                }
            }
        }
        catch(SocketException _e)
        {
            _e.printStackTrace();
        }
    }
}
```

图 5.7　NetAddress2.java 的编译、执行和输出结果

在 Windows 操作系统中，ipconfig 是获取网络信息的强大命令。可以执行 ipconfig 命令来获取计算机网络信息的摘要，并且可以执行 ipconfig /all 命令来获取更详细的网络信息。例 5.4 展示了如何在 Java 中执行 ipconfig 命令以使用系统调用获取网络信息(就像第 3 章那样)。图 5.8 给出了编译、执行和输出结果。

例 5.4　IPConfig1.java

```java
//Example 5.4 IPConfig1.java program
import java.io.*;

public class IPConfig1 {
    public static void main(String args[]) {
        try {
            String line;
            String[] cmd = {"cmd.exe", "/c", "ipconfig /all"};
            Process p = Runtime.getRuntime().exec(cmd);
            BufferedReader input = new BufferedReader(new
                InputStreamReader(p.getInputStream()));

            while ((line = input.readLine()) != null) {
                System.out.println(line);
            }
            input.close();
        }
        catch (Exception e) {
            e.printStackTrace();
        }
    }
}
```

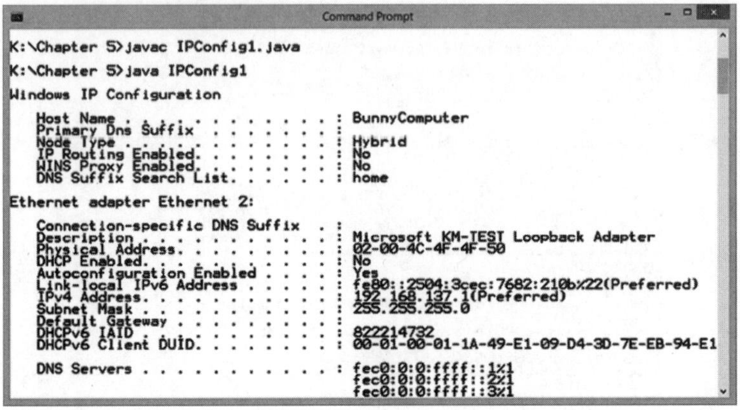

图 5.8　IPConfig1.java 的编译、执行和输出结果

例 5.5 是例 5.4 的改进版本。例 5.5 使用 Java 的 Pattern 和 Match 类在 ipconfig /all 命令的输出中搜索 DHCP 服务器。Java 模式匹配是在给定文本中搜索特定内容的有效方法。以下代码展示了使用模式匹配的步骤：

```
Pattern pattern = Pattern.compile("DHCP Server");   //Create a pattern
Matcher matcher = pattern.matcher("");              //Create a text
matcher.reset(line);                                //Reset the text
matcher.find()                    //Search the pattern in the text
```

图 5.9 给出了例 5.5 的编译、执行和输出结果。

例 5.5　IPConfig2.java

```java
//Example 5.5 IPConfig2.java program
import java.util.*;
import java.util.regex.*;
import java.io.*;
public class IPConfig2 {
    public static void main(String args[]) {
        try {
            String line;
            String[] cmd = {"cmd.exe", "/c", "ipconfig /all"};
            Process p = Runtime.getRuntime().exec(cmd);

            BufferedReader input = new BufferedReader(new
                InputStreamReader(p.getInputStream()));

            Pattern pattern = Pattern.compile("DHCP Server");
            Matcher matcher = pattern.matcher("");
            while ((line = input.readLine()) != null) {
                matcher.reset(line);
                if (matcher.find()) {
                    System.out.println(line);
                }
            }
            input.close();
        } catch (Exception e) {
```

```
            e.printStackTrace();
        }
    }
}
```

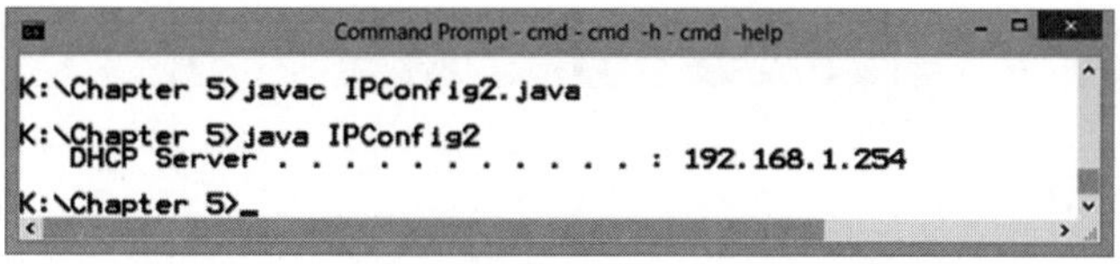

```
Command Prompt - cmd - cmd  -h - cmd -help
K:\Chapter 5>javac IPConfig2.java
K:\Chapter 5>java IPConfig2
   DHCP Server . . . . . . . . . . . : 192.168.1.254
K:\Chapter 5>_
```

图 5.9　IPConfig2.java 的编译、执行和输出结果

练习 5.3　修改例 5.5 中的 Java 程序，使其可以从 ipconfig /all 命令的输出中查找和显示其他信息，例如 DNS 服务器、默认网关、IPv4 地址、子网掩码、物理地址和主机名。

练习 5.4　完成练习 5.3 后，修改 Java 程序，使其可以从 Java 命令行参数中获取 DNS 服务器、默认网关等信息，然后从 ipconfig /all 命令的输出中搜索并显示它们。

5.3　Java 套接字编程

在计算机网络中，套接字是 Internet 上两台计算机之间双向通信通道的端点。套接字由 IP 地址和端口号组成。Java 的 Socket API 支持两种类型的套接字：UDP 套接字和 TCP 套接字。使用 UDP 套接字可以开发 Java 网络应用，例如音频或视频聊天应用，这些应用需要简单、能够进行快速的数据传输并具有最小的延迟。使用 TCP 套接字可以开发 Java 网络应用，例如 Web 应用、电子邮件应用和文件传输应用，它们需要可靠的数据传输和流量控制。

5.3.1　Java UDP 客户端-服务器编程

UDP 是传输层协议。UDP 提供了无连接、不可靠但简单且又快速的通信服务。因此，UDP 主要用于时间关键型应用，例如 Internet 电话、Web 广播或在线视频，在这些应用中，无延迟发送数据比无错误发送数据更重要。例 5.6(分为例 5.6A 和例 5.6B)展示了一个简单的 UDP 客户端-服务器程序。其中，例 5.6A 中的 UDPServer1 是服务器程序，例 5.6B 中的 UDPClient1.java 是客户端程序。在这里，服务器程序需要先运行，等待客户端的连接，并且客户端程序需要在服务器程序运行之后运行，以便可以连接到服务器。在客户端程序中，用户输入将要发送到服务器的单行句子。当服务器接收到这些数据时，就将它们回显给客户端。最后，客户端打印出服务器的响应结果。

以下伪代码展示了 UDPServer1.java 的执行逻辑：

```
begin
    create a Server Datagram Socket on port 3301
    create a Datagram packet for receiving data
    receive data from a client
    print client's IP address and port number
```

```
    create another Datagram packet containing client's data
    send packet to client
    close Socket
end
```

例 5.6A UDPServer1.java

```java
//Example 5.6A UDPServer1.java program
import java.net.*;

class UDPServer1 {
   public static void main(String argv[]) throws Exception{
      byte[] buf = new byte[256];
      String strData;
      int PORT = 3301;
      DatagramSocket serverSocket = new DatagramSocket(PORT);
      DatagramPacket packet = new DatagramPacket(buf, buf.length);

      serverSocket.receive(packet);
      strData = new String(packet.getData());
      InetAddress IPAddress = packet.getAddress();
      PORT = packet.getPort();
         System.out.println("Datagram received: "+IPAddress+":"+PORT);

      buf = strData.getBytes();
      packet = new DatagramPacket(buf, buf.length, IPAddress, PORT);
      serverSocket.send(packet);
      serverSocket.close();
   }
}
```

以下伪代码展示了 UDPClient1.java 的执行逻辑:

```
begin
    create a BufferReader associating with keyboard
    create a client Datagram Socket
    read in a line from keyboard
    create a Datagram packet containing that line
    send packet to server
    create another Datagram packet for receiving data from server
    receive data from server
    print server's response
    close Socket
end
```

例 5.6B UDPClient1.java

```java
// Example 5.6B UDPClient1.java program
import java.io.*;
import java.net.*;
class UDPClient1 {
   public static void main(String argv[]) throws Exception {
       int PORT = 3301;
       String HOST="localhost";
       String strData;
```

```
        System.out.println("Please enter your text:");
        BufferedReader inputLine =
        new BufferedReader(new InputStreamReader(System.in));

        DatagramSocket clientSocket = new DatagramSocket();
        InetAddress IPAddress = InetAddress.getByName(HOST);
        byte[] buf = new byte[256];
        strData = inputLine.readLine();
        buf = strData.getBytes();

        DatagramPacket packet =
        new DatagramPacket(buf, buf.length, IPAddress, PORT);
        clientSocket.send(packet);

        packet = new DatagramPacket(buf, buf.length);
        clientSocket.receive(packet);
        strData = new String(packet.getData());
        System.out.println("FROM SERVER:" + strData.toUpperCase());
        clientSocket.close();
    }
}
```

图 5.10 给出了 UDPServer1.java 和 UDPClient1.java 的编译、执行和输出结果。在这个示例中，服务器在本地主机(127.0.0.1)的 3301 端口上运行，客户端在本地主机(127.0.0.1)的 3024 端口上运行。请注意，这里需要使用两个单独的控制台窗口来运行 UDPServer1.java 和 UDPClient1.java。

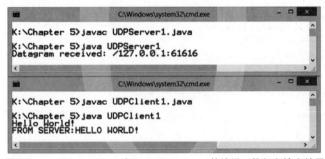

图 5.10 UDPServer1.java 和 UDPClient1.java 的编译、执行和输出结果

练习 5.5 修改例 5.6A 中的 UDP 服务器程序，使其在从客户端接收数据之后将这些数据更改为大写形式，然后回显这些数据。

练习 5.6 修改例 5.6A 中的 UDP 客户端程序，使其使用 for 循环将数据发送到服务器 10 次，并相应地修改 UDP 服务器程序。

5.3.2 Java TCP 客户端-服务器编程

TCP 也是传输层协议。与 UDP 不同，TCP 提供面向连接的、可靠的流服务。由于具有错误检测、错误纠正和流量控制功能，TCP 现在已被广泛应用于 Web、电子邮件、FTP 下载和 Telnet 服务中。在数据的传输过程中，数据的完整性具有最高的优先级。

例 5.7(分为例 5.7A 和例 5.7B)展示了一个简单的 TCP 回显程序。其中，例 5.7A 中的 TCPServer1.java 是提供回显服务的服务器程序；例 5.7B 中的 TCPClient1.java 是客户端程序，用于将一行数据发送到服务器并在屏幕上打印服务器的回显应答。以下伪代码展示了 TCPServer1.java 的执行逻辑：

```
begin
    create a server TCP Socket on port 3301
    create a client Socket waiting for client's connection
    print client's IP address and port number
    create a BufferReader for receiving data from client
    create a DataOutputStream for sending data to client
    send data to client
    close client Socket
    close server Socket
end
```

例 5.7A TCPServer1.java

```java
// Example 5.7A TCPServer1.java program
import java.io.*;
import java.net.*;

class TCPServer1 {
  public static void main(String argv[]) throws Exception {
      ServerSocket serverSocket = null;
      Socket clientSocket = null;
      //You can use any other port numbers (1024 - 65535), if 3301 is
      //not available
      int PORT = 3301;
      serverSocket = new ServerSocket(PORT);
      clientSocket = serverSocket.accept();
      System.out.println("Connected from:"
                    + clientSocket.getInetAddress() + ":"
                    + clientSocket.getPort());
      BufferedReader inputLine =
         new BufferedReader(new InputStreamReader(clientSocket.
            getInputStream()));
      DataOutputStream outputLine =
         new DataOutputStream(clientSocket.getOutputStream());
      outputLine.writeBytes(inputLine.readLine());

      clientSocket.close();
      serverSocket.close();
   }
}
```

以下伪代码展示了 TCPClient1.java 的执行逻辑：

```
begin
    create a BufferReader associating with keyboard
    create a client TCP Socket associating with a TCP server
    print server's IP address and port number
    create a DataOutputStream for sending data to server
    create a BufferReader for receiving data from server
```

```
            read in a line from keyboard and send it to server
            create a Datagram packet containing that line
            receive data from server and print server's response
            close Socket
end
```

例 5.7B　TCPClient.java

```
// Example 5.7B TCPClient1.java program
import java.io.*;
import java.net.*;
class TCPClient1 {
    public static void main(String argv[]) throws Exception {
        Socket clientSocket = null;
        int PORT = 3301;
        String HOST = "localhost";

        BufferedReader inputLine =
            new BufferedReader(new InputStreamReader(System.in));

        clientSocket = new Socket(HOST, PORT);
        System.out.println("Connected to: " + clientSocket.
            getInetAddress() + ":" + clientSocket.getPort());
        DataOutputStream outputLine =
            new DataOutputStream(clientSocket.getOutputStream());
        BufferedReader replyLine =
            new BufferedReader(new InputStreamReader(clientSocket.
                            getInputStream()));

        outputLine.writeBytes(inputLine.readLine() + '\n');
        System.out.println("FROM SERVER: " + replyLine.readLine().
            toUpperCase());

        clientSocket.close();
    }
}
```

图 5.11 给出了 TCPServer1.java 和 TCPClient1.java 的编译、执行和输出结果。同样，你需要首先在一个控制台窗口中运行 TCPServer1.java，然后在另一个控制台窗口中运行 TCPClient1.java。在这里，TCP 服务器在本地主机(127.0.0.1)的 3301 端口上运行，而 TCP 客户端在本地主机(127.0.0.1)的 3020 端口上运行。

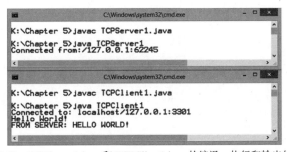

图 5.11　TCPServer1.java 和 TCPClient1.java 的编译、执行和输出结果

练习 5.7 修改例 5.7A 中的 TCPServer1.java 程序,使其可以无限循环地运行。当接收到 date 一词时,把日期信息发送给客户端;当接收到 time 一词时,把时间信息发送给客户端;当接收到 bye 一词,向客户端发送 good-bye 消息,然后终止循环。

练习 5.8 修改例 5.7B 中的 TCPClient1.java 程序,使其也可以无限循环地运行,并且能够将诸如 date、time 和 bye 的文本发送到服务器。

5.3.3　Java 多线程回显服务器编程

前面提到的回显服务器是在单个线程上运行的,因此一次只能接收一个客户端发送过来的数据。例 5.8A 展示的多线程回显服务器允许回显服务器同时接收并处理多个客户端发送过来的数据,并分别回显来自每个客户端的消息。

例 5.8A　多线程回显服务器

```java
//Example 5.8A Multithreaded Echo Server
import java.io.*;
import java.net.*;

public class EchoServer2 {
    public static void main(String [] args){
        ServerSocket echoServer;
        int id=0;
        try{
            echoServer=new ServerSocket(9999);
            while(true){
                Socket clientSocket = echoServer.accept();

                ChatThread cliThread = new ChatThread(clientSocket,id++);
                cliThread.start();
            }
        }
        catch (IOException e)
        {
            System.out.println(e);
        }
    }
}
class ChatThread extends Thread{
    Socket clientSocket;
    int id;

    BufferedReader br;
    PrintWriter os;
    String line;

    ChatThread ( Socket clientSocket, int id ) {
        this.clientSocket = clientSocket;
        this.id = id;
    }
    public void run(){
        try{
```

```java
            br = new BufferedReader(new InputStreamReader(clientSocket
                              .getInputStream()));
            os = new PrintWriter(clientSocket.getOutputStream(),true);

            while ((line = br.readLine())!= null){
                System.out.println(id+"-Received: "+line);
                os.println(line);
            }
        }
        catch (IOException e){
            System.out.println(e);
        }
        finally{
            try{
                br.close();
                os.close();
                clientSocket.close();
                System.out.println(id+"...Stopped");
            }
            catch(Exception e){
                e.printStackTrace();
            }
        }
        //System.out.println(message);
    }
}
```

例 5.8B 展示了相应的回显客户端。

例 5.8B 回显客户端

```java
//Example 5.8B Echo Client
import java.io.*;
import java.net.*;

public class EchoClient{

    public static void main(String[] args){
        Socket echoSocket;
        PrintWriter out;
        BufferedReader in;
        try
        {
            echoSocket = new Socket("localhost", 9999);
            out = new PrintWriter(echoSocket.getOutputStream(),true);
            in = new BufferedReader(new InputStreamReader(
                echoSocket.getInputStream()));
            BufferedReader stdIn = new BufferedReader(
                new InputStreamReader(System.in));

            String userInput;
            while ((userInput = stdIn.readLine()) != null)
            {
                out.println(userInput);
                System.out.println("echo: " + in.readLine());
```

```
            }
                out.close();
                in.close();
                stdIn.close();
                echoSocket.close();
            }
            catch(Exception e)
            {
                System.out.println("Error: "+e.toString());
                System.exit(-1);
            }
        }
    }
```

练习 5.9 修改例 5.8A 中的多线程回显服务器,使其能够从命令行参数(通过使用 args)获取端口号。

练习 5.10 修改例 5.8B 中的回显客户端,使其能够从命令行参数(通过使用 args)获取服务器的 IP 地址和端口号。

有关 Java 套接字编程的更多信息,请参见以下资源:

https://docs.oracle.com/javase/tutorial/networking/sockets/index.html
https://www.javaworld.com/article/2077322/core-java/core-java-sockets-programming-in-java-a-tutorial.html
https://o7planning.org/en/10393/java-socket-programming-tutorial

5.4 Java HTTP 编程

HTTP 是 Web 客户端和服务器之间的通信协议。HTTP 以请求和响应的模式进行工作,其中 Web 客户端(Web 浏览器)将请求发送到 Web 服务器,然后 Web 服务器以网页的形式进行响应。HTTP 是万维网背后的推动力,毫无疑问,万维网无疑是当今 Internet 上最受欢迎的服务。利用 Java 可以轻松地使用 HTTP 开发 Web 应用。

5.4.1 Java HTTP/HTTPS 客户端

例 5.9(分为例 5.9A、例 5.9B 和例 5.9C)展示了一系列 Java HTTP/HTTPS 客户端。首先,例 5.9A 展示了使用 URL 和 URLConnection 类的简单客户端。在例 5.9A 中,我们将打开由统一资源定位符(Uniform Resource Locator,URL)指定的网页,在本例中为 www.google.com/。图 5.12 给出了编译、执行和输出结果。

例 5.9A 一个 HTTP 客户端

```
//Example 5.9A HTTP Client
import java.net.*;
import java.io.*;
```

```
public class HTTPClient {
    public static void main(String[] args) throws Exception {
        URL u = new URL("http://www.google.com/");
        URLConnection uc = u.openConnection();
        BufferedReader in = new BufferedReader(new InputStreamReader(
                            uc.getInputStream()));
        String inputLine;
        while ((inputLine = in.readLine()) != null)
           System.out.println(inputLine);
        in.close();
    }
}
```

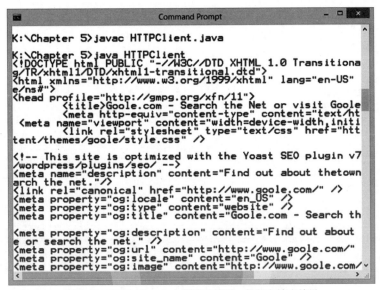

图 5.12　HTTPClient.java 程序的编译、执行和输出结果

练习 5.11　修改例 5.9A 中的 Java 程序，使其可以从命令行参数(通过使用 args)获取网站的 URL。

例 5.9B 是例 5.9A 的另一个版本，例 5.9B 借助 HTTPS 协议并使用 HttpsURLConnection() 来连接到安全的网站。

例 5.9B　一个 HTTPS 客户端

```
//Example 5.9B: HTTPS Client
import java.net.*;
import java.io.*;
import javax.net.ssl.HttpsURLConnection;

public class HTTPSClient {
    public static void main(String[] args) throws Exception {
        URL u = new URL("https://www.google.com/");
        HttpsURLConnection uc = (HttpsURLConnection) u.openConnection();
        BufferedReader in = new BufferedReader(new InputStreamReader(
                            uc.getInputStream()));
```

```
        String inputLine;
        while ((inputLine = in.readLine()) != null)
            System.out.println(inputLine);
        in.close();
    }
}
```

例 5.9C 是例 5.9B 的另一个版本，例 5.9C 也从 HTTPS 服务器打印数字证书信息。

例 5.9C　另一个 HTTPS 客户端

```
//Example 5.9c HTTP Client 2
import java.net.*;
import java.io.*;
import java.security.cert.Certificate;
import javax.net.ssl.HttpsURLConnection;
public class HTTPSClient2 {
    public static void main(String[] args) throws Exception {
        String urltxt = "https://www.google.com/";
        URL u = new URL(urltxt);
        HttpsURLConnection uc = (HttpsURLConnection) u.openConnection();
        try {
            System.out.println("Response Code : " + uc.getResponseCode());
            System.out.println("Cipher Suite : " + uc.getCipherSuite());
            System.out.println("\n");

            Certificate[] certs = uc.getServerCertificates();
            for(Certificate cert : certs){
                System.out.println("Cert Type : " + cert.getType());
                System.out.println("Cert Hash Code : " + cert.hashCode());
                System.out.println("Cert Public Key Algorithm : "
                                   + cert.getPublicKey().getAlgorithm());
                System.out.println("Cert Public Key Format : "
                                   + cert.getPublicKey().getFormat());
                System.out.println("\n");
            }
        } catch (IOException e){
            e.printStackTrace();
        }

        BufferedReader in = new BufferedReader(new InputStreamReader(
                              uc.getInputStream()));
        String inputLine;
        while ((inputLine = in.readLine()) != null)
            System.out.println(inputLine);
        in.close();
    }
}
```

HTTP/HTTPS 客户端可以使用 GET 或 POST 方法与 HTTP/HTTPS 服务器进行通信。一般默认使用的是 GET 方法，GET 方法用于从服务器检索信息。POST 方法用于向服务器发送信息，例如发送登录详情、提交表单、搜索关键字等。例 5.10 展示了一个简单的 Java HTTP/HTTPS 客户端程序，它使用 GET 方法从 HTTP/HTTPS 服务器获取信息，与之前的示例类似。例 5.10

将打开由 URL 指定的网页，在这里为 https://docs.oracle.com/javase/tutorial/。图 5.13 给出了编译、执行和输出结果。

例 5.10 HTTPGET1.Java

```java
//Example 5.10 The HTTPGet1.java program
import java.io.*;
import java.net.*;

public class HTTPGet1 {
    // HTTP GET request
    public static void getHTML(String website) throws Exception {
        URL url = new URL(website);
        HttpURLConnection conn = (HttpURLConnection) url.openConnection();
        conn.setRequestMethod("GET");
        conn.setRequestProperty("User-Agent", "Chrome/51.0.2704.63");
        conn.setRequestProperty("Accept-Language", "en-US,en");
        int responseCode = conn.getResponseCode();
        System.out.println("\nSending 'GET' request to URL : " + website);
        System.out.println("Response Code : " + responseCode);

        BufferedReader br = new BufferedReader(
                new InputStreamReader(conn.getInputStream()));
        StringBuilder response = new StringBuilder();
        String line;
        while ((line = br.readLine()) != null) {
            response.append(line);
        }
        br.close();
        System.out.println( response.toString());
    }
    public static void main(String[] args) throws Exception
    {
        getHTML("https://docs.oracle.com/javase/tutorial/");
    }
}
```

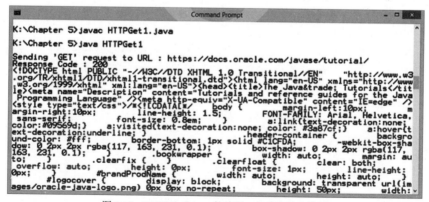

图 5.13 HTTPGet1.java 的编译、执行和输出结果

练习 5.12 修改例 5.10 中的 Java 程序，使其可以使用 args 从命令行参数获取网站的 URL。

练习 5.13 在例 5.10 所示 Java 程序的基础上，对 Certificate 类进行研究学习，并编写一个 Java 示例程序以使用 Certificate 类定义的其他方法，例如 verify(PublicKey key)、hashCode()和 toString()。

例 5.11 展示了一个简单的 Java HTTP 客户端程序，该客户端程序使用 POST 方法将搜索结果发送到服务器。在这个示例中，我们将打开网页 https://www.amazon.co.uk/s/ref=nb_sb_noss_2? 并搜索关键字 java。图 5.14 给出了编译、执行和输出结果。由于输出长度受限，这里只打印前 1000 个字符。你还可以使用其他网站(例如 Google)进行搜索，只需要取消注释代码中的以下两行即可：

```
//String website="https://www.google.co.uk/search?";
//String urlParameters="q=java";
```

例 5.11　HTTPPost1.java

```
//Example 5.11 HTTPPost1.java program
import java.io.*;
import java.net.*;
import javax.net.ssl.HttpsURLConnection;

public class HTTPPost1 {
    // HTTP POST request
    private static void postHTML(String website, String urlParameters)
throws Exception {
        URL url = new URL(website);
        HttpsURLConnection conn = (HttpsURLConnection) url.openConnection();
        conn.setRequestMethod("POST");
        conn.setRequestProperty("User-Agent", "Chrome/51.0.2704.63");
        conn.setRequestProperty("Accept-Language", "en-US,en");
        // Send post request
        conn.setDoOutput(true);
        DataOutputStream wr = new DataOutputStream(conn.getOutputStream());
        wr.writeBytes(urlParameters);
        wr.flush();
        wr.close();

        int responseCode = conn.getResponseCode();
        System.out.println("\nSending 'POST' request to URL : " + website);
        System.out.println("Post parameters : " + urlParameters);
        System.out.println("Response Code : " + responseCode);

        BufferedReader in = new BufferedReader(
            new InputStreamReader(conn.getInputStream()));
        String inputLine;
        StringBuffer response = new StringBuffer();

        while ((inputLine = in.readLine()) != null) {
            response.append(inputLine);
        }
        in.close();

        //print result (first 1000 characters)
```

```
            System.out.println(response.toString().substring(0,1000));
    }
    public static void main(String[] args) throws Exception
    {
        //Search java in Amazon
        String website = "https://www.amazon.co.uk/s/ref=nb_sb_noss_2?";
        String urlParameters = "url=search-alias%3Daps&field-keywords=java";
        //Search java in Google
        //String website = "https://www.google.co.uk/search?";
        //String urlParameters = "q=java";

        postHTML(website, urlParameters);
    }
}
```

图 5.14　HTTPPost1.java 的编译、执行和输出结果

练习 5.14　修改例 5.11 中的 Java 程序，使其可以使用 args 从命令行参数获取网站的 URL 和 URL 参数。

例 5.12 展示了另一个使用 URL 类的示例程序。例 5.12 通过解析网页的 URL 来显示相关的信息，例如协议、主机、端口、路径和查询。图 5.15 给出了编译、执行和输出结果。

例 5.12　URL 解析程序

```
//Example 5.12 URL Parsing
import java.net.*;
public class URLExample1{
    public static void main(String[] args) {
        try {
            URL url = new URL("http://www.google.come:80/search?q=java");
            System.out.println("URL created: " + url);
            System.out.println("protocol: " + url.getProtocol());
            System.out.println("host: " + url.getHost());
            System.out.println("port: " + url.getPort());
            System.out.println("path: " + url.getPath());
            System.out.println("query: " + url.getQuery());
        }
        catch (MalformedURLException e) {
```

```
            System.out.println("Malformed URL: " + e.getMessage());
        }
    }
}
```

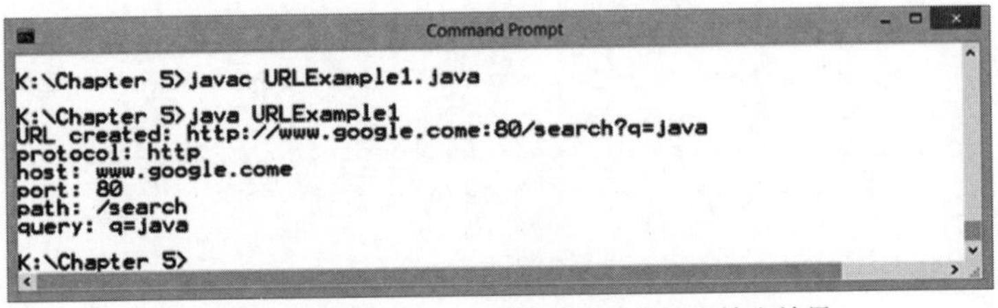

```
Command Prompt
K:\Chapter 5>javac URLExample1.java
K:\Chapter 5>java URLExample1
URL created: http://www.google.come:80/search?q=java
protocol: http
host: www.google.come
port: 80
path: /search
query: q=java
K:\Chapter 5>
```

图 5.15　URLExample1.java 的编译、执行和输出结果

5.4.2　Java HTTP 服务器

例 5.13 展示了一个使用 ServerSocket 类的 Java HTTP 服务器。它运行在 8088 端口上，并且在连接 Web 浏览器客户端时，将在浏览器中显示 Hello World！图 5.16 的上半部给出了编译、执行和输出结果。要查看网页，请打开 Web 浏览器，输入计算机的 IP 地址和端口号 8088，然后按 Enter 键，效果如图 5.16 的下半部所示。

例 5.13　一个 Java HTTP 服务器

```java
//Example 5.13 HTTP Server
import java.io.*;
import java.net.*;

public class HTTPServer {
  protected void start() {
    ServerSocket s;

    System.out.println("Starting up HTTP Server...");
    try {
      s = new ServerSocket(8088);
    } catch (Exception e) {
      System.out.println("Error: " + e);
      return;
    }

    System.out.println("Waiting... ");
    for (;;) {
      try {
        Socket remote = s.accept();
        System.out.println("Connected....");
        BufferedReader in = new BufferedReader(new InputStreamReader(
            remote.getInputStream()));
        PrintWriter out = new PrintWriter(remote.getOutputStream());

        String str = ".";
        while (!str.equals("")){
          str = in.readLine();
```

```
            System.out.println(str);
        }

        out.println("HTTP/1.1 200 OK");
        out.println("Content-Type: text/html");
        out.println("Server: Java HTTP Server");
        out.println("");
        // Send the HTML page
        out.println("<H1>Hello World!</H1>");
        out.flush();
        remote.close();
    } catch (Exception e) {
        System.out.println("Error: " + e);
    }
  }
}

    public static void main(String args[]) {
        HTTPServer sr = new HTTPServer();
        sr.start();
    }
}
```

图 5.16　HTTPServer.java 的编译、执行和输出结果(上半部)以及测试结果(下半部)

练习 5.15 修改例 5.13 中的 Java 程序，使其能够在网页上打印日期和时间信息以及客户端的 IP 地址。

5.4.3 Java 多线程 HTTP 服务器

例 5.14A 是例 5.13 中的 Java HTTP 服务器的多线程版本。在这个版本中，我们首先在 8088 端口上创建一个 ServerSocket 对象，然后使用 while 循环侦听客户端请求。每当接到客户端连接请求时，就生成一个单独的线程来处理连接。线程内部的 run() 方法用于从客户端读取 HTTP 请求消息的所有行，并回送 Hello Multithreaded HTTP Server! 信息。

例 5.14A 一个 Java 多线程 HTTP 服务器

```java
//Example 5.14A Multithreaded HTTP Server
import java.io.*;
import java.net.*;

public class HTTPServer2 {
    public static void main(String args[]) {
        int port =8088;
        ServerSocket web;
        try {
            web = new ServerSocket(port);
        } catch (Exception e) {
            System.out.println("Error: " + e);
            return;
        }

        while(true){
            System.out.println("MultiThreaded Web Server Running on port:" + port);
            try{
                Socket client = web.accept();
                System.out.println("Accepted Client : " + client);
                HTTPServer2Thread s = new HTTPServer2Thread(client);
                s.start();
            }
            catch(Exception e){
                System.out.println("Error: " + e);
                return;
            }
        }
    }
}
class HTTPServer2Thread extends Thread{
        Socket client;
        HTTPServer2Thread(Socket client){
            this.client = client;
        }
    public void run(){
        try {
            BufferedReader in = new BufferedReader(new
                InputStreamReader(client.getInputStream()));
```

```
            PrintWriter out = new PrintWriter(client.getOutputStream());

            String str = ".";
            while (!str.equals("")){
                str = in.readLine();
                System.out.println(str);
            }
            out.println("HTTP/1.1 200 OK");
            out.println("Content-Type: text/html");
            out.println("Server: Java HTTP Server");
            out.println("");
            // Send the HTML page
            out.println("<H1>Hello Multithreaded HTTP Server!</H1>");
            out.flush();
            out.close();
            client.close();
        } catch (Exception e) {
            System.out.println("Error: " + e);
        }
    }
}
```

> **练习 5.16** 修改例 5.14A 中的 Java 程序，让它使用主类继承 Thread 类，而不是使用单独的 Thread 类。

例 5.14B 是例 5.13 的另一个多线程版本。在这个版本中，我们找出客户端请求的文件并打开，逐行读取，然后逐行发送给客户端。为了测试服务器，只需要打开 Web 浏览器，然后输入计算机的 IP 地址和 8088 端口，显示的结果与图 5.16 完全相同。

例 5.14B　一个 Java 多线程 HTTP 文件服务器

```
//Example 5.14B Multithreaded HTTP File Server
import java.io.*;
import java.net.*;

public class HTTPServer3 {
    public static void main(String args[]) {
        int port =8088;
        ServerSocket web;
        try {
            web = new ServerSocket(port);
        } catch (Exception e) {
            System.out.println("Error: " + e);
            return;
        }

        while(true){
            System.out.println("MultiThreaded Web Server Running on port:" + port);
            try{
                Socket client = web.accept();
                System.out.println("Accepted Client : " + client);
                HTTPServer3Thread s = new HTTPServer3Thread(client);
```

```java
                    s.start();
                }
                catch(Exception e){
                    System.out.println("Error: " + e);
                    return;
                }
            }
        }
    }
}
class HTTPServer3Thread extends Thread{
    Socket client;
    HTTPServer3Thread(Socket client){
        this.client = client;
    }
    public void run(){
        try {
            BufferedReader in = new BufferedReader(new
                InputStreamReader(client.getInputStream()));
            PrintWriter out = new PrintWriter(client.getOutputStream());
            String reqMeth = "";
            String reqURL = "";
            String reqProto = "";
            String str = ".";
            int i=0;
            while (!str.equals("")){
                str = in.readLine();
                System.out.println(i+": " +str);
                if (i==0){
                    reqMeth = str.substring(0, 3);
                    reqURL = str.substring(5, (str.lastIndexOf("HTTP/1.")));
                    reqProto = str.substring(str.indexOf("HTTP/1."));
                }
                i++;
            }
            //Send HTTP header
            out.println("HTTP/1.1 200 OK");
            out.println("Content-Type: text/html");
            out.println("Server: Java HTTP Server");
            out.println("");
            // Open the file and send out line by line
            FileReader file=new FileReader(reqURL.trim());
            BufferedReader filebuff= new BufferedReader(file);
            boolean endof=false;
            String line;
            while (!endof) {
                line=filebuff.readLine();
                if (line == null){ endof=true; break;}
                out.println(line);
            }
            file.close();

            //Close all the streams
            out.flush();
            out.close();
            client.close();
        } catch (Exception e) {
```

```
            System.out.println("Error: " + e);
        }
    }
}
```

练习 5.17　修改例 5.14B 中的 Java 程序，让它使用主类继承 Thread 类，而不是使用单独的 Thread 类。

5.5　Java 电子邮件 SMTP 编程

简单邮件传输协议(Simple Mail Transfer Protocol，SMTP)用于通过 Internet 传输电子邮件。例 5.15 展示了一个 Java SMTP 示例程序。在这里，我们连接到运行 SMTP 的位于端口 25 的电子邮件服务器，并使用 HELO、MAIL、RCPT TO、DATA 和 QUIT 向接收者发送电子邮件。

例 5.15　一个 Java SMTP 示例程序

```
//Example 5.15 SMTP Example
import java.io.*;
import java.net.*;

public class smtpClient {
    public static void main(String[] args) {
        Socket smtpSocket = null;
        DataOutputStream os = null;

        BufferedReader br = null;
        //Please use your own email settings for the following parameters
        String client = "Your Name";
        String server = " yourmailserver.com";
        String sender = "youremail@yourmailserver.com";
        String receiver = "receiver@receiverserver.com";

        try {
            smtpSocket = new Socket(server, 25);
            os = new DataOutputStream(smtpSocket.getOutputStream());

            br = new BufferedReader(new InputStreamReader(smtpSocket.
                    getInputStream()));
        } catch (UnknownHostException e) {
           System.err.println("Don't know about host: "+server);
        } catch (IOException e) {
           System.err.println("Couldn't get I/O for the connection to:
                        hostname");
        }

        if (smtpSocket != null && os != null && br != null) {
            try {
                os.writeBytes("HELO "+ client +"\n");
                os.writeBytes("MAIL From: "+ sender +"\n");
```

```
            os.writeBytes("RCPT To: "+ receiver +"\n");
            os.writeBytes("DATA\n");
            os.writeBytes("From: "+ sender +"\n");
            os.writeBytes("Subject: testing\n");
            os.writeBytes("Hi there 1\n"); // message body
            os.writeBytes("\n.\n");
            os.writeBytes("QUIT\n");

            String responseLine;
            while ((responseLine = br.readLine()) != null) {
               System.out.println("Server: " + responseLine);
               if (responseLine.indexOf("Ok") != -1) {
                  break;
               }
            }

            os.close();
            smtpSocket.close();
        } catch (UnknownHostException e) {
            System.err.println("Trying to connect to unknown host: "+ e);
        } catch (IOException e) {
            System.err.println("IOException: " + e);
        }
    }
}
```

发送和检查电子邮件的一种更优雅但也更复杂的方法是使用 JavaMail API。为了使用 JavaMail API，需要下载两个 JAR 文件。其中一个名为 javax.mail.jar，这个 JAR 文件可以通过网址 https://javaee.github.io/javamail/ 获取。另一个名为 activation.jar，这个 JAR 文件可以通过网址 https://www.oracle.com/technetwork/java/javase/downloads/index-135046.html 获取，进而使用 JAF(JavaBeans Activation Framework)。

将下载的 javax.mail.jar 和 Activation.jar 文件复制到当前的 Java 目录中。例 5.16A 所示的 Java 程序使用 JavaMail API 通过安全传输层协议(Transport Layer Security，TLS)发送电子邮件。请相应地修改 SMTP 服务器的名称、用户名、密码、电子邮件地址以及收件人的电子邮件地址。注意，如果使用双因素或多因素身份验证来设置电子邮件，那么例 5.16A 中的代码将不起作用。

例 5.16A　JavaMail API 示例程序

```
//Example 5.16A JavaMail API Example
import java.util.*;
import javax.mail.*;
import javax.mail.internet.*;

public class JavaMail1 {
    public static void main(String[] args) {
        final String username = "youremail@gmail.com";
        final String password = "yourpassword";
        Properties props = new Properties();
        props.put("mail.smtp.auth", "true");
        props.put("mail.smtp.starttls.enable", "true");
```

```
        props.put("mail.smtp.host", "smtp.gmail.com");
        props.put("mail.smtp.port", "587"); //SSL/TLS port :587 normal:465

        Session session = Session.getInstance(props,
          new javax.mail.Authenticator() {
            protected PasswordAuthentication getPasswordAuthentication()
            {
               return new PasswordAuthentication(username, password);
            }
          });

        try {
            Message message = new MimeMessage(session);
            message.setFrom(new InternetAddress("youremail@gmail.com"));
            message.setRecipients(Message.RecipientType.TO,
               InternetAddress.parse("someone@somewhere.com"));
            message.setSubject("Your Research Paper");
            message.setText("Dear Perry,"
               + "\n\n Could please send me your GLCM paper?"
               + "\n\n Cheers!"
               + "\n\n Tom");

            Transport.send(message);
            System.out.println("Done");

        } catch (MessagingException e) {
            throw new RuntimeException(e);
        }
    }
}
```

为了编译和执行例 5.14A 中的 Java 程序，需要在类路径中包括 javax.mail.jar 和 activation.jar 文件，如下所示：

```
javac -classpath .;javax.mail.jar;activation.jar JavaMail1.java
java -classpath .;javax.mail.jar;activation.jar JavaMail1
```

例 5.16B 展示了另一个简单的 Java 应用，用于使用 POP3(Post Office Protocol 3)协议检查电子邮件。同样，这里也使用 JavaMail API 通过 TLS 建立连接。请根据实际情况相应地修改 POP3 服务器的主机名、用户名和密码。

例 5.16B　JavaMail API POP3 示例程序

```
//Example 5.16B JavaMail API POP3 Example
import java.util.*;
import javax.mail.*;

public class JavaMail2 {
    public static void main(String[] args) {

        String host = "pop.gmail.com";
        String username = "yourname@gmail.com";
```

```java
            String password = "yourpassword";

            try {
                //create properties field
                Properties properties = new Properties();
                properties.put("mail.pop3.host", host);
                properties.put("mail.pop3.port", "995");
                properties.put("mail.pop3.starttls.enable", "true");
                Session emailSession = Session.getDefaultInstance(properties);

                //create the POP3 store object and connect with the server
                Store store = emailSession.getStore("pop3s");
                store.connect(host, username, password);

                //create the folder object and open it
                Folder emailFolder = store.getFolder("INBOX");
                emailFolder.open(Folder.READ_ONLY);

                // retrieve the messages from the folder in an array and print it
                Message[] messages = emailFolder.getMessages();
            for (int i = 0, n = messages.length; i < n; i++) {
                Message message = messages[i];
                System.out.println("---------------------------------");
                System.out.println("Email Number " + (i + 1));
                System.out.println("Subject: " + message.getSubject());
                System.out.println("From: " + message.getFrom()[0]);
                System.out.println("Text: " + message.getContent().toString());

                }
                //close the store and folder objects
                emailFolder.close(false);
                store.close();
            } catch (Exception e) {
                e.printStackTrace();
            }
        }
    }
```

为了编译并执行上面的 Java 程序，只需要输入以下内容：

```
javac -classpath .;javax.mail.jar;activation.jar JavaMail2.java
java -classpath .;javax.mail.jar;activation.jar JavaMail2
```

有关 SMTP 和 JavaMail API 的更多信息，请访问以下链接：

```
https://en.wikipedia.org/wiki/Simple_Mail_Transfer_Protocol
https://www.oracle.com/technetwork/java/javamail/index.html
https://www.tutorialspoint.com/javamail_api/index.htm
```

https://www.journaldev.com/2532/javamail-example-send-mail-in-java-smtp
https://www.javatpoint.com/java-mail-api-tutorial

5.6 Java RMI 客户端-服务器编程

远程方法调用(Remote Method Invocation，RMI)是一种分布式系统技术，这种技术允许一个 Java JVM(Java Virtual Machine，虚拟机)调用运行在网络上其他位置的另一个 JVM 上的对象方法，这有点类似于 20 世纪 80 年代开发的远程过程调用(Remote Procedure Calls，RPC)，RPC 允许以任意语言编写的过程式程序调用驻留在另一台计算机上的函数。但是 RPC 仅支持有限的简单数据类型，并且不支持对象。RMI 尽管当前仅支持 Java 语言，但也允许传递和返回 Java 对象，这使得 RMI 相比 RPC 更强大。RMI 对于大规模系统的开发很有用，因为可以在多台机器上分配资源和处理负载。

RMI 应用分为两种：RMI 服务器和 RMI 客户端。RMI 服务器创建远程对象并将其注册到查找服务，以允许 RMI 客户端找到它们。RMI 客户端使用了 RMI 服务器中一个或多个远程对象的远程引用，然后调用对象的方法。

为了开发 RMI 应用，需要创建如下四个程序：
- 远程接口程序
- 远程接口实施程序
- RMI 服务器程序
- RMI 客户端程序

远程接口程序是一个 Java 接口，它定义了 RMI 客户端可以调用的方法。远程接口实施程序实现了远程接口程序中定义的方法。RMI 服务器程序用于提供服务，供 RMI 客户端程序使用。远程接口程序、远程接口实施程序和 RMI 服务器程序在服务器计算机上运行，而 RMI 客户端程序在客户端计算机上运行。

例 5.17 展示的 RMI 应用可以计算一系列双精度浮点数的平均值，其中包含四个文件：Analysor1.java、Analysor1Impl.java、Analysor1Server.java 和 Analysor1Client.java。文件 Analysor1.java 是仅仅定义了 Average()方法的远程接口程序。Analysor1Impl.java 实现了 Average()方法。Analysor1Server.java 提供了名为 AnalysorService 的 RMI 服务，其中包含 Average()方法。默认情况下，RMI 服务在端口 1099 上运行。以下是 RMI 服务器程序中的两条关键语句：

```
Analysor1 c = new Analysor1Impl();
Naming.rebind("rmi://localhost:1099/AnalysorService", c);
```

Analysor1Client.java 是 RMI 客户端程序，用于查找 AnalysorService 服务并调用 Average() 方法。以下是 RMI 客户端程序中的两条关键语句：

```
Analysor1 c = (Analysor1) Naming.lookup("rmi://localhost:1099/AnalysorService");
System.out.println( "The average of array d: "+ c.Average(d));
```

例 5.17A 展示了 Analysor1.java 文件。

例 5.17A　Analysor1.java 远程接口程序

```java
// Example 5.17a Analysor1.java RMI interface program
public interface Analysor1 extends java.rmi.Remote {
    public double Average(double d[])
        throws java.rmi.RemoteException;
}
```

例 5.17B 展示了 Analysor1Impl.java 文件。

例 5.17B　Analysor1Impl.java 远程接口实施程序

```java
// Example 5.17B Analysor1Impl.java RMI interface implementation program
public class Analysor1Impl
    extends java.rmi.server.UnicastRemoteObject implements Analysor1 {
    public Analysor1Impl() throws java.rmi.RemoteException {
        super();
    }

    public double Average(double d[])
        throws java.rmi.RemoteException {
        double r=0.0;
        for (int i=0;i<d.length;i++){
            r+=d[i];
        }
        r/=d.length;
        return r;
    }
}
```

例 5.17C 展示了 Analysor1Server.java 文件。

例 5.17C　Analysor1Server.java RMI 服务器程序

```java
// Example 5.17c Analysor1Server.java RMI server program
import java.rmi.Naming;

public class Analysor1Server {

  public Analysor1Server() {
    try {
      Analysor1 c = new Analysor1Impl();
      Naming.rebind("rmi://localhost:1099/AnalysorService", c);
    } catch (Exception e) {
      System.out.println("Trouble: " + e);
    }
  }

  public static void main(String args[]) {
    new Analysor1Server();
  }
}
```

例 5.17D 展示了 Analysor1Client.java 文件。

例 5.17D　Analysor1Client. java RMI 客户端程序

```java
// Example 5.17d Analysor1Client.java RMI client program
import java.rmi.Naming;
import java.rmi.RemoteException;
import java.net.MalformedURLException;
import java.rmi.NotBoundException;

public class Analysor1Client {

    public static void main(String[] args) {

        double d[]=new double[10];
        for (int i=0;i<d.length;i++){
            d[i]=i;
        }

        try {
            Analysor1 c = (Analysor1) Naming.lookup("rmi://localhost:1099/
                AnalysorService");
            System.out.println( "The average of array d: "+c.Average(d));
        }
        catch (MalformedURLException murle) {
            System.out.println();
            System.out.println("MalformedURLException");
            System.out.println(murle);
        }
        catch (RemoteException re) {
            System.out.println();
            System.out.println("RemoteException");
            System.out.println(re);
        }
        catch (NotBoundException nbe) {
            System.out.println();
            System.out.println("NotBoundException");
            System.out.println(nbe);
        }
        catch (java.lang.ArithmeticException ae) {
            System.out.println();
            System.out.println("java.lang.ArithmeticException");
            System.out.println(ae);
        }
    }
}
```

为了运行 RMI 应用，请执行下列步骤：

(1) 启动 RMI 注册表，以便 RMI 服务器可以注册服务。

(2) 启动 RMI 服务器。

(3) 启动 RMI 客户端。

详细信息如图 5.17 所示。

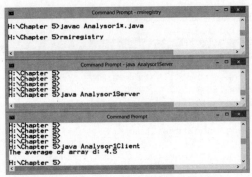

图 5.17 所有 RMI Java 程序的编译以及 RMI 注册表的启动结果(顶部);Analysor1Server.java RMI 服务器的启动结果(中间);Analysor1Client.java RMI 客户端的启动结果(底部)

5.7 SDN 入门

本章在前面探讨了软件定义网络(SDN)的概念。SDN 有着巨大的潜力,无疑是计算机网络的未来发展方向。如果想要开始使用 SDN,以下将要介绍的一些教程是不错的起点。在开始使用 SDN 之前,这些教程将会向你介绍一些最重要的 SDN 工具:OpenFlow、Floodlight 和 OpenDaylight。

5.7.1 OpenFlow 入门

OpenFlow 的最简单入门方法是使用 Mininet。图 5.18 展示的 GitHub 网站(https://github.com/mininet/openflow-tutorial/wiki)介绍了如何下载和设置相关软件,并给出了一些示例,涉及学习开发工具、创建学习交换机、如何控制真实网络的一部分、路由器练习、高级拓扑以及创建防火墙。

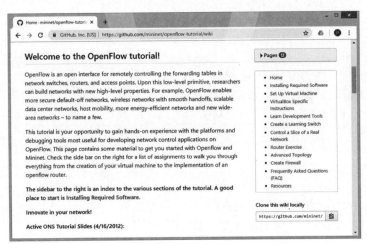

图 5.18 OpenFlow 教程中介绍的 GitHub 网站

我们首先需要下载以下软件。

- Mininet VM 镜像:https://github.com/mininet/mininet/wiki/Mininet-VM-Images。

- VirtualBox：https://www.virtualbox.org/wiki/Downloads。
- Xming：https://sourceforge.net/projects/xming/files/Xming/6.9.0.31/Xming-6-9-0-31- setup.exe/Downloads。
- Putty.exe：http://the.earth.li/~sgtatham/putty/latest/x86/putty.exe。

然后参考以下链接将 Mininet 虚拟机文件导入 VirtualBox 软件：

```
https://github.com/mininet/openflow-tutorial/wiki/Set-up-Virtual-Machine
http://mininet.org/vm-setup-notes/
```

图 5.19 展示了 VirtualBox 中的 Mininet 虚拟机，图 5.20 展示了 Mininet 虚拟机的登录界面。启动 Mininet 虚拟机，使用 mininet 作为用户名和密码进行登录。Mininet 虚拟机基于 Ubuntu Linux 操作系统。

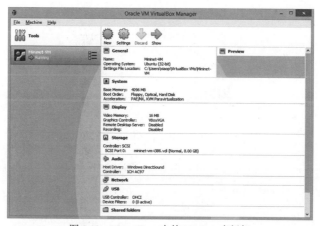

图 5.19　VirtualBox 中的 Mininet 虚拟机

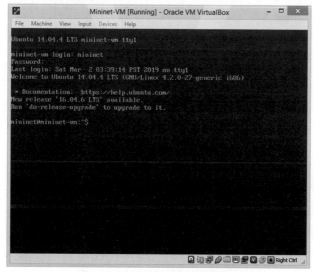

图 5.20　Mininet 虚拟机的登录界面

登录后，在 Linux shell 提示符($)中，可以通过输入以下命令来启动 Mininet 虚拟机并进入命令行界面(CLI)模式。sudo 命令表示以超级用户或管理员身份运行 Mininet 虚拟机。

```
$ sudo mn
```

这将启动具有最小拓扑的 Mininet 模拟网络，里面包括一个控制器(c0)、一台交换机(s1)和两台主机(h1 和 h2)，如图 5.21 所示。

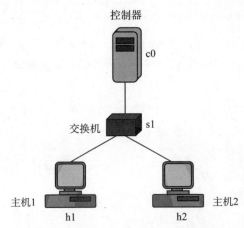

图 5.21　具有最小拓扑的 Mininet 模拟网络

也可以使用--topo 选项从其他拓扑启动 Mininet 模拟网络。例如，下面的命令将使用一台交换机和四台主机启动 Mininet 模拟网络：

```
$ sudo mn --topo single,4
```

可以使用下面的命令启动另一个 Mininet 模拟网络，这个 Mininet 模拟网络会将三台交换机连接在一起，其中的每台交换机都有一台主机：

```
$ sudo mn --topo linear,3
```

可以使用下面的命令来显示 Mininet 虚拟机的更多启动选项：

```
$ sudo mn -h
```

启动 Mininet CLI 后，在 mininet>提示符下输入以下命令以显示 Mininet 模拟网络中的所有节点：

```
mininet> nodes
```

下面的命令用来显示 Mininet 模拟网络中的网络信息：

```
mininet> net
```

下面的命令用来显示 Mininet 模拟网络中所有链接的信息：

```
mininet> links
```

下面的命令用来显示 Mininet 模拟网络中所有节点的详细信息，包括每个节点的 IP 地址：

```
mininet> dump
```

下面的命令用来显示所有 Mininet CLI 命令的帮助信息：

mininet> help

图 5.22 展示了上述部分命令的输出结果。

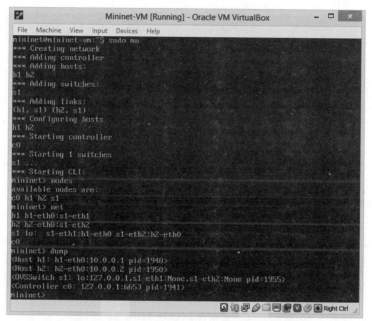

图 5.22　以最小拓扑显示 Mininet 模拟网络的信息

ifconfig 是非常有用的 Linux 命令，用于显示网络信息和配置网络接口。ifconfig 类似于我们先前使用的 Windows ipconfig 命令。可以在 h1 节点上执行 ifconfig 命令，方法是输入以下命令以显示 h1 节点的网络信息：

mininet> h1 ifconfig

在这种情况下，所有节点都使用 IP 地址 10.0.0.0。Mininet 模拟网络可能使用不同的 IP 地址，例如 192.168.0.0，具体取决于计算机网络设置。

练习 5.18　修改上面的命令以显示 h2 节点的网络信息，然后对 s1 和 c0 节点执行相同的操作。

同样，Linux 命令 ps 用于显示有关系统中运行的进程的信息。ps -a 命令将显示所有进程。可以通过输入以下命令在 h1 节点上执行 ps -a 命令：

mininet> h1 ps -a

练习 5.19　修改上面的命令以显示 h2 节点上的所有进程，然后对 s1 节点和 c0 节点执行相同的操作。

Linux 命令 ping 用于检查两个节点之间的连接状态。可以通过输入以下命令来实现从主机

1 ping 主机 2。ping -c 10 命令表示 ping 十次。图 5.23 给出了结果。

```
mininet> h1 ping -c 10 h2
```

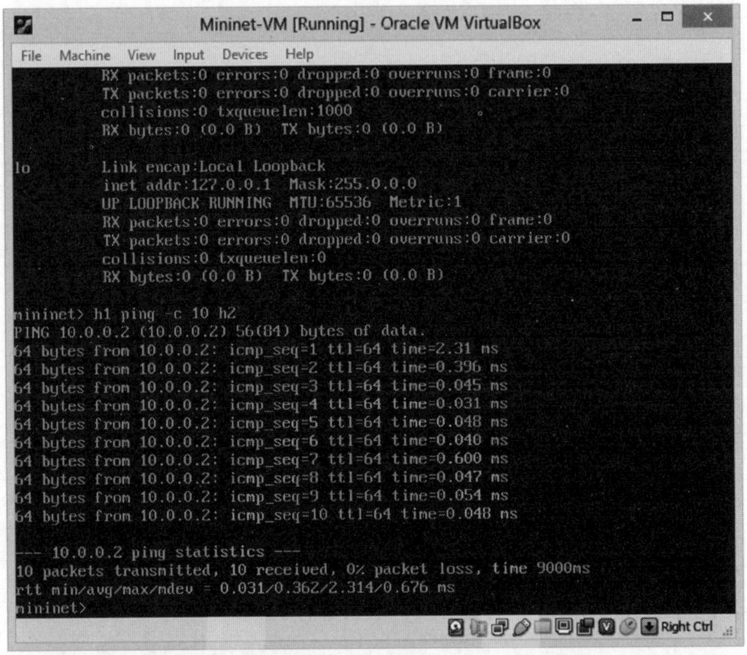

图 5.23　使用 ping 命令从主机 1 ping 主机 2 十次

还可以通过输入以下命令来 ping 所有节点：

```
mininet> pingall
```

可以通过输入以下命令来测试 h1 节点和 h2 节点之间的 TCP 带宽：

```
mininet> iperf
```

可以通过输入以下命令在 h1 节点上启动终端：

```
mininet> xterm h1
```

可以通过输入以下命令在 h1 节点上启动简单的 Python Web 服务器。其中的 80 表示在端口 80 上运行服务器，& 表示在后台运行服务器。

```
mininet> h1 python -m SimpleHTTPServer 80 &
```

可以通过输入以下命令在 h2 节点上以文本模式查看 Web 服务器内容：

```
mininet> h2 wget -O - h1
```

可以通过输入以下命令在 h1 节点上停止 Python Web 服务器：

```
mininet> h1 kill %python
```

你还可以将新节点添加到网络中。例如，可输入以下命令，将名为 h3 的节点添加到网络中：

`mininet> py net.addHost ('h3')`

可输入以下命令，在 s1 节点和 h3 节点之间添加新的链接：

`mininet> py net.addLink (s1, net.get('h3'))`

可输入以下命令，为 s1 节点附加新的以太网接口：

`mininet> py net.attach ('s1-eth3')`

可输入以下命令，配置 h3 节点的 IP 地址：

`mininet> py net.get ('h3').cmd('ifconfig h3-eth0 10.0.0.3')`

可输入以下命令，在 h3 节点上显示网络信息：

`mininet> h3 ifconfig`

图 5.24 给出了上述命令的输出结果。如你所见，还可以从其他节点 ping h3 节点。

可以使用以下链接提供的 Mininet Topology Visualizer 工具来可视化 Mininet 拓扑：
http://demo.spear.narmox.com/app/?apiurl=demo#!/mininet。

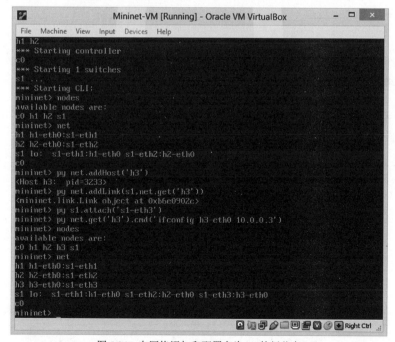

图 5.24　向网络添加和配置名为 h3 的新节点

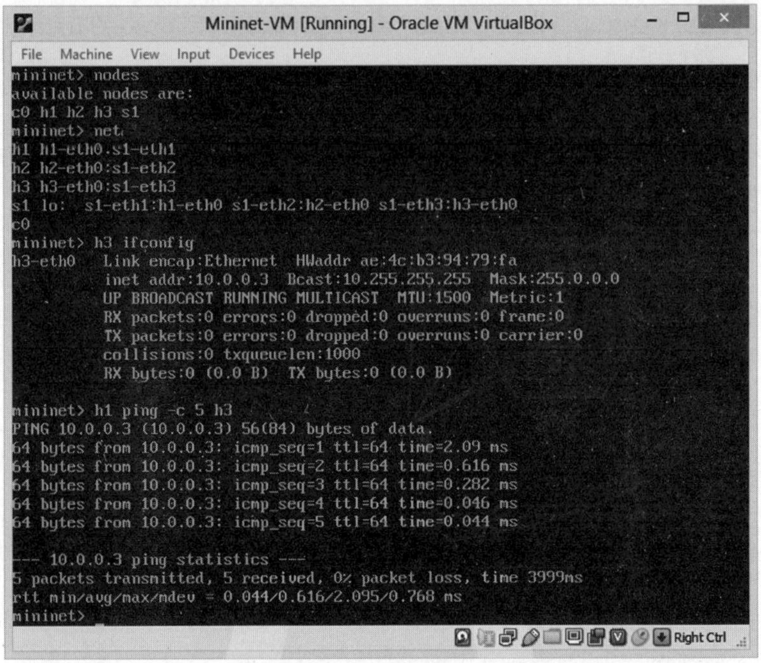

图 5.24(续)

只需要将 mininet> links 和 mininet> dump 命令的结果复制并粘贴到页面中,然后单击 Render Graph 按钮即可生成网络拓扑,如图 5.25 所示。

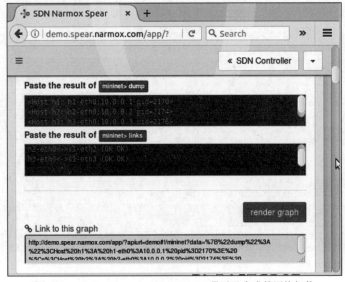

图 5.25 Mininet Topology Visualizer 工具以及生成的网络拓扑

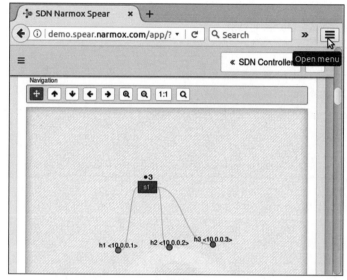

图 5.25(续)

最后，可以通过输入以下命令来退出 Mininet 模拟网络：

```
mininet> exit
```

可以通过输入以下命令来清理 Mininet 模拟网络：

```
$ sudo mn -c
```

有关 Mininet OpenFlow 的更多信息和高级用法，例如配置路由器和交换机，以及使用 Xming 服务器、Putty.exe 等，请访问以下链接：

```
https://github.com/mininet/openflow-tutorial/wiki
http://mininet.org/
https://github.com/mininet/mininet/wiki/Introduction-to-Mininet
https://academy.gns3.com/p/sdn-and-openflow-introduction
https://www.cisco.com/c/en_uk/solutions/software-defined-networking/
overview.html
```

5.7.2　Floodlight 入门

Floodlight 官方网站 http://www.projectfloodlight.org/getting-started/ 提供了详细的入门指南，还提供了有关如何下载和安装 Floodlight 的说明。

5.7.3　OpenDaylight 入门

OpenDaylight 的 WiKi 网站 https://wiki.opendaylight.org/view/Main_Page 介绍了如何下载 OpenDaylight 软件，还介绍了如何通过设置开发环境和获取代码来入门。

5.8　Java 网络编程资源

以下是一些有趣的 Java 网络教程。
- Tutorialspoint：https://www.tutorialspoint.com/java/java_networking.htm。
- Java2S：http://www.java2s.com/Tutorial。
- GeeksforGeeks：https://www.geeksforgeeks.org/socket-programming-in-java/。

5.9　小结

本章介绍了计算机网络的基本概念，包括最新的软件定义网络。本章还提供了一些简单的示例，这些示例使用 Java 来获取网络信息、UDP 套接字、TCP 套接字、HTTP 和 HTTPS 客户端、HTTP 服务器、SMTP 客户端、Java RMI 客户端和 RMI 服务器。本章最后介绍了 SDN。第 7 章将提供一些更复杂的 Java 示例，它们基于 REST 设计模型和最新的 MQTT 协议。

5.10　本章复习题

1. 什么是 Internet？什么是 TCP/IP？
2. 什么是局域网和广域网？
3. 思科的三层网络架构是什么？
4. 交换机和路由器有什么区别？
5. 什么是 DNS 和 DHCP？
6. 什么是 SDN？什么是 OpenFlow、Floodlight 和 OpenDaylight？
7. 什么是 IP 地址、端口号和 MAC 地址？
8. IPv4 地址和 IPv6 地址之间有什么区别？
9. 什么是套接字？
10. UDP 和 TCP 有什么区别？
11. 什么是 HTTP？HTTP 有几个版本？HTTP 和 HTTPS 之间的区别是什么？
12. 什么是 RMI？RMI 是如何工作的？

第 6 章

面向移动应用的Java编程

"学习知识，只有融会贯通，才算真正学到手。"

——John Wooden

6.1 引言
6.2 Android Studio
6.3 Hello World 应用
6.4 Button 和 TextView 组件的应用
6.5 传感器应用
6.6 部署 Android 应用
6.7 Android 应用中 activity 的生命周期
6.8 MIT App Inventor
6.9 5G
6.10 小结
6.11 本章复习题

6.1 引言

自 2007 年苹果公司首次推出 iPhone 以来，我们对智能手机的热情变得越来越浓厚。智能手机销量猛增。根据 Statista(https://www.statista.com/)提供的数据，2019 年全球约有 25 亿部智能手机。智能手机是现代技术的典范，它彻底改变了我们的生活。借助智能手机，只需要滑动手指即可搜索 Internet、预订机票和酒店、订购食物和打出租车。在许多地方，人们只需要带上手机就可以进行移动支付，而不必准备现金或信用卡！用了大约十年的时间，智能手机已真正从奢侈品变成日用品。

市场上通常有两种主要的智能手机类型：谷歌的 Android 手机和苹果的 iPhone 手机。我们估计世界上有 50 亿移动手机用户。Android 手机拥有全球最大的手机市场份额，占比超过 80%。顶级的 Android 手机制造商是三星、华为、谷歌、HTC、LG、中兴、小米、OPPO、VIVO 等。但最大的单一手机公司仍然是苹果，拥有的市场份额约为 15%。

本章首先介绍如何使用 Android Studio 开发用于 Android 手机的移动应用，然后介绍 MIT App Inventor，这是开发 Java Android 应用的另一种常用方式。MIT App Inventor 是基于 Web 的可视化编程工具，它让用户可以使用可视化对象构建 Java Android 应用。MIT App Inventor 在初学者中特别受欢迎。本章最后将介绍 5G，这是最受关注且研究最多的下一代移动技术。5G 将显著改变我们的通信方式，因此了解 5G 是什么以及 5G 如何工作将非常有用。

6.2 Android Studio

Android 是谷歌基于 Linux 操作系统开发的针对移动设备的开源移动操作系统。Android Studio 是用于开发 Android 应用的官方 IDE。Android Studio 实际上是基于 IntelliJ IDEA 进行工作的，因此，除了使用 Android Studio 之外，还可以使用 IntelliJ IDEA 开发 Android 应用。

要想开发 Android 应用，需要满足以下条件：

- 拥有 Java JDK 5 或以上版本。
- 拥有 Android Studio 或 IntelliJ IDEA。

因为你在第 2 章中已经学习并安装了 Java JDK，所以这里只需要安装 Android Studio 即可。图 6.1 显示的是 Android 开发人员网站(https://developer.android.com/studio/)，可以从中下载并安装 Android Studio。Android Studio 在安装过程中，还将安装用于编译程序的 Android SDK 和用于模拟程序运行的 Android 虚拟设备(Android Virtual Device，AVD)。

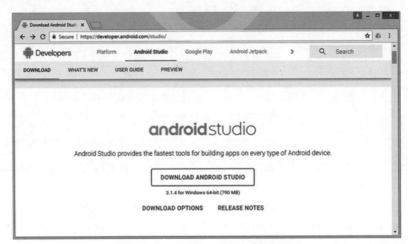

图 6.1 Android Studio 开发人员网站

不过，安装和配置 Android Studio 可能不会一帆风顺。有几个棘手的地方可能需要进行多次尝试才能成功。图 6.2 展示了来自 Tutorialspoint(https://www.tutorialspoint.com/android/android_studio.htm)的出色的 Android Studio 安装教程。可供参考的还有 YouTube 教程系列(https://www.youtube.com/watchv=LN8fBh7LH9k&list=PLt72zDbwBnAW5TU96U HUbLtnivjviIKks)，如图 6.3 所示。

图 6.2　来自 Tutorialspoint 的 Android Studio 教程

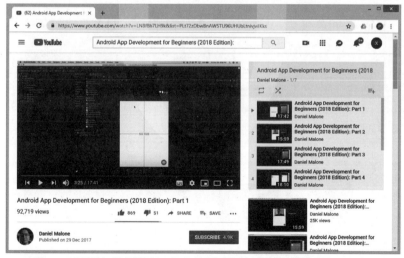

图 6.3　适用于初学者学习 Android 应用开发的 YouTube 教程系列(2018 版)

6.3　Hello World 应用

成功安装 Android Studio 后，即可开始开发自己的 Android 应用。在第一个示例中，我们将创建一个简单的 Android 应用，用于在屏幕上显示 Hello World！。

下面使用 Android Studio 创建一个 Android 项目。首次运行 Android Studio 时，将出现欢迎界面，可以从中单击 Start a new Android Studio project，如图 6.4 所示。从出现的项目选择界面中选择默认的 Empty Activity，然后单击 Next 按钮，如图 6.5 所示。在出现的项目配置界面中，设置项目名称，默认使用的名称是 MyApplication，如图 6.6 所示，单击 Finish 按钮以完成项目的创建。

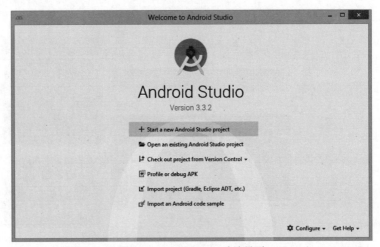

图 6.4　Android Studio 欢迎界面

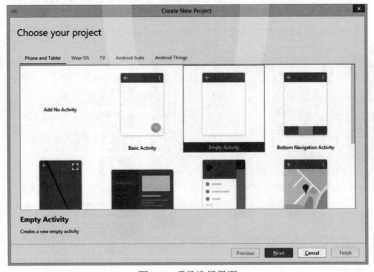

图 6.5　项目选择界面

上面创建的是一个空白的 Android 项目。Android Studio 的左侧面板将显示项目的结构，右侧面板将显示文件的内容。Android Studio 项目有几个关键文件，可以使用以下顺序从左侧面板中找到这些文件：

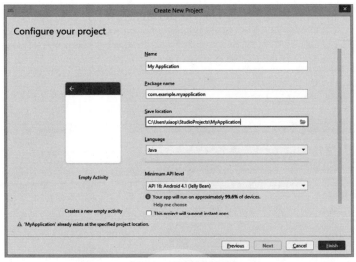

图 6.6　项目配置界面

- app | java | com.example.myapplication | MainActivity.java
- app | res | layout | activity_main.xml
- app | manifest | AndroidManifest.xml
- Gradle Scripts 文件

MainActivity.java 是 Android Studio 项目中最重要的 Java 程序，双击它就可以在右侧面板中查看里面的内容，如图 6.7 所示。在 MainActivity 类的内部，首次创建 MainActivity 类时将调用 onCreate()方法。setContentView()方法提供了一些有关布局资源的信息，这些信息定义在 activity_main.xml 文件中。

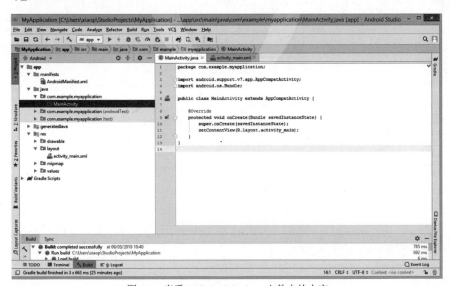

图 6.7　查看 MainActivity.java 文件中的内容

activity_main.xml 文件定义了 Java 应用的 UI 布局,双击即可在右侧面板的另一个选项卡中查看里面的内容。在 activity_main.xml 选项卡的底部,还有两个子选项卡:Design 和 Text。Design 子选项卡用来显示 UI 布局的外观(参见图 6.8),Text 子选项卡用来显示 XML 代码(参见图 6.9)。在这种情况下,UI 布局中仅包含一个 TextView 组件,用于显示 Hello World!文本。可以简单地修改 XML 代码以显示其他文本。

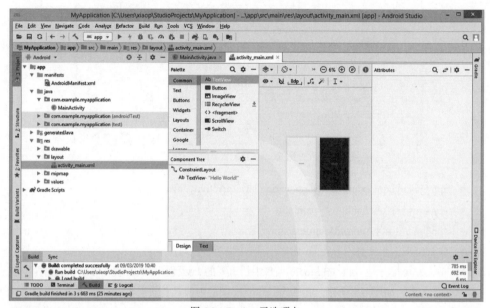

图 6.8　Design 子选项卡

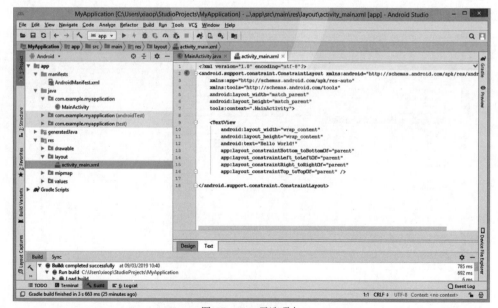

图 6.9　Text 子选项卡

AndroidManifest.xml 文件描述了 Java 应用的基本特征，双击后可以在右侧面板的另一个选项卡中查看里面的内容，如图 6.10 所示。

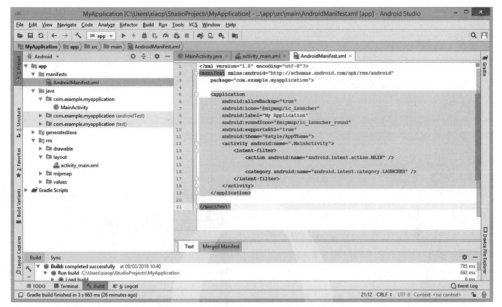

图 6.10 AndroidManifest.xml 文件的内容

Gradle Scripts 文件用于构建项目。

例 6.1A 展示了 MainActivity.java 文件中的代码，例 6.1B 展示了 activity_main.xml 文件中的代码。

例 6.1A MainActivity.java 文件中的代码

```
package com.example.myapplication;

import android.support.v7.app.AppCompatActivity;
import android.os.Bundle;

public class MainActivity extends AppCompatActivity {

    @Override
    protected void onCreate(Bundle savedInstanceState) {
        super.onCreate(savedInstanceState);
        setContentView(R.layout.activity_main);
    }
}
```

例 6.1B activity_main.xml 文件中的代码

```
<?xml version="1.0" encoding="utf-8"?>
<android.support.constraint.ConstraintLayout
    xmlns:android="http://
    schemas.android.com/apk/res/android"
```

```
            xmlns:app="http://schemas.android.com/apk/res-auto"
            xmlns:tools="http://schemas.android.com/tools"
            android:layout_width="match_parent"
            android:layout_height="match_parent"
            tools:context=".MainActivity">

    <TextView
        android:layout_width="wrap_content"
        android:layout_height="wrap_content"
        android:text="Hello World!"
        app:layout_constraintBottom_toBottomOf="parent"
        app:layout_constraintLeft_toLeftOf="parent"
        app:layout_constraintRight_toRightOf="parent"
        app:layout_constraintTop_toTopOf="parent"/>

</android.support.constraint.ConstraintLayout>
```

要想运行项目，只需要在 Android Studio 中单击 Run 按钮或按 Shift + F10 组合键即可，之后项目将在 Android 虚拟设备(Android Virtual Device，AVD)上运行，AVD 是 Android 手机的模拟器。注意，Android Studio 要求在首次进行设置时下载 AVD 相关的磁盘镜像。最终效果是在空白屏幕的中间显示 Hello World!，如图 6.11 所示。第一次运行时，AVD 会花费一些时间，因为需要模仿手机的启动过程。

恭喜！你已经成功开发了第一个 Android 应用。

练习 6.1 修改 MyApplication 项目的 XML 代码，使其可以显示 Hello XXX!，其中的 XXX 代表你的姓名。

练习 6.2 修改 MyApplication 项目的 XML 代码，使其可以显示日期和时间。

也可以直接在智能手机上运行项目，这样就

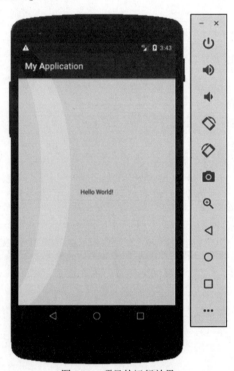

图 6.11　项目的运行效果

可以直接通过 Android 调试桥(Android Debug Bridge，ADB)连接智能手机以测试和调试 Android 程序。在这种情况下，首先需要更改手机的设置并使用 ADB 运行 Android 程序。步骤如下：

(1) 在手机上打开 Settings 菜单，选择 Developer Options，然后启用 USB Debugging。

(2) 使用 USB 电缆将手机连接到计算机。

(3) 单击 Android Studio 中的 Run 按钮，在手机上生成并运行 Android 程序。

有关开发 Android 应用的更多信息，请访问以下链接：

https://developer.android.com/training/basics/firstapp/
https://www.youtube.com/playlist?list=PLS1QulWo1RIbb1cYyzZpLFCKvdYV_yJ-E
https://www.javatpoint.com/android-tutorial

6.4 Button 和 TextView 组件的应用

本节将创建一个 Android 应用，其中包含一个 Button 组件和一个 TextView 组件。我们还将向这个 Button 组件添加一些动作，使得当单击 Button 组件时可以查看 TextView 组件中的文本。

下面使用与 6.3 节完全相同的方法在 Android Studio 中创建另一个空白的 Android 项目，然后将项目的名称改为 MyApplication2。在 activity_main.xml 选项卡中，选择底部的 Design 子选项卡，然后将 Button 组件拖动到视图中，所以视图中现在有两个组件——一个 TextView 组件和一个 Button 组件，如图 6.12 所示。选择 Text 子选项卡，注意 Android Studio 已经为 Button 组件自动生成了相应的 XML 代码。将 TextView 组件的 ID 设置为 textview，并将 Button 组件的 ID 设置为 button，如图 6.13 所示。

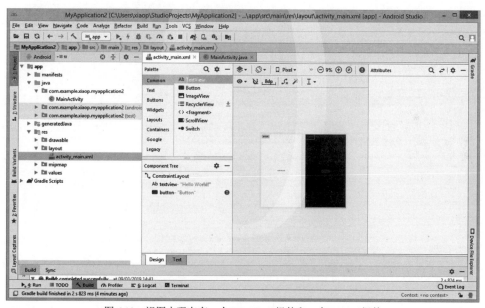

图 6.12　视图中现在有一个 TextView 组件和一个 Button 组件

现在修改 MainActivity.java 文件中的代码以向 Button 组件添加一些动作，如图 6.14 所示。在这种情况下，当单击按钮时，将把 TextView 组件中的文本更改为 Button has been clicked！并将字体大小更改为 25。图 6.15 展示了输出结果。

例 6.2A 展示了 MainActivity.java 文件中的代码，例 6.2B 展示了 activity_main.xml 文件中的代码。

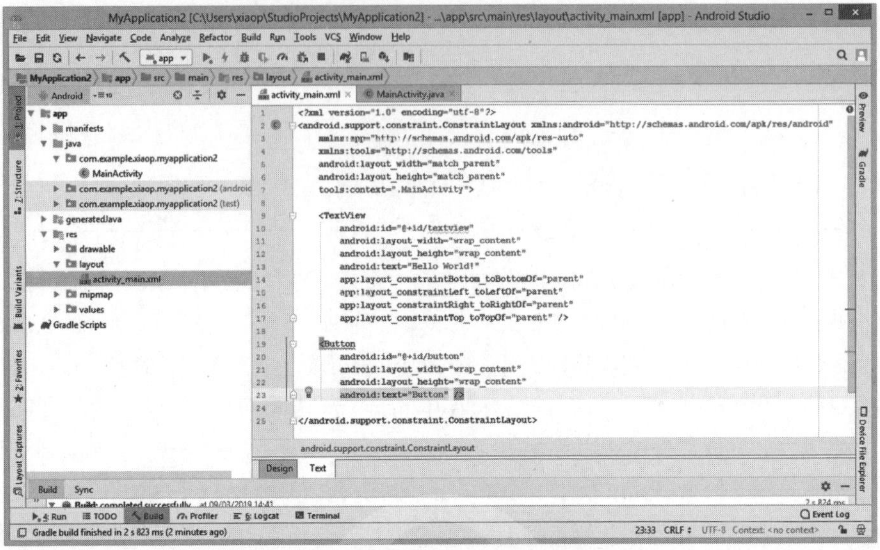

图 6.13　修改 TextView 和 Button 组件的 ID

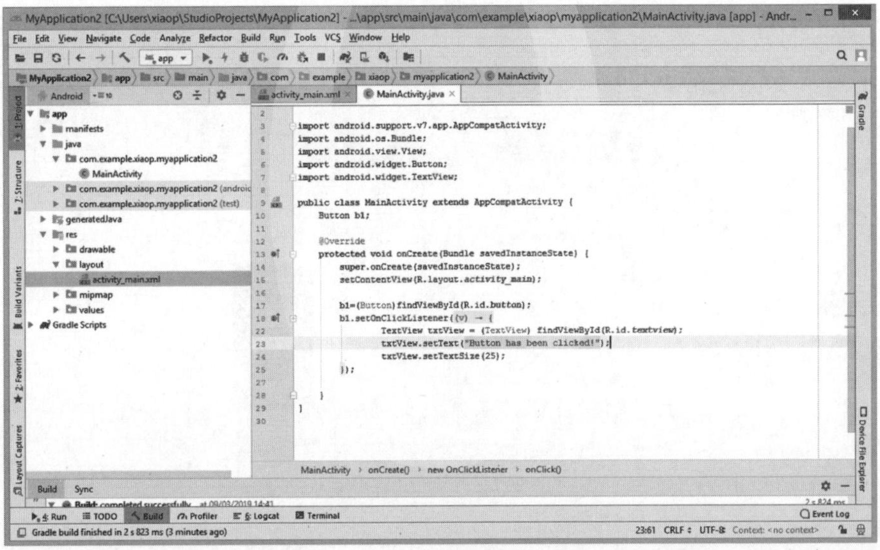

图 6.14　向 Button 组件添加一些动作

例 6.2A　MainActivity.java 文件中的代码

```
import android.support.v7.app.AppCompatActivity;
import android.os.Bundle;
import android.view.View;
import android.widget.Button;
import android.widget.TextView;

public class MainActivity extends AppCompatActivity {
    Button b1;
```

```java
@Override
protected void onCreate(Bundle savedInstanceState) {
    super.onCreate(savedInstanceState);
    setContentView(R.layout.activity_main);
    b1=(Button)findViewById(R.id.button);
    b1.setOnClickListener(new View.OnClickListener() {

        @Override
        public void onClick(View v) {
            TextView txtView = (TextView) findViewById(R.id.textview);
            txtView.setText("Button has been clicked!");
            txtView.setTextSize(25);
        }
    });
}
```

图 6.15　项目的输出结果

例 6.2B　activity_main.xml 文件中的代码

```xml
<?xml version="1.0" encoding="utf-8"?>
<android.support.constraint.ConstraintLayout
    xmlns:android="http://
    schemas.android.com/apk/res/android"
    xmlns:app="http://schemas.android.com/apk/res-auto"
    xmlns:tools="http://schemas.android.com/tools"
    android:layout_width="match_parent"
    android:layout_height="match_parent"

    tools:context=".MainActivity">
    <TextView
```

```
        android:id="@+id/textview"
        android:layout_width="wrap_content"
        android:layout_height="wrap_content"
        android:text="Hello World!"
        app:layout_constraintBottom_toBottomOf="parent"
        app:layout_constraintLeft_toLeftOf="parent"
        app:layout_constraintRight_toRightOf="parent"
        app:layout_constraintTop_toTopOf="parent"/>

    <Button
        android:id="@+id/button"
        android:layout_width="wrap_content"
        android:layout_height="wrap_content"
        android:text="Button"/>

</android.support.constraint.ConstraintLayout>
```

练习 6.3 基于 MyApplication2 项目,在视图中添加一个 EditText 组件,每当单击按钮时,就将 EditText 组件中的文本复制到 TextView 组件中。

6.5 传感器应用

本节将创建一个 Android 应用,用于在屏幕上显示来自传感器的信息。

使用与 6.3 节完全相同的方法在 Android Studio 中创建另一个空白的 Android 项目,将项目的名称更改为 MyApplication3。在 activity_main.xml 选项卡中,选择 Text 子选项卡,将 TextView 组件的 ID 设置为 textview,如图 6.16 所示。

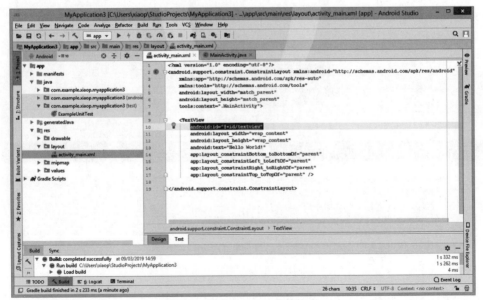

图 6.16 修改 TextView 组件的 ID

现在，修改 MainActivity.java 文件以添加一些动作来获取所有传感器信息，如图 6.17 所示。在这种情况下，SensorManager 类用于处理传感器服务，List 类用于存储所有传感器，TextView 组件用于显示信息。for 循环用于遍历所有传感器并获取它们的名称、供应商和版本，然后在 TextView 组件中显示这些信息。

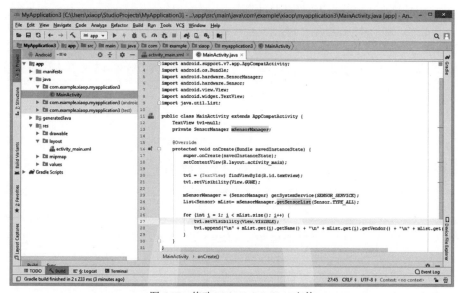

图 6.17　修改 MainActivity.java 文件

图 6.18 展示了 AVD 模拟器中的输出结果。

图 6.18　AVD 模拟器中的输出结果

例 6.3A 展示了 MainActivity.java 文件中的代码,例 6.3B 展示了 activity_main.xml 文件中的代码。

例 6.3A　MainActivity.java 文件中的代码

```java
package com.example.xiaop.myapplication3;

import android.support.v7.app.AppCompatActivity;
import android.os.Bundle;
import android.hardware.SensorManager;
import android.hardware.Sensor;
import android.view.View;
import android.widget.TextView;
import java.util.List;

public class MainActivity extends AppCompatActivity {
    TextView tv1=null;
    private SensorManager mSensorManager;

    @Override
    protected void onCreate(Bundle savedInstanceState) {
        super.onCreate(savedInstanceState);
        setContentView(R.layout.activity_main);

        tv1 = (TextView) findViewById(R.id.textview);
        tv1.setVisibility(View.GONE);

        mSensorManager = (SensorManager) getSystemService(SENSOR_SERVICE);
        List<Sensor> mList= mSensorManager.getSensorList(Sensor.TYPE_ALL);

        for (int i = 1; i < mList.size(); i++) {
            tv1.setVisibility(View.VISIBLE);
            tv1.append("\n" + mList.get(i).getName() + "\n" + mList
                .get(i).getVendor() + "\n" + mList.get(i).getVersion());
        }
    }
}
```

例 6.3B　Activity_Main.xml 文件中的代码

```xml
<?xml version="1.0" encoding="utf-8"?>
<android.support.constraint.ConstraintLayout
    xmlns:android="http://
    schemas.android.com/apk/res/android"
    xmlns:app="http://schemas.android.com/apk/res-auto"
    xmlns:tools="http://schemas.android.com/tools"
    android:layout_width="match_parent"
    android:layout_height="match_parent"
    tools:context=".MainActivity">

    <TextView
        android:id="@+id/textview"
        android:layout_width="wrap_content"
        android:layout_height="wrap_content"
        android:text="Hello World!"
```

```
            app:layout_constraintBottom_toBottomOf="parent"
            app:layout_constraintLeft_toLeftOf="parent"
            app:layout_constraintRight_toRightOf="parent"
            app:layout_constraintTop_toTopOf="parent"/>

</android.support.constraint.ConstraintLayout>
```

练习 6.4 基于 MyApplication3 项目，添加两个 Button 组件，通过修改代码，实现当单击第一个 Button 组件时，在 TextView 组件中显示所有传感器信息；而当单击第二个 Button 组件时，清除所有 TextView 组件中的信息。

练习 6.5 基于 MyApplication3 项目，添加两个 Button 组件，通过修改代码，实现当单击第一个 Button 组件时检索温度传感器的值，并在 TextView 组件中显示温度；而当单击第二个 Button 组件时，清除 TextView 组件中的所有信息。

6.6 部署 Android 应用

Android 应用在完成开发之后，在准备分发给他人或在 Google Play 商店中发布之前，需要使用 Android Studio 构建签名的 APK(Android Package Kit)文件。APK 是一种文件格式，由 Android 操作系统用于分发和安装移动应用的软件包，类似于 Windows 中的 EXE 文件格式。APK 文件通常包含已编译程序的代码、资源、证书和清单文件。当想要对应用进行签名时，需要生成上传密钥和存储密钥。在 Java 中，密钥库用于存储授权证书或公共密钥证书，并受密钥库密码的保护。在基于 Java 的应用中，密钥库通常用于加密和身份验证。有关安全性、加密和身份验证的更多详细信息，请参见第 7 章。

举个例子，当想要为 MyApplication3 项目构建 APK 文件时，首先可以从 Android Studio 菜单中选择 Build | Generate Signed Bundle/APK，这将显示 Generate Signed Bundle or APK 对话框，可以从中选择用来生成签名的签名包或 APK，如图 6.19 所示。选择 APK，然后单击 Next 按钮。在打开的下一个对话框中，在 Key store path 文本框的下方单击 Create new 按钮，如图 6.20 所示。在打开的下一个对话框中填写信息：对于密钥存储路径，请为 Java KeyStore (JKS) 文件选择路径和名称；对于密码，可以自行选择；如图 6.21 所示。之后单击 OK 按钮。在打开的下一个对话框中选择签名版本，这里选择的是 V2(Full APK Signature)，如图 6.22 所示。最后单击 Finish 按钮。

APK 文件可以在 Android 项目文件夹\MyApplication3\app\release\app-release.apk 中找到。之后，可以重命名 APK 文件并将其复制到智能手机以进行安装或在 Google Play 商店中发布。

有关用于生成签名的 APK 文件和发布 Android 应用的更多详细信息，请参见以下资源：

```
http://www.androiddocs.com/tools/publishing/app-signing.html
https://developer.android.com/studio/publish/
```

图 6.19　选择 APK　　　　　图 6.20　单击 Create new 按钮

图 6.21　设置密钥库　　　　　图 6.22　生成签名的 APK

6.7　Android 应用中 activity 的生命周期

在 Android 系统中，activity 是用户可以执行的单个任务。每个 Android 应用都可以包含一个或多个 activity。图 6.23 展示了 Android 应用中 activity 的生命周期图，这幅图来自 Android 开发者网站(https://developer.android.com/guide/components/activities/activity-lifecycle)。Android 应用中 activity 的生命周期分为 6 个阶段，它们可以通过如下不同的回调函数来触发：onCreate()、onStart()、onResume()、onPause()、onStop()、onRestart() 和 onDestroy()。

首次创建 activity 时会调用 onCreate() 回调函数，它类似于标准 Java 程序中的 main() 方法。相反，onDestroy() 回调函数在 activity 被系统关闭之前被调用。当 activity 对用户可见时，将调用 onStart() 回调函数。当 activity 与用户进行交互时，将调用 onResume() 回调函数。当用户使用应用时，应用保持在这种状态，直到用户执行其他操作以将注意力从应用转移开。当 activity

暂停时(例如，在多屏操作期间或在顶部使用透明应用时)，将调用 onPause()回调函数。暂停后，再次调用 onResume()回调函数。当 activity 不再可见时，将调用 onStop()回调函数。当 activity 停止后重新启动时，将调用 onRestart()回调函数。

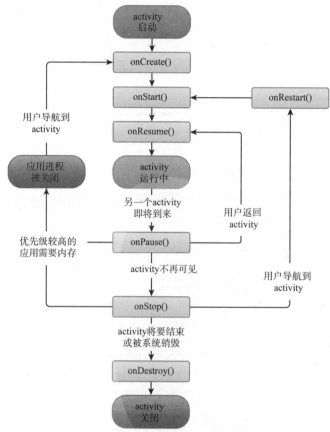

图 6.23　Android 应用中 activity 的生命周期

6.8　MIT App Inventor

除了 Android Studio 或类似的 IDE 之外，开发 Android 应用的方法还有很多，例如使用 MIT App Inventor。MIT App Inventor 提供了一种开源的、基于 Web 的在线 Android 开发环境，由麻省理工学院(Massachusetts Institute of Technology，MIT)和 Google 联合开发。MIT App Inventor 旨在教导新手在无须编写 Java 代码的情况下，使用 Open Blocks 可视化程序开发 Android 应用。MIT App Inventor 有两个版本，当前版本是 MIT App Inventor 2。

1．语音识别应用

下面将使用 MIT App Inventor 构建一个带有标签和按钮的语音识别应用。当单击按钮时，语音识别引擎将启动。然后，你可以说出一条指令，例如"打开 Google 地图""打开 BBC 天气""打开 YouTube"或"打开 Facebook"，这个应用将识别发出的指令并在手机上执行相应的操作。要使用 MIT App Inventor 2，请首先访问 http://ai2.appinventor.mit.edu/，然后使用 Google 账户登录，创建一个名为 PerryAsk 的新项目。MIT App Inventor 的右上角有两个按钮：Designer 和 Blocks。Designer 按钮用于显示应用的前端或外观。Blocks 按钮用于显示应用的后端。在这里，可以使用功能模块编写程序。MIT App Inventor 的开发理念就基于这种前端和后端概念，这种概念与另一种流行的可视化编程工具 LabVIEW 相同。

在设计视图中，可以显示一个或多个屏幕(默认显示一个屏幕)。要查看屏幕的工作原理，请首先从左侧的工具面板中将 Label 和 Button 组件拖动到屏幕中，然后再拖入如下 5 个不可见的组件：TextToSpeech、SpeechRecognizer、OrientationSensor、LocationSensor 和 ActivityStarter。这 5 个组件虽然不会出现在屏幕上，但是它们会出现在设计视图中屏幕的下方，如图 6.24 所示。接下来单击 Blocks 按钮，编辑模块代码，如图 6.25 所示。

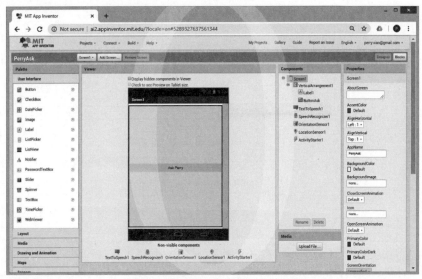

图 6.24　将组件拖入设计视图中的屏幕

对于这个项目，本书提供完整的源代码，它们被压缩在名为 PerryAsk.aia 的文件中。

屏幕初始化模块用于启用 LocationSensor，如图 6.26 所示。该模块也可以用于初始化程序中使用的其他变量。

按钮模块则通过调用 SpeechRecognizer 从语音中获取文本，如图 6.27 所示。这将触发 SpeechRecognizer 模块的运行，如图 6.28 所示。

SpeechRecognizer 模块的内部是一个 do 循环，该循环首先在标签中显示 SpeechRecognizer 文本，然后根据 SpeechRecognizer 文本使用一系列 if … else if 语句执行不同的操作。例如，如果你说"打开地图"，那么这个语音识别应用将使用 ActivityStarter 启动 Google 地图，然后使用 LocationSensor 获取当前地址并显示在地图上。

第 6 章 面向移动应用的 Java 编程　173

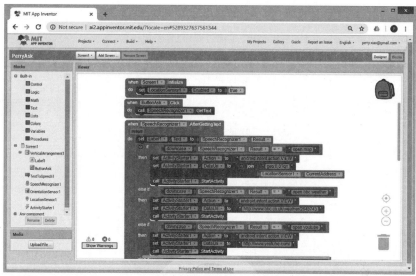

图 6.25　编辑模块代码

图 6.26　屏幕初始化模块

图 6.27　按钮模块

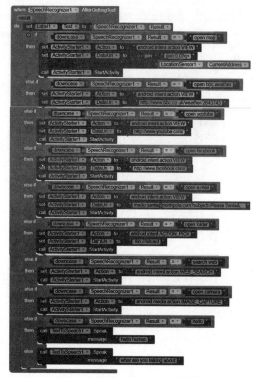

图 6.28　SpeechRecognizer 模块

这个语音识别应用还可以搜索网络、打开相机，当你说 hello 时，它将使用 TextToSpeech 模块，通过人工语音向你说 Hello,human。要在手机上部署和运行这个语音识别应用，请从菜单中选择 Build | App(provide QR code for .apk)。这将会编译程序并生成二维码，如图 6.29 所示。可使用手机扫描这个二维码以下载并安装该应用。

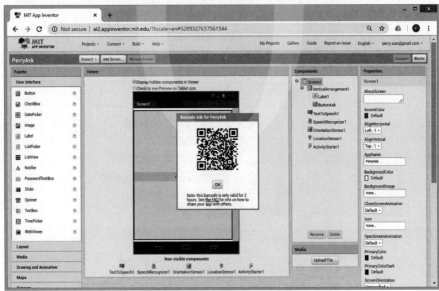

图 6.29　编译程序并生成二维码

练习 6.6　修改 MIT App Inventor 程序，当你说"文本消息"时，使其向你的手机联系人之一发送 hello 短信。

练习 6.7 修改 MIT App Inventor 程序，使其可以进行一些简单的对话，如下所示。

你：Hello!

MIT App Inventor 程序：Hello, Human!

你：What is your name?

MIT App Inventor 程序：My name is Android.

你：Where do you live?

MIT App Inventor 程序：Inside this silly phone case.

你：Bye!

MIT App Inventor 程序：Bye!

2. 翻译应用

MIT App Inventor 2 还带有名为 YandexTranslate 的令人印象深刻的语言翻译引擎。接下来，我们将构建一个用于执行语音操作的应用。可借助这个应用，让语音识别和语言翻译同时进行。按下一个按钮，然后用英语说些什么，这个应用就会识别你所说的话，并翻译成中文，然后对你用中文回复。而当按下另一个按钮并说中文时，这个应用同样可以识别你所说的话，并翻译为英文，然后对你用英文回复。

使用 MIT App Inventor 2 创建一个名为 PerryTranslate 的新项目，如图 6.30 所示。在设计视图中，将一些组件拖动到屏幕中，它们分别是用于英文的 TextBox、Label 和 Button 组件，以及相应的用于中文的 TextBox、Label 和 Button 组件。你还需要将以下三个不可视组件拖动到屏幕中：TextToSpeech、SpeechRecognizer 和 YandexTranslate 组件。同样，这个项目的源代码已被压缩到 PerryTranslate.aia 文件中。

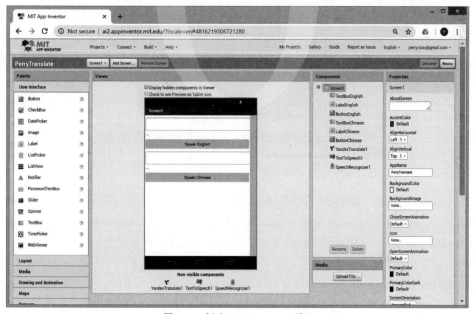

图 6.30 创建 PerryTranslate 项目

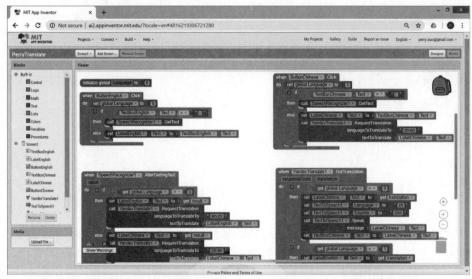

图 6.30(续)

单击 Blocks 按钮以显示模块编辑器。屏幕初始化模块已将 Language 变量设置为 0。在这种情况下，0 代表英文，1 代表中文。ButtonEnglish 模块则将 Language 变量的值重置为 0。如果 TextBoxEnglish 为空，就启动 SpeechRecognizer 并从英文语音中获取文本，这将触发 SpeechRecognizer 模块的运行。如果 TextBoxEnglish 不为空，就将文本复制到 LabelEnglish 中，如图 6.31 所示。ButtonChinese 模块会将 Language 变量的值重置为 1。如果 TextBoxChinese 为空，就启动 SpeechRecognizer 并从中文语音中获取文本，这将触发 SpeechRecognizer 模块的运行。如果 TextBoxChinese 不为空，就将文本复制到 LabelChinese 中，如图 6.32 所示。

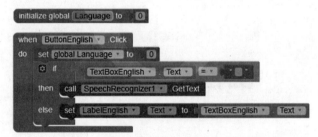

图 6.31　屏幕初始化模块和 ButtonEnglish 模块

SpeechRecognizer 模块包含一个 do 循环，如图 6.33 所示。在这个 do 循环内，如果 Language 变量为 0，就把 LabelEnglish 设置为 SpeechRecognizer 中的文本，然后调用 YandexTranslate 引擎以将 LabelEnglish 中的文本翻译成中文。YandexTranslate 引擎在生成译文之后，将触发 GotTranslation 模块，从而将 LabelChinese 中的文本设置为译文，调用 TextSpeech 以读出译文，然后将 TextBoxChinese 中的文本设置为与 LabelChinese 中的文本相同。在 do 循环内，如果 Language 变量的值为 1，就执行与上面相同的翻译过程，但不同的是从中文翻译为英文。

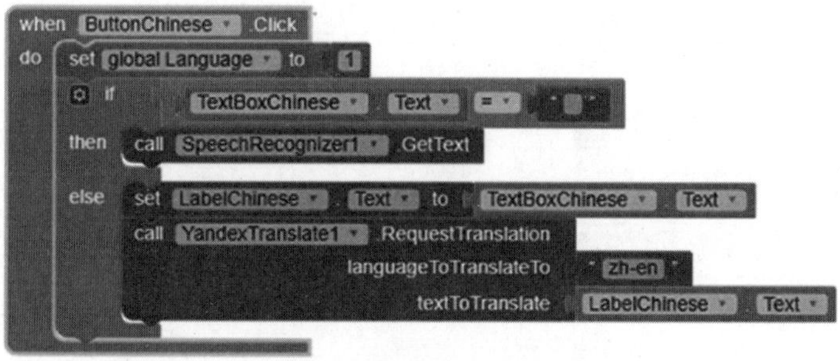

图 6.32　ButtonChinese 模块

图 6.33　设置 AfterGettingText 和 GotTranslation 模块中的 do 循环

与前面一样，要在手机上运行应用，请从菜单中选择 Build | App(provide QR code for .apk) 以编译程序并生成二维码，如图 6.34 所示。然后使用手机扫描二维码以下载并安装这个翻译应用。

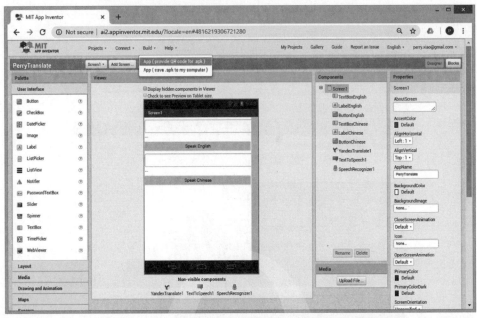

图 6.34 编译程序并生成二维码

练习 6.8 修改 MIT App Inventor 程序，使其可以将英文翻译成另一种语言，例如西班牙语、法语或德语。

练习 6.9 修改 MIT App Inventor 程序，添加一个多选组合框以包含多种语言。当按下按钮时，就把你所说的话转换为文本，然后翻译为这个多选组合框指定的另一种语言。

练习6.10 修改 MIT App Inventor 程序，添加两个多选组合框，其中每个多选组合框都包含多种语言。第一个多选组合框指定的是原始语言，第二个多选组合框指定的是目标语言。当单击第一个多选组合框时，就将你用原始语言所说的话转换为文本，然后将转换后的文本翻译为目标语言。

有关 MIT App Inventor 的更多信息，请参见以下资源：

```
http://appinventor.mit.edu/explore/ai2/tutorials.html
http://www.appinventor.org/book2
```

6.9　5G

在谈论移动应用时，5G 是一个绕不开的话题，因为这是目前最受关注且研究最多的技术之一。随着所有大的智能手机厂商开始发布支持 5G 的智能手机，通过了解 5G 的确切含义来结束本章不失为一种有趣的方式。

5G 是指第五代移动通信技术，是 1G、2G、3G 和 4G 技术的延续。这些技术的主要区别在于频谱，因为所有无线移动技术都依靠电磁波发送和接收信号。与前几代产品不同，5G 使用更高的频率范围，通常为 3 GHz～86 GHz。5G 使用的是"基于蜂窝的移动网络"，其中每个蜂窝里的手机都将连接到基站(Base Station，BS)。每个基站都连接到移动交换中心(Mobile Switch Center，MSC)，MSC 则连接到有线网络。因此，移动网络也称为蜂窝网络。图 6.35 展示了蜂窝网络的体系结构以及移动用户之间如何通信。1G 的蜂窝大小最大，2G、3G、4G 的蜂窝大小与 1G 相比小得多，而 5G 的蜂窝大小甚至更小。图 6.36 展示了 5G、4G 等不同移动通信技术的频率范围和蜂窝大小。

5G 的主要优点是速度快、延迟低且连接性更好。这使 5G 成为未来无人驾驶汽车、虚拟现实、增强现实、互联网游戏和物联网(IoT)应用的关键技术。

速度快　5G 网络将比以前的网络快得多。表 6.1 比较了播放时长为两小时的高清电影的下载时间。对于 3G 网络需要 24 小时，对于 4G 网络需要 7 分钟，而对于 5G 网络则可能只需要几秒。即使大多数用户体验到的速度只是潜在最高速度的一小部分，这也仍然会给人留下深刻的印象。

表 6.1　两小时播放时长的高清电影在不同网络中的下载速度

网络	最高下载速度	最短时间
3G 网络	384 Kbps	24 小时
4G 网络	100 Mbps	7 分钟
5G 网络	1 Gbps～10 Gbps 或更快	只需要几秒

低延迟　延迟时间是指从你将数据发送到 Internet 到收到反馈数据为止的时间。与上一代网络相比，5G 网络的延迟时间非常短，如表 6.2 所示。

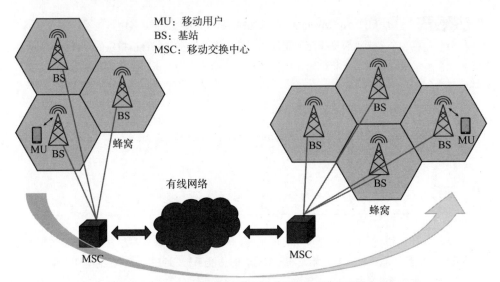

图 6.35　蜂窝网络的体系架构以及移动用户之间的通信方式

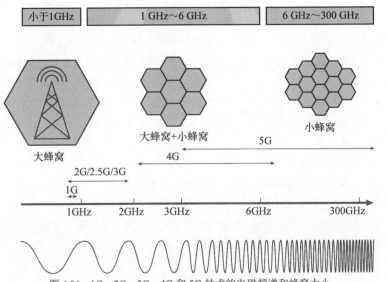

图 6.36　1G、2G、3G、4G 和 5G 技术的电磁频谱和蜂窝大小

表 6.2　不同网络的典型延迟时间

网络	延迟时间/ms
3G	100
4G	50
5G	1

更好的连接性　5G 不仅可以连接更多的人和设备,而且可以提供更可靠的连接,停机时间却非常短。据称,5G 具有保持在线的连接特性。表 6.3 对 1G、2G/2.5G/2.75G、3G、4G 和 5G

技术做了比较。

表 6.3　1G、2G/2.5G/2.75G、3G、4G 和 5G 技术的比较

	1G	2G/2.5G/2.75G	3G	4G	5G
时期	1970—1990 年	1990—2000 年	2000—2010 年	2010—2020 年	2020—2030 年
带宽	30 kHz	200 kHz	20 MHz	20 MHz	100 MHz
频率	800 MHz (模拟信号)	800 MHz~1.8 GHz (数字信号)	1.6 GHz~2.1 GHz	2 GHz~8 GHz	3 GHz~300 GHz
数据速率	2 Kbps	9.6 Kbps (2G) 64 Kbps~114 Kbps(2.5G) 384 Kbps(2.75G)	384 Kbps(移动) 2 Mbps(静止)	100 Mbps	大于 1 Gbps
标准	MTS、AMTS 和 IMTS	2G：GSM 2.5G：GPRS 2.75G：EDGE	IMT-2000 3.5G-HSDPA 3.75G-HSUPA	统一标准	统一标准
服务	模拟语音 (无数据服务)	数字语音、SMS、MMS、网页浏览(2.5G)	音频、视频、流媒体、网页浏览、IPTV	增强的音频/视频流、网页浏览、高清电视	动态信息访问以及频流、网页浏览、具有 AI 功能的可穿戴设备
多路复用	FDMA	TDMA 和 CDMA	CDMA	CDMA	CDMA
核心网络	PSTN	PSTN	分组网络	Internet	Internet
交换	电路	电路和数据包	数据包	所有数据包	所有数据包
特点	第一代无线通信	数字信号	数字带宽	数字带宽、高速率、基于 IP	数字带宽、超高速率
技术	模拟蜂窝	数字蜂窝	CDMA、UMTS和EDGE	LTE、WiFi、WiMAX	WWW和Unified IP

6.9.1　毫米波

5G 有两个频率范围：频率范围 1(小于 6 GHz)和频率范围 2(24 GHz~86 GHz)。频率范围 1 的带宽为 100 MHz，频率范围 2 的带宽为 400 MHz。在物理学中，因为频率为 30 GHz~300 GHz 的电磁波的波长范围为 1~10 mm，所以它们被称为毫米波。这与过去用于上一代移动通信技术的 6 GHz 以下的电磁波不同，后者的波长往往为几十厘米。因为毫米波很容易被建筑物或障碍物阻挡，抑或被树叶和雨水吸收，所以它们在空间中不易传播，这就是 5G 网络使用更小蜂窝的原因。同样，由于频率差异，5G 网络无法使用现有的 4G/3G/2G 基站提供的天线来发送和接收信号。5G 网络需要自己的天线，因此也需要自己的基础设施。以上所有这些因素造就了 5G 网络的小型蜂窝技术。

6.9.2　小蜂窝

5G 网络的小型蜂窝通常只有几百米，而 4G/3G/2G 网络的蜂窝则有几千米。因为小蜂窝中的 5G 天线需要发射微小的毫米波，因此它们的尺寸也比传统天线小。人们通常将 5G 天线小型化，并使用更少的功率进行操作，从而更易于安装在灯柱和建筑物的顶部。这种新的蜂窝结构

能提供更有针对性且更有效的频谱。但是，这也意味着在农村地区，5G 网络很难普及建设。

6.9.3 大规模 MIMO

多输入多输出(Multiple-Input Multiple-Output，MIMO)描述的是一种使用两个或多个发射机和接收机一次性发送和接收更多数据的无线系统。MIMO 已成功应用于当今的 4G 网络，在 4G 网络中，基站通常有 12 个天线端口：8 个发射机端口和 4 个接收机端口。5G 基站可以支持大约 100 个端口,这被称为大规模 MIMO。大规模 MIMO 通过同时允许更多用户发送和接收信号，显著提高了移动网络的容量。

6.9.4 波束成形

大规模 MIMO 存在的最大问题是如何减少干扰，通俗地讲，就是如何一次从更多天线发射更多信息。波束成形是专为基站设计的，用于识别到达特定用户的最有效的数据传输路径，并减少数据传输过程中对附近用户产生的干扰。波束成形使基站能够以精确协调的模式在不同的方向上发送单独的数据包，从而使大规模 MIMO 阵列中的大量用户和天线能够一次性交换更多信息。波束成形还可以将信号聚集到仅指向用户方向的集成波束中，以避开障碍并减少对其他所有人产生的干扰。

6.9.5 全双工

在当前的移动网络中，基站和手机的收发器(发送器和接收器)必须轮流以相同的频率发送和接收信息，或者同时使用不同的频率发送和接收信号。借助 5G，收发器将能够以相同的频率同时发送和接收数据。这种全双工技术可以使无线网络的容量增加一倍。全双工技术面临的主要挑战是需要设计一种能够路由输入输出信号的电路，以避免在天线同时发送和接收数据时发生冲突。有利就有弊，全双工技术也会产生更多的信号干扰。

6.9.6 未来的 6G 和 7G

目前，有关 6G 和 7G 的研究已经开始了。6G 不会取代 5G，而是将 5G 与现有的卫星网络集成以实现全球覆盖。6G 将提供超快的 Internet 访问，并将用于智能家居和智能城市。

7G 将走得更远，7G 会将 6G 和 5G 集成在一起,提供空间漫游，并使整个世界完全无线化。

有关 5G 的更多信息，请访问以下资源：

https://spectrum.ieee.org/video/telecom/wireless/everything-you-need-to-know-about-5g
https://5g.co.uk/guides/how-fast-is-5g/
https://5g.co.uk/guides/5g-frequencies-in-theuk-what-you-need-toknow/
https://www.qorvo.com/design-hub/blog/small-cell-networks-and-the-evolution-of-5g
http://www.emfexplained.info/?ID=25916
https://www.digitaltrends.com/mobile/what-is-5g/

6.10 小结

本章介绍了用于移动应用的 Java 编程。我们首先介绍了移动应用开发工具 Android Studio，然后提供了三个示例，最后介绍了移动应用的部署。本章还介绍了 MIT App Inventor，这是开发 Android 应用的另一种流行方式。本章在末尾介绍了下一代的 5G 移动通信技术，并简要概述了 5G 的工作原理及用途。

6.11 本章复习题

1. 什么是 Android？
2. 比较 Android 手机、苹果的 iPhone 手机以及其他智能手机的市场份额。
3. 什么是 Android SDK 和 Android AVD？
4. 简述下载和安装 Android Studio 的步骤。
5. 简述使用 Android Studio 创建 Android 项目的步骤。
6. APK(Android Package Kit)文件是什么？
7. JKS(Java KeyStore)文件是什么？
8. 简述部署 Android 应用的步骤。
9. 简述 Android 应用中 activity 的生命周期。
10. 什么是 MIT App Inventor？当前版本是多少？
11. MIT App Inventor 中的 Designer 视图(设计视图)和 Block 视图(代码视图)有什么区别？
12. 什么是 5G？5G 的主要优势是什么？
13. 5G 使用了哪些关键技术？
14. 什么是 6G 和 7G？

第 III 部分

第 7 章：面向物联网应用的 Java 编程
第 8 章：面向人工智能应用的 Java 编程
第 9 章：面向网络安全应用的 Java 编程
第 10 章：面向区块链应用的 Java 编程
第 11 章：面向大数据应用的 Java 编程

第 7 章

面向物联网应用的Java编程

"简单是可靠的先决条件。"
——Edsger Dijkstra

7.1 什么是物联网

7.2 物联网通信协议

7.3 物联网平台

7.4 物联网安全

7.5 为什么使用 Java

7.6 使用树莓派的 Java 物联网

7.7 Oracle Java ME 嵌入式客户端

7.8 适用于 Java 的物联网平台

7.9 小结

7.10 本章复习题

7.1 什么是物联网

毋庸置疑,物联网(Internet of Things,IoT)是当前最热门的话题之一。与物联网紧密相关的概念是工业 4.0。那么,什么是物联网?什么是工业 4.0?

从广义上讲,物联网指的是互联的设备或"物"的全球网络。物联网是 Internet 的未来,Internet 是目前互联的计算机(包括智能手机和平板电脑)的全球网络。物联网中的"物"是指日常的物理设备,它们不是计算机,而是具有嵌入式计算硬件(微控制器)的设备,例如电视、冰箱、炊具、水壶、灯、汽车、门、椅子等，如图7.1所示。

图7.1 物联网

为了将这些日常设备连接到Internet，需要使用如下几种技术。

地址 物联网中的每一台设备都有唯一的地址，这样就可以对设备进行唯一的标识，因此需要很多地址。IPv6可提供多达 2^{128} 个不同的地址，这足以为地球上的每一粒沙子分配不同的IP地址。

通信协议 物联网中的每一台设备都需要与其他设备进行通信，它们使用与Internet中的计算机不同的协议进行通信。

传感器和执行器 物联网中的每一台设备都需要有传感器，传感器可以提供一些信息，比如电视是开着还是关着、冰箱是满的还是空的、椅子是空的还是有人坐着、窗户是开着还是关着，等等。物联网中的每一台设备可能还需要有执行器，如电机或压电设备。有了执行器，就可以打开门、关上窗户，等等。

微控制器 物联网中的每一台设备都需要有微控制器来读取传感器数据，与其他设备通信并执行某些任务。只有当微控制器足够小、足够便宜并且具有足够低的功耗时，才可能嵌入设备。

云计算 随着数以亿计的设备连接到物联网，物联网将产生大量的数据，这些数据无法用常规方法进行存储和分析。这就是为什么需要使用云计算来存储、分析和显示数据的原因。

据电子产品全球分销商法内尔有限公司(Farnell Ltd)介绍，物联网系统通常可分为四个组成部分(https://uk.farnell.com/internet-of-things)，如图7.2所示。

- **收集模块**：边缘设备(传感器和执行器)。
- **连接模块**：无线或有线方式。
- **控制模块**：网关设备。
- **云模块**：数据中心。

图 7.2　物联网系统的组成部分

收集模块是查找边缘设备的地方，边缘设备与传感器和执行器集成在一起。传感器用于测量，包括温度传感器、湿度传感器、烟雾传感器、火焰传感器、有毒气体传感器、距离传感器、水位传感器、运动传感器、光传感器、磁传感器等。执行器用于执行某些操作，例如打开和关闭阀门或门、移动组件等。收集模块是物联网系统的前端。

连接模块是设备与设备之间以及设备与用户之间进行通信的地方。可以通过无线或有线连接的方式进行通信。对于无线通信，有一系列技术，例如近场通信(Near Field Communication，NFC)、射频识别(Radio Frequency Identification，RFID)、蓝牙、Zigbee、Z-Wave(WPAN)、WiFi、LoRa(LPWAN)、Sigfox(LPWAN)以及最受关注的 5G 蜂窝技术。

在控制模块中，可以找到桥接收集模块和云模块之间间隙的网关设备。网关设备接收来自物联网设备传感器或系统中其他设备的数据，并将这些数据发送到云端。网关设备将是任何物联网系统的关键部分，因为它们还负责协议转换、数据处理、数据存储、数据过滤和数据安全。

云模块是存储、分析和显示数据的地方。物联网将产生大量的数据，传统的个人计算机甚至本地服务器由于存储空间和计算能力有限，都无法存储和分析这些数据。存储和分析这些海量数据的最佳地点是云端，这是一种通过互联网提供的计算服务。借助云计算，可以享用服务器、存储、数据库、网络、软件、分析、智能等服务。许多技术巨头公司，如 IBM、Google、Microsoft 和 Amazon，都提供了自己的云计算服务。例如，使用 IBM 的 Watson 物联网云，可以轻松地连接物联网设备，存储和分析数据，并在 Web、计算机和移动设备上显示结果。Watson 提供了许多复杂的分析软件，包括人工智能软件，可以充分利用提供的数据。云模块是物联网系统的大脑。

借助物联网，企业可以提高效率、生产力和客户参与度，增强数据收集和数据分析能力并减少浪费。但是，使用物联网还会出现一些问题，例如安全性、隐私、复杂性和合规性。安全性始终是重中之重。有关安全性的更多详细信息参见本章后面的内容以及第 9 章。

到目前为止，连接到物联网的设备数量每年以 30%的速度增长，2017 年已经有大约 84 亿台物联网设备。2020 年预计将有 300 亿台物联网设备，全球市场将达到 7 万亿美元。物联网在很大程度上被视为下一个大事件，一些公司在物联网研发方面投入了数十亿美元。

物联网的工业版被称为工业 4.0——第四次工业革命。第一次工业革命发生在 18 世纪，当时蒸汽机使工业生产发生了革命。第二次工业革命发生在 19 世纪，当时电力和装配线使大规模生产成为可能。第三次工业革命也称为数字革命，发生在 20 世纪，电子和计算机使自动化生产成为可能。前三次工业革命都显著提高了生产力。

第四次工业革命即将到来。在这场革命中，供应商、物流、工厂和客户将紧密结合在一起，这将更加显著地提高生产率。机器人技术和人工智能将得到广泛应用。所有数据将存储在云中，人工智能将用于分析数据、预测趋势和识别故障。虚拟生产还将使制造商能够在虚拟环境中设计、创建和测试产品。客户将能够在线查看和定制产品，并将越来越关注可定制的智能产品。这种趋势也被称为智能制造(intelligent manufacturing)，制造出来的产品被称为智能制造产品。图 7.3 阐述了上述概念并与前几次工业革命做了对比。

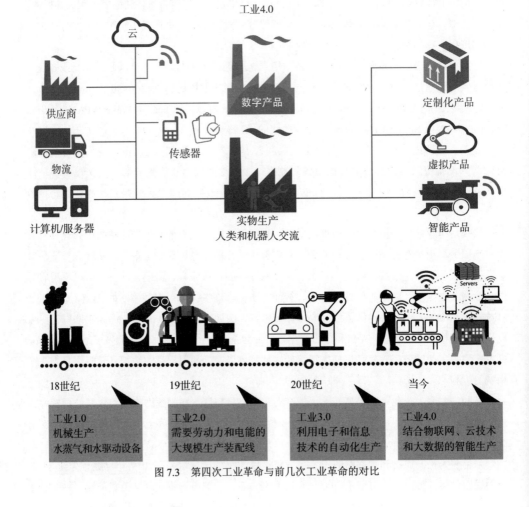

图 7.3　第四次工业革命与前几次工业革命的对比

7.2　物联网通信协议

物联网中的设备与互联网中的计算机相比，通信方式不同，后者往往使用大量带宽来传输大量数据,因此对于后者而言,时机的掌握并不总是很重要。最常用的互联网协议是超文本传输协议(Hypertext Transfer Protocol,HTTP)。物联网中的设备往往只有有限的电源、计算能力和带宽,它们通常需要间歇性地传输少量数据,但定时通常很重要。有几种通信协议专门为这种通信而设计。下面介绍一些最重要的物联网协议。

7.2.1 MQTT

消息队列遥测传输(Message Queuing Telemetry Transport,MQTT)是 IBM 于 1999 年开发的一种发布-订阅消息传递协议。MQTT 的核心是 MQTT 代理,用于接收并发布消息。图 7.4 展示了 MQTT 的工作原理。在图 7.4 中,我们首先在 MQTT 代理上创建名为 temperature 的主题,接下来温度传感器、便携式计算机和移动设备都订阅了这个主题。温度传感器将不时进行测量,并将数据发布到 MQTT 代理上的主题。然后,MQTT 代理将数据中继到便携式计算机或移动设备。与 HTTP 协议相比,MQTT 需要的计算能力、带宽和功耗更少。

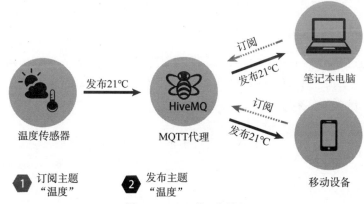

图 7.4 MQTT 的工作原理

MQTT 提供如下三个级别的服务质量(Quality of Service,QoS)。
- 0:最多一次。消息只发送一次。
- 1:至少一次。在收到确认信号之前,多次发送消息。
- 2:正好一次。只发送消息的一个副本(由两级握手保证)。

有关 MQTT 的更多信息,请参阅以下资源:

```
http://mqtt.org/
https://www.hivemq.com/blog/how-to-get-started-with-mqtt
```

7.2.2 CoAP

约束应用协议(Constrained Application Protocol,CoAP)是一种专门用于受限物联网设备(具有有限的计算能力和功耗)和受限物联网网络(具有有限的带宽)的 Web 传输协议。CoAP 专为机器对机器(Machine to Machine,M2M)应用而设计,如智能能源和楼宇自动化。CoAP 基于请求和响应消息，类似于 HTTP，但 CoAP 使用的是用户数据报协议(UDP)而不是传输控制协议(TCP)。因此，CoAP 的报头更小、速度更快。

有关 CoAP 的更多信息，请参阅以下资源：

http://coap.technology/

7.2.3 XMPP

可扩展消息和状态协议(Extensible Messaging and Presence Protocol，XMPP)是基于可扩展标记语言(Extensible Markup Language，XML)的开放标准通信协议。XMPP 使用结构化、可扩展的消息实现了物联网设备之间的实时通信。XMPP 拥有十分广泛的应用，包括即时消息、状态和协作。

有关 XMPP 的更多信息，请参阅以下资源：

https://xmpp.org/about/

7.2.4 SOAP

简单对象访问协议(Simple Object Access Protocol，SOAP)是另一种基于 XML 的消息传递协议，由微软设计，用于通过 Internet 在计算机之间交换信息。对于 XML 消息，SOAP 是高度可扩展的，可以用于 Web 服务。

有关 SOAP 的更多信息，请参阅以下资源：

https://www.w3schools.com/xml/xml_soap.asp

7.2.5 REST

代表性状态转移(Representational State Transfer，REST)是 Web 服务的轻量级体系结构样式。REST 基于统一资源标识符(Uniform Resource Identifier，URI)和超文本传输协议(Hypertext Transfer Protocol，HTTP)，并且使用 JavaScript 对象符号(JavaScript Object Notation，JSON)表示数据格式。REST 与浏览器完全兼容。

有关 REST 的更多信息，请参阅以下资源：

http://rest.elkstein.org/

7.3 物联网平台

物联网平台是运行在 Internet 中的软件，它们使用物联网协议(如 MQTT、CoAP、XMPP 等)连接传感器和用户设备。物联网平台还被用于存储、分析和显示数据。人工智能通常是在物联网平台上实现的，以提供复杂的分析。

一些十分受欢迎的物联网平台如下。

- **Eclipse 物联网**：https://iot.eclipse.org/。
- **Oracle 物联网云**：https://cloud.oracle.com/iot。
- **IBM Watson**：https://www.ibm.com/watson/。
- **亚马逊 AWS**：https://aws.amazon.com/。
- **微软 Azure**：https://azure.microsoft.com/en-us/。
- **谷歌云**：https://cloud.google.com/。
- **Salesforce**：https://www.salesforce.com/uk/products/iot-cloud/overview/。

7.4 物联网安全

安全对于物联网至关重要。随着越来越多的设备连接到物联网，黑客攻击以及发生入侵和数据盗窃的风险越来越高。对于许多企业来说，物联网安全是采用物联网技术时的重中之重。

为了保护物联网系统和设备，物联网制造商应该从一开始就确保硬件安全，使其能够防篡改，确保安全升级并提供固件更新/补丁。物联网软件开发人员应专注于安全软件开发和安全软件集成。对于那些部署物联网系统的人来说，重点应该放在硬件的安全性和用户的身份验证上。最后，物联网运营商应保持系统最新，并通过减少恶意软件、执行审核、保护基础设施和保护凭据来防止系统安全受到损害。

7.5 为什么使用 Java

Java 因具有丰富的网络功能和高度的安全性，可以在许多物联网应用中使用。根据 Oracle 的说法，Java 可以用于机器对机器(Machine to Machine，M2M)设备、无线模块、工业控制、智能仪表/传感器和电子医疗保健/远程医疗保健。例如，Java 可以用于许多小型的嵌入式系统，比如树莓派(Raspberry Pi)，以读取传感器数据、将数据传输到云端并进行控制。

有了 Java，设备可以变得更加集成，在信息交换方面更加高效，从而提供更好的体验。设备还具有自我升级和自我管理的能力，从而延长了产品的生命周期、降低了产品支持成本。Java 是一种易于学习和使用的语言，世界上有 900 多万 Java 开发人员。通过可重用模块和平台独立性，Java 可以提高市场覆盖率。最后，使用 Java 能使设备更安全可靠。

7.6 使用树莓派的 Java 物联网

为了在嵌入式设备中编译 Java 程序，需要有完整的操作系统。以下是三种最受欢迎的单片机，它们运行在完整的 Linux 操作系统中并完全支持 Java(对于所有这些设备，显示的规格都基于撰写本书时的可用型号，它们在你阅读本书时可能已经升级）。

树莓派 四核 ARM Cortex CPU，Broadcom GPU，RAM，四个 USB 端口，10/100 Mbps 以太网，802.11n 无线局域网，蓝牙 4.0，HDMI 端口和 GPIO 引脚。价格：35$。https://www.raspberrypi.org/。

BeagleBone ARM Cortex CPU，图形引擎，RAM，4 GB 板载闪存，两个 PRU 32 位微控制器，两个 USB 端口，10/100 Mbps 以太网，HDMI 和 LCD 接口，串行端口，ADC，I2C，SPI，PWM 引脚。价格：67.99$。https://beagleboard.org/。

Odroid 三星 Cortex CPU，Mali GPU，RAM，eMMC 闪存，两个 USB 3.0 端口，一个 USB 2.0 端口，千兆以太网，HDMI 端口和 GPIO 引脚。价格：59$。https://www.hardkernel.com/。

这里我们只关注树莓派，因为这是当前最流行的嵌入式设备。树莓派是一款信用卡大小的单片机，是为促进计算机科学教学而开发的。自 2005 年引入以来，树莓派的受欢迎程度一直稳步增长。截至 2018 年 3 月，全球共售出 1900 万台树莓派设备。树莓派有多个版本，最新的是树莓派 3 代，内置有 WiFi 和蓝牙。图 7.5 展示了树莓派 3 代 B 型板及其关键部件。图 7.6 展示了树莓派 3 代 B 型板的 40 GPIO 引脚排列图（https://pi4j.com/1.2/pins/Model-3b-rev1.html）。

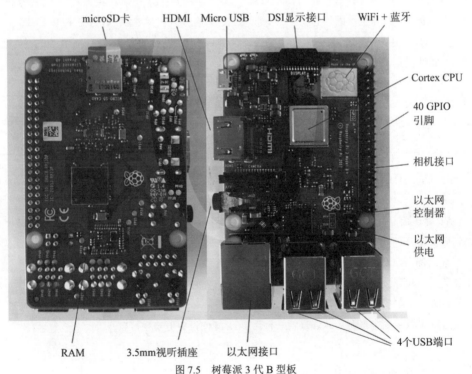

图 7.5　树莓派 3 代 B 型板

引脚#	名称			名称	引脚#
	3.3V直流电源	1	2	5.0V直流电源	
8	引脚8 SDA1(I2C)	3	4	5.0V直流电源	
9	引脚9 SCL1(I2C)	5	6	地线	
7	引脚7 GPCLK0	7	8	引脚15 TxD(UART)	15
	地线	9	10	引脚16 RxD(UART)	16
0	引脚0	11	12	引脚1 PCM_CLK/PWM0	1
2	引脚2	13	14	地线	
3	引脚3	15	16	引脚4	4
	3.3V直流电源	17	18	引脚5	5
12	引脚12 MOSI(SPI)	19	20	地线	
13	引脚13 MISO(SPI)	21	22	引脚6	6
14	引脚14 SCLK(SPI)	23	24	引脚10 CE0(SPI)	10
	地线	25	26	引脚11 CE1(SPI)	11
30	引脚14 (I2C ID EEPROM)	27	28	SCL0 (I2C ID EEPROM)	31
21	引脚21 GPCLK1	29	30	地线	
22	引脚22 GPCLK2	31	32	引脚26 PWM0	26
23	引脚23 PWM1	33	34	地线	
24	引脚24 PCM_FS/PWM1	35	36	引脚27	27
25	引脚25	37	38	引脚28 PCM_DIN	28
	地线	39	40	引脚29 PCM_DOUT	29

Raspberry Pi 3 Model B (J8 Header)

图 7.6 树莓派 3 代 B 型板的 40 GPIO 引脚排列图

7.6.1 设置树莓派

为了在树莓派上运行 Java，首先需要有树莓派硬件。可以访问树莓派网站，找到最近的分销商，也可以直接从亚马逊网站购买。

除树莓派外，还需要 8 GB 或更大容量的 microSD 卡来存储 Raspbian 操作系统——一种 GNU Debian Linux 操作系统。可以通过以下方式获得预安装的 Raspbian microSD 卡。

- RS(搜索树莓派 microSD)：https://uk.rs-online.com/。
- Pi Hut：https://thepihut.com/products/raspbian-preinstalled-sd-card。
- 从 https://www.raspberrypi.org/downloads/下载操作镜像。

图 7.7 展示了树莓派软件指南网站(https://www.raspberrypi.org/learning/software-guide/quickstart/)，上面提供了有关 Raspbian 软件如何下载和安装以及如何设置树莓派以运行的全面指南。

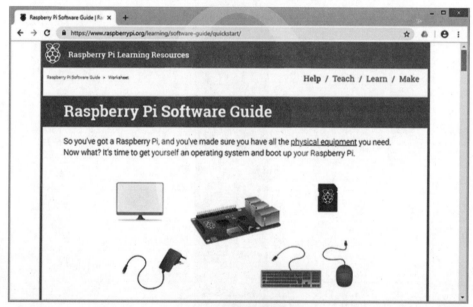

图 7.7 树莓派软件指南网站

为了与树莓派交互，需要有 USB 键盘、USB 鼠标以及带有 HDMI 端口的电视或带有 HDMI-VGA 转换器的标准计算机显示器。

也可以使用 SSH 协议或 Windows 远程桌面连接功能远程连接树莓派。为了使用 SSH 协议，需要在树莓派上启用 SSH 服务。可以通过以下三种方式之一启用 SSH 服务。

- 在树莓派桌面上，从 Preferences 菜单启动 Raspberry Pi Configuration，导航到 Interfaces 选项卡，选中 SSH 旁边的 Enabled 复选框，然后单击 OK 按钮。
- 在树莓派终端，输入 sudo raspi-config。从 configuration 菜单中选择 Interfacing 选项，导航到并选择 SSH，选择 Yes，单击 OK 按钮，然后单击 Finish 按钮。这里，sudo 的意

思是以超级用户(也就是管理员)的身份执行命令。

- 在树莓派终端，输入如下命令以启用并启动 SSH 服务：

```
$ sudo systemctl enable ssh
$ sudo systemctl start ssh
```

笔者最喜欢的方式是使用 Windows 远程桌面连接功能连接树莓派，因为 SSH 仅提供文本模式的连接。为此，首先需要使用以下命令在树莓派上安装 xrdp 和 tightvncserver 软件包：

```
$ sudo apt-get remove xrdp vnc4server tightvncserver
$ sudo apt-get install tightvncserver
$ sudo apt-get install xrdp
```

接下来，使用以太网电缆或 WiFi 将树莓派连接到 Internet。为了使用 Windows 远程桌面连接协议连接到树莓派，你需要知道树莓派的 IP 地址。如果没有带 HDMI 端口或 HDMI-VGA 转换器的电视，可以使用名为 Advanced IP Scanner(https://www.advanced-ip-scanner.com/)的免费程序来扫描树莓派的 IP 地址，如图 7.8 所示。

找到 IP 地址后，可使用 Windows 远程桌面连接功能远程登录树莓派，如图 7.9 所示。输入用户名和密码(默认为 pi 和 raspberry)并登录后，你应该会看到树莓派桌面，如图 7.10 所示。

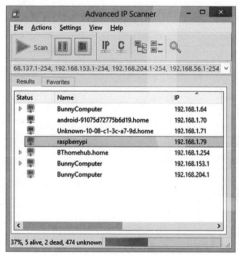

图 7.8　Advanced IP Scanner

图 7.9　远程登录树莓派

树莓派启动并运行后，你将需要 Java JDK，Java JDK 应随标准 Raspbian 操作系统(包括 BlueJ Java IDE)一起提供。如果想安装其他版本的 Java JDK，那么可以在 Raspbian 终端简单地运行以下命令：

```
$ sudo apt-get update
$ sudo apt-get install oracle-java8-jdk
```

这些命令可将 Java 8 安装到树莓派上，但 Java 9 和更高版本不容易获得。为了读写树莓派的通用输入输出(GPIO)引脚，还需要下载并安装 Pi4J 库(http://pi4j.com/)，如图 7.11 所示。下载和安装 Pi4J 库的方法有多种。笔者更喜欢从 GitHub 下载整个源代码：https://github.com/Pi4J/pi4j。

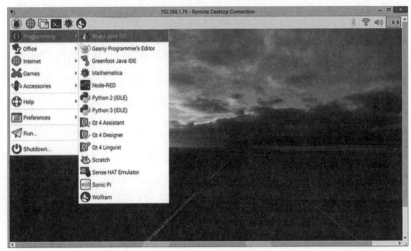

图 7.10　树梅派桌面

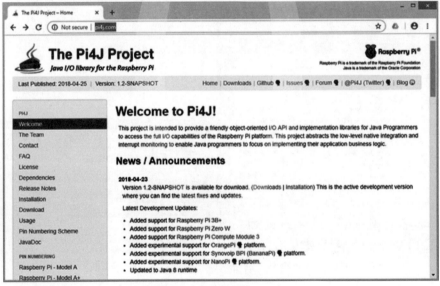

图 7.11　下载并安装 Pi4J 库

7.6.2　Java GPIO 示例

下面开始构建第一个树莓派 Java 应用：使用 GPIO 引脚控制闪烁的 LED。

首先，我们要为 Java 程序创建一个专用文件夹。笔者选择在标准的 Documents 文件夹下创建一个名为 IoT Java 的文件夹，如图 7.12 所示。在 Pi4J 库的下载文件夹中，找到名为 pi4j-core.jar 的文件，然后复制到 IoT Java 文件夹中。另外，找到名为 ControlGpioExample.java 的示例文件，也复制到 IoT Java 文件夹中。然后，在终端窗口中，使用 cd 命令导航到 IoT Java 文件夹。可以使用命令 nano ControlGpioExample.java 查看和修改代码。

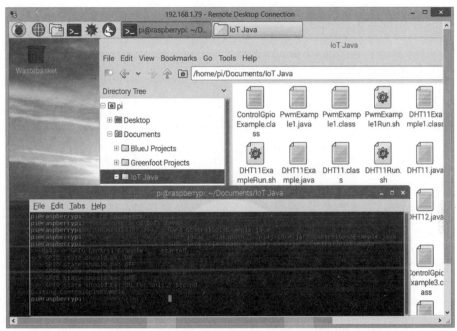

图 7.12　ControlGpioExample.java 程序的编译和执行

以下是完整的代码，作用主要是使用 GpioFactory.getInstance()创建 GPIO 对象，并使用 provisionDigitalOutputPin()将 GPIO 引脚设置为数字输出引脚。在这种情况下，可选择将 GPIO 引脚 01 作为数字输出引脚，然后切换为高电平，等待 5000 ms 后，再切换为低电平，等待另一个 5000 ms，最终将 GPIO 引脚 01 打开和关闭。如果将 LED 连接到 GPIO 引脚 01，那么将相应地打开和关闭 LED。

```java
import com.pi4j.io.gpio.GpioController;
import com.pi4j.io.gpio.GpioFactory;
import com.pi4j.io.gpio.GpioPinDigitalOutput;
import com.pi4j.io.gpio.PinState;
import com.pi4j.io.gpio.RaspiPin;
/**
 *
 * This example code demonstrates how to perform simple state
 * control of a GPIO pin on the Raspberry Pi.
 *
 * @author Robert Savage
 */
public class ControlGpioExample {
    public static void main(String[] args) throws InterruptedException {
        System.out.println("<--Pi4J--> GPIO Control Example ... started.");
        // create gpio controller
        final GpioController gpio = GpioFactory.getInstance();
        // provision gpio pin #01 as an output pin and turn on
```

```
            final GpioPinDigitalOutput pin = gpio.provisionDigitalOutputPin(R
                    aspiPin.GPIO_01, "MyLED", PinState.HIGH);
            // set shutdown state for this pin
            pin.setShutdownOptions(true, PinState.LOW);
            System.out.println("--> GPIO state should be: ON");
            Thread.sleep(5000);

            // turn off gpio pin #01
            pin.low();
            System.out.println("--> GPIO state should be: OFF");
            Thread.sleep(5000);

            // toggle the current state of gpio pin #01 (should turn on)
            pin.toggle();
            System.out.println("--> GPIO state should be: ON");
            Thread.sleep(5000);

            // toggle the current state of gpio pin #01 (should turn off)
            pin.toggle();
            System.out.println("--> GPIO state should be: OFF");
            Thread.sleep(5000);

            // turn on gpio pin #01 for 1 second and then off
            System.out.println("--> GPIO state should be: ON for only 1 second");
            pin.pulse(1000, true); // set second argument to 'true' use a
                                   // blocking call
            // stop all GPIO activity/threads by shutting down the GPIO controller
            // (this method will forcefully shutdown all GPIO monitoring
            // threads and scheduled tasks)
            gpio.shutdown();
            System.out.println("Exiting ControlGpioExample");
        }
    }
```

为了编译和运行程序,请输入以下命令。编译和运行程序时,在类路径中包含 pi4j-core.jar 文件很重要,以超级用户身份运行程序也很重要。

```
$ javac -classpath ".:pi4j-core.jar" ControlGpioExample.java
$ sudo java -classpath ".:pi4j-core.jar" ControlGpioExample
```

如果不想在每次编译和运行程序时重新输入所有冗长的类路径的详细信息,那么还可以创建一个能自动执行上述操作的 shell 脚本,如图 7.13 所示。在本例中,笔者创建了一个名为 ControlGpioExample.sh 的 Bash shell 脚本,内容如下:

```
#!/bin/Bash
javac -classpath ".:pi4j-core.jar" ControlGpioExample.java
sudo java -classpath ".:pi4j-core.jar" ControlGpioExample
```

哈希感叹号(#!)被称为 shebang,后跟 shell 脚本的路径。这里使用的是 Bash shell 脚本,但是还有许多其他类型的 shell 脚本可用。shebang 必须出现在脚本文件的第一行,并且#和!之间也没有空格。为了运行 shell 脚本,只需要输入以下命令:

```
$ base ControlGpioExample.sh
```

基于 ControlGpioExample.java 程序，可以创建一个简单地使用 GPIO 引脚控制闪烁的 LED 演示程序，如例 7.1 所示。例 7.1 将使用 GPIO 引脚 01 作为数字输出引脚，并每隔 500 ms 切换一次高低电压。如果将 LED 连接到这个引脚，如图 7.14 所示，LED 将相应地闪烁。

图 7.13　用于编译和执行 ControlGpioExample.java 程序的 Linux shell 脚本

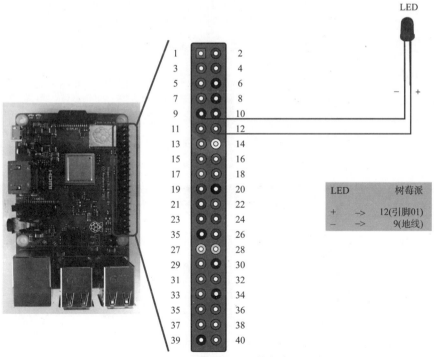

图 7.14　树莓派和 LED 的电路

例 7.1 一个简单地使用 GPIO 引脚控制闪烁的 LED 演示程序

```java
// Example 7.1 ControlGpioExample1.java
import com.pi4j.io.gpio.GpioController;
import com.pi4j.io.gpio.GpioFactory;
import com.pi4j.io.gpio.GpioPinDigitalOutput;
import com.pi4j.io.gpio.PinState;
import com.pi4j.io.gpio.RaspiPin;

public class ControlGpioExample1 {
    // create gpio controller
    final static GpioController gpio = GpioFactory.getInstance();
    // provision gpio pin #01 as an output pin and turn on
    final static GpioPinDigitalOutput pin = gpio.provisionDigitalOutputPin(
        RaspiPin.GPIO_01, "MyLED", PinState.HIGH);

    public static void main(String[] args) throws InterruptedException {
        // set shutdown state for this pin
        pin.setShutdownOptions(true, PinState.LOW);

        pin.high();
        Thread.sleep(500);
        pin.low();
        Thread.sleep(500);
        pin.high();
        Thread.sleep(500);

        gpio.shutdown();
    }
}
```

练习 7.1 修改例 7.1 中的 Java 程序，使其在不确定的循环中设置引脚电压的高低。

练习 7.2 修改例 7.1 中的 Java 程序，使其在 for 循环中将 Thread.sleep() 的引脚电压设置为高电平和低电平，范围为 0~1000，步长为 100。

下面创建一个简单的莫尔斯电码演示程序，如例 7.2 所示。该程序使用 GPIO 引脚 01 作为数字输出引脚，并根据字母 A 的莫尔斯电码(点、破折号、短空格)将引脚电压设置为高电平和低电平。同样，如果将 LED 连接到这个引脚，那么将相应地闪烁莫尔斯电码。

例 7.2 一个简单的莫尔斯电码演示程序

```java
// Example 7.2 Morse Code demo program – MorseExample1.java
import com.pi4j.io.gpio.GpioController;
import com.pi4j.io.gpio.GpioFactory;
import com.pi4j.io.gpio.GpioPinDigitalOutput;
import com.pi4j.io.gpio.PinState;
import com.pi4j.io.gpio.RaspiPin;

public class MorseExample1 {
    // create gpio controller
    final static GpioController gpio = GpioFactory.getInstance();
    // provision gpio pin #01 as an output pin and turn on
```

```
    final static GpioPinDigitalOutput pin = gpio.provisionDigitalOutputPin(
        RaspiPin.GPIO_01, "MyLED", PinState.HIGH);
    public static void main(String[] args) throws InterruptedException {
        // set shutdown state for this pin
        pin.setShutdownOptions(true, PinState.LOW);

        System.out.println("Morse Code 'A' - dot, dash, shortspace");
        dot();
        dash();
        shortspace();
        gpio.shutdown();
    }
    static void dot() throws InterruptedException {
        pin.high();
        Thread.sleep(300);
        pin.low();
        Thread.sleep(300);
    }
    static void dash() throws InterruptedException {
        pin.high();
        Thread.sleep(900);
        pin.low();
        Thread.sleep(300);
    }
    static void shortspace() throws InterruptedException {
        Thread.sleep(600);
    }
}
```

练习 7.3 修改例 7.2 中的 Java 程序，使其在 LED 上显示 SOS。在莫尔斯电码中，字母 S 为点、点、点、短空格，字母 O 为破折号、破折号、破折号、短空格。

例 7.3 使用 GPIO 引脚 01 作为数字输出引脚，在 for 循环中使用 toggle() 每 1000 毫秒切换一次高低电平。

例 7.3　第三个 GPIO 示例程序

```
//Example 7.3
import com.pi4j.io.gpio.*;

public class ControlGpioExample2 {
    // create gpio controller
    final static GpioController gpio = GpioFactory.getInstance();
    // provision gpio pin #01 as an output pin and turn on
    final static GpioPinDigitalOutput pin = gpio.provisionDigitalOutputPin(
        RaspiPin.GPIO_01, "MyLED", PinState.HIGH);
    public static void main(String[] args) throws InterruptedException {
        // set shutdown state for this pin
        pin.setShutdownOptions(true, PinState.LOW);

        // toggle the current state of gpio pin #01
        for (int i=0; i<10; i++){
            pin.toggle();
            Thread.sleep(1000);
        }
```

```
        gpio.shutdown();
    }
}
```

GPIO 引脚也可以设置为数字输入引脚。示例 7.4 展示了如何将 GPIO 引脚 29 设置为数字输入引脚，并在 for 循环中使用 getState() 显示引脚状态 10 次。可以将按钮连接到 GPIO 引脚 29，如图 7.15 所示。电阻仅用于保护电路板。在运行程序时，如果按下按钮，GPIO 引脚 29 的状态将显示为低电平，否则将显示为高电平。

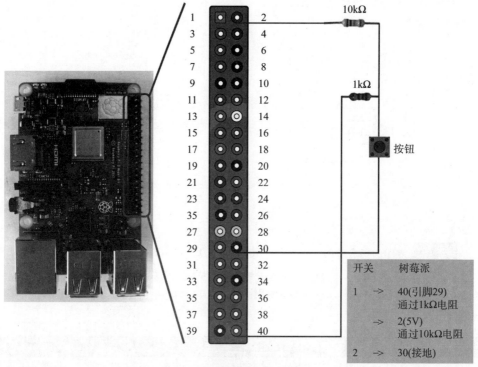

图 7.15　树莓派和按钮的电路

例 7.4　使用 GPIO 引脚进行输入

```
//Example 7.4
import com.pi4j.io.gpio.*;
public class ControlGpioExample3 {

    public static void main(String[] args) throws InterruptedException {

        // create gpio controller
        final GpioController gpio = GpioFactory.getInstance();

        // provision gpio pin #29 as an input pin
        final GpioPinDigitalInput Inp = gpio.provisionDigitalInputPin(
            RaspiPin.GPIO_29, PinPullResistance.PULL_DOWN);
```

```
    for (int i =0; i<10; i++){
        //Get the state of the pin, High or Low
        System.out.println(Inp.getState());
        Thread.sleep(1000);
    }

    // this method will forcefully shutdown all GPIO monitoring
    // threads and scheduled tasks
    gpio.shutdown();

    }
}
```

练习 7.4 修改例 7.4 中的 Java 程序，使其使用不同的 GPIO 引脚作为数字输入引脚。

7.6.3 从 Java 程序中调用 Python 程序

如第 3 章所述，Java 还可以运行系统程序和命令。在树莓派中，Python 是默认的编程语言，拥有本机支持的所有函数。因此，如果在 Java 中有一些不太好做的事情，那么可以从 Java 程序中调用 Python 程序。

例如，下面这个简单的 Python GPIO 程序将每隔 0.5 秒就打开和关闭一次 GPIO 引脚 01。这里使用 Python GPIO Zero 库处理 GPIO 引脚。假设这个程序名为 led.py，位于/home/pi/文件夹中。请注意，Python 使用#作为注释符。

```
#Example 7.4 Python GPIO example - led.py
#Modified from https://gpiozero.readthedocs.io/en/stable/
from gpiozero import LED    #import the GPIO Zero library
from time import sleep

led = LED(18) #GPIO_01 is 18 in GPIO Zero

while True:
    led.on()
    sleep(0.5)
    led.off()
    sleep(0.5
```

有关 Python GPIO Zero 库及其引脚编号的更多信息，请访问以下资源：

```
https://gpiozero.readthedocs.io/en/stable/
https://gpiozero.readthedocs.io/en/stable/recipes.html
```

例 7.5 演示了如何在 Java 程序中进行系统调用以运行 led.py。由于 led.py 运行在一个无限循环中，因此需要按 Ctrl+C 组合键才能停止程序的运行。

例 7.5 从 Java 程序中调用 Python 程序

```
//Example 7.5 Java example to call Python program
import java.io.BufferedReader;
import java.io.InputStreamReader;
```

```
public class PythonExample {

    public static void main(String[] args) throws Exception {
        System.out.println("Running led.py, press CTRL + C to stop... ");
        Runtime rt= Runtime.getRuntime();
        Process p=rt.exec("python led.py"); [
    }
}
```

练习 7.5 修改例 7.5 中的 Java 程序，使其可以通过命令行参数获取 Python 程序的名称及路径，进而运行任何 Python 程序。

练习 7.6 修改例 7.5 中的 Java 程序，通过提示用户输入要运行的 Python 程序的名称，使其可以运行任何 Python 程序。

7.6.4 Java PWM 示例

脉冲宽度调制(Pulse Width Modulation，PWM)是一种非常有用的嵌入式技术。PWM 可用于控制 LED 的亮度或控制电机。在 Pi4J 库中，还有一个 PWM 示例程序，名为 PwmExample.java。我们可以在这个 PWM 示例程序的基础上，创建一个简单的 Java PWM 示例程序，如例 7.6 所示。该程序使用 CommandArgumentParser.getPin()设置默认参数，如 PWM 引脚(在本例中为 GPIO 引脚 01)，并使用 provisionPwmOutputPin()创建 PWM 对象。这里使用 pwmSetMode()设置 PWM 模式(PWM_MODE_BAL 或 PWM_MODE_MS)，使用 pwmSetRange()设置 PWM 范围(默认为 1024，此处设置为 1000)，并使用 pwmSetClock()设置 PWM 时钟的除数以指定 PWM 信号的频率(参见表 7.1)：

表 7.1 PWM 时钟的除数以及对应的信号频率

除数	频率	说明
2048	9.375 kHz	
1024	18.75 kHz	
512	37.5 kHz	
256	75 kHz	
128	150 kHz	
64	300 kHz	
32	600 kHz	
16	1.2 MHz	
8	2.4 MHz	
4	4.8 MHz	
2	9.6 MHz	最快可用
1	4.6875 kHz	与除数为 4096 时相同

最后，可以使用 setPwm()设置 PWM 速率，这里设置为 500，表示 50%的 PWM 占空比。在 PWM 中，占空比是指在一个时间间隔或一段时间内，脉宽调制引脚接通的时间百分比。因此，50%意味着 PWM 引脚一半的时间是打开的，而另一半的时间是关闭的。

例 7.6 一个简单的 Java PWM 程序

```java
//Example 7.6
import com.pi4j.io.gpio.*;
import com.pi4j.util.CommandArgumentParser;
public class PwmExample1
    public static void main(String[] args) throws InterruptedException {
        // create GPIO controller instance
        GpioController gpio = GpioFactory.getInstance();

        Pin pin = CommandArgumentParser.getPin(
                RaspiPin.class,      // pin provider class to obtain pin
                                     // instance from
                RaspiPin.GPIO_01,    // default pin if no pin argument found
                args);               // argument array to search in

        GpioPinPwmOutput pwm = gpio.provisionPwmOutputPin(pin);

        com.pi4j.wiringpi.Gpio.pwmSetMode(com.pi4j.wiringpi.Gpio.PWM_MODE_MS);
        com.pi4j.wiringpi.Gpio.pwmSetRange(1000);
        com.pi4j.wiringpi.Gpio.pwmSetClock(500);

        // set the PWM rate to 500
        pwm.setPwm(500);
        System.out.println("PWM rate is: " + pwm.getPwm());

        Thread.sleep(10000); //wait for 10 seconds
        pwm.setPwm(0);
        System.out.println("PWM rate is: " + pwm.getPwm());
        // stop all GPIO activity/threads
        gpio.shutdown();
    }
}
```

以下命令展示了如何编译和运行程序。同样，以超级用户身份运行程序也是很重要的。

```
$ javac -classpath ".:pi4j-core.jar" PwmExample1.java
$ sudo java -classpath ".:pi4j-core.jar" PwmExample1
```

练习 7.7 修改例 7.6 中的 Java 程序，将 setPwm()的值从 100、200 或 300 改为 1000，再将 LED 连接到 GPIO 引脚 01 并观察效果。

练习 7.8 修改例 7.6 中的 Java 程序，将 pwmSetClock()改为不同的值，例如 1、2、4、8、16、32、64、128、256、512、1024、2048 和 4096，再将数字示波器连接到 GPIO 引脚 01，并观察使用不同值的影响。

例 7.6A 展示了使用 Pi4J 库的 connectionPi.SoftPwm 类运行 PWM 的另一种方式：使用 Gpio.wiringPiSetup()初始化 wiring 库，并使用 SoftPwm.softPwmCreate()设置 PWM 引脚和 PWM 范

围。在这种情况下,使用的是 GPIO 引脚 01,范围是 0~100。然后使用 SoftPwm.softPwmWrite() 将值写入 PWM 引脚(100 中的 50),这意味着 50%的 PWM 占空比。

例 7.6A　使用 wiringpi.SoftPwm 类运行 PWM

```
//Example 7.6A
import com.pi4j.wiringpi.Gpio;
import com.pi4j.wiringpi.SoftPwm;
public class SoftPWM {
    private static int PIN_NUMBER = 1;
    public static void main(String[] args) throws InterruptedException {
        // initialize wiringPi library, this is needed for PWM
        Gpio.wiringPiSetup();

        // the range is set like (min=0 ; max=100)
        SoftPwm.softPwmCreate(PIN_NUMBER, 0, 100);
        int counter = 0;
        while (counter<10) {
            // softPwmWrite(int pin, int value)
            SoftPwm.softPwmWrite(PIN_NUMBER, 50);
            Thread.sleep(500);
            counter++;
        }
    }
}
```

7.6.5　Java PIR 和 LED 示例

基于前面的示例,可以使用无源红外(PIR)传感器和 LED 创建简单的树莓派和 Java 智能照明系统,电路图如图 7.16 所示。在这种情况下,我们使用 PIR 传感器检测人员的存在并相应地打开和关闭 LED。例 7.7 展示了用到的 Java 代码。

例 7.7　一个简单的树莓派和 Java 智能照明系统

```
//Example 7.7
import com.pi4j.io.gpio.*;

public class PIRGpioExample {
    public static void main(String[] args) throws InterruptedException {

        // create gpio controller
        final GpioController gpio = GpioFactory.getInstance();

        // provision gpio pin #01 (GPIO_01) as an output pin and turn on
        final GpioPinDigitalOutput LEDpin = gpio.provisionDigitalOutputPin
            (RaspiPin.GPIO_01, "MyLED", PinState.HIGH);

        // set shutdown state for this pin
        LEDpin.setShutdownOptions(true, PinState.LOW);

        // provision gpio pin #40 (GPIO_29) as an PIR input pin
        final GpioPinDigitalInput PIRpin = gpio.provisionDigitalInputPin
            (RaspiPin.GPIO_29, PinPullResistance.PULL_DOWN);
```

```
for (int i =0; i<10; i++){
    //Get the state of the pin, High or Low
    System.out.println(PIRpin.getState());

    if (PIRpin.getState()==PinState.HIGH){
        LEDpin.high();
    }
    else{
        LEDpin.low();
    }
    Thread.sleep(1000);
}

// this method will forcefully shutdown all GPIO monitoring threads
// and scheduled tasks
gpio.shutdown();

    }
}
```

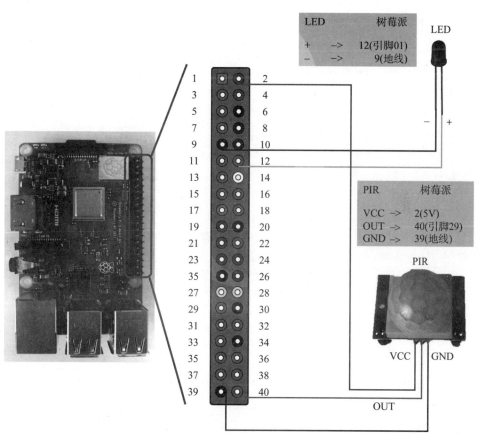

图 7.16　使用了 PIR 传感器和 LED 的树莓派以及 Java 智能照明系统的电路图

练习 7.9　修改例 7.7 中的 Java 程序，使用另一个 GPIO 引脚作为数字输入引脚。当这个 GPIO 引脚打开时，LED 应该一直亮着；而当这个 GPIO 引脚关闭时，LED 应该根据 PIR 引脚的状态亮起或灭掉。将按钮的打开/关闭开关连接到相应的 GPIO 引脚并观察效果。

例 7.8 基于例 7.7，演示了如何使用 Gpio.pinMode()将 GPIO 引脚设置为输入或输出引脚。

例 7.8　使用 Gpio.pinMode()设置 GPIO 引脚

```
//Example 7.8
import com.pi4j.io.gpio.*;
import com.pi4j.wiringpi.Gpio;
import com.pi4j.wiringpi.GpioUtil;

public class PIRGpioExample2 {
    public static void main(String[] args) throws InterruptedException {
        // create gpio controller
        final GpioController gpio = GpioFactory.getInstance();

        // provision gpio pin #01(GPIO_01) as an output pin and turn on
        final GpioPinDigitalOutput LEDpin = gpio.provisionDigitalOutputPin(
            RaspiPin.GPIO_01, "MyLED", PinState.HIGH);

        // set shutdown state for this pin
        LEDpin.setShutdownOptions(true, PinState.LOW);

        // provision gpio pin #40 (GPIO_29) as an PIR input pin
        Gpio.pinMode(40, Gpio.INPUT);

        for (int i =0; i<10; i++){
            if (Gpio.digitalRead(30)==1){
                LEDpin.high();
                System.out.println("High");
            }
            else{
                LEDpin.low();
                System.out.println("Low");
            }
            Thread.sleep(1000);
        }
        // this method will forcefully shutdown all GPIO monitoring threads and
        // scheduled tasks
        gpio.shutdown();
    }
}
```

7.6.6　Java I2C 示例

集成电路总线(I2C)是由飞利浦半导体公司(现为 NXP Semiconductors)于 1982 年发明的一种流行的串行通信总线和协议。I2C 是两线接口，用于连接低速设备，例如微控制器、数字传感器、EEPROM、A/D 和 D/A 转换器以及嵌入式系统中类似的外设。树莓派支持 I2C，并且具有两个默认的 I2C 引脚：用于 I2C SDA 的引脚 3 和用于 I2C SCL 的引脚 5。为了将 I2C 与树莓派结合起来使用，首先需要使用树莓派配置软件输入以下命令：

```
$ sudo raspi-config
```

raspi-config 界面在不同的 Raspbian 版本中略有不同。图 7.17 展示了 Raspbian GNU / Linux 8 (Jessie) 中的 raspi-config 界面。可以使用以下命令查看 Raspbian 操作系统的版本：

```
$ cat /etc/os-release
```

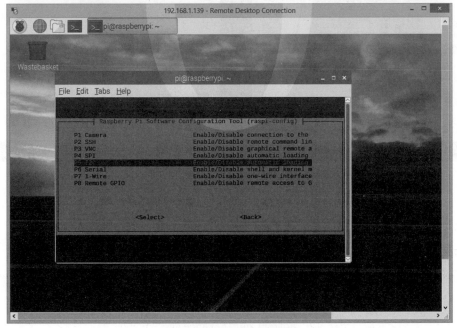

图 7.17　树莓派中 I2C 的配置

可以使用以下命令升级 Raspbian 操作系统：

```
$ sudo apt-get update
$ sudo apt-get dist-upgrade
```

在 raspi-config 界面中，使用向下箭头选择 Interface Options。然后选择 P5 I2C，按 Enter 键，选择 Yes 以完成配置程序，最后重新启动树莓派。

MAX30205 是一种流行的、精确的、基于 I2C 的数字体温传感器。图 7.18 展示了带有 MAX30205 体温传感器的简单树莓派的 I2C 电路图。例 7.9 给出了相应的 I2C Java 应用代码。这里首先使用 I2CFactory.getInstance()创建一个 I2C 对象并使用 i2c.getDevice()获取位于特定地址的设备，然后使用 write()向设备写入字节，使用 read()从设备读取字节。

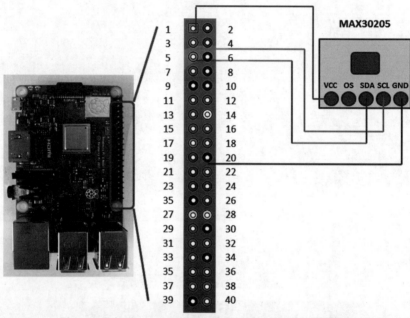

图 7.18 带有 MAX30205 体温传感器的简单树莓派的 I2C 电路图

例 7.9 一个简单的 I2C Java 应用

```java
//Example 7.9
import java.io.IOException;
import com.pi4j.io.i2c.I2CBus;
import com.pi4j.io.i2c.I2CDevice;
import com.pi4j.io.i2c.I2CFactory;
import com.pi4j.io.i2c.I2CFactory.UnsupportedBusNumberException;
public class I2CExample {
    public static void main(String args[]) throws InterruptedException, UnsupportedBusNumberException, IOException {
        int addr = 0x48; // MAX30205 I2C address
        System.out.println("I2C Example");
        I2CBus i2c = I2CFactory.getInstance(I2CBus.BUS_1);
        I2CDevice device = i2c.getDevice(addr);
```

```
        while(true){
            device.write(addr, (byte) 0x00);
            System.out.println(device.read(addr));
            Thread.sleep(2000);
        }
    }
}
```

有关 MAX30205 体温传感器的更多信息，请参阅以下资源：

https://www.maximintegrated.com/en/products/sensors/MAX30205.html
https://www.tindie.com/products/closedcube/max30205-01degc-human-body-temperature-sensor/
https://github.com/closedcube/ClosedCube_MAX30205_Arduino

例 7.10 展示了另一个 I2C Java 应用，该应用扫描从 1 到 128 的所有地址，以查看是否连接了任何 I2C 设备。

例 7.10 另一个 I2C Java 应用

```
//Example 7.10
import java.util.*;
import com.pi4j.io.i2c.*;

public class I2CScan {

    public static void main(String[] args) throws Exception {
        final I2CBus bus = I2CFactory.getInstance(I2CBus.BUS_1);
        for (int i = 1; i < 128; i++) {
          try {
                I2CDevice device = bus.getDevice(i);
                device.write((byte)0);
                System.out.println("Found Address: " + Integer.toHexString(i));
            } catch (Exception ignore) { }
        }
    }
}
```

还有更多的 I2C Java 应用示例，请参阅以下资源：

https://learn.sparkfun.com/tutorials/raspberry-pi-spi-and-i2c-tutorial/all
https://github.com/Pi4J/pi4j/blob/master/pi4j-example/src/main/java/bananapro/I2CExample.java
https://github.com/oksbwn/MCP23017-_Raspberry-Pi

7.6.7　Java ADC 示例

树莓派没有模数转换器(ADC)引脚，因此无法直接从模拟传感器读取数据。但是，由于树莓派支持 I2C，因此可以使用基于 I2C 的 ADC 模块(例如 16 位的 ADS1115(https://www.adafruit.com/product/1085))连接 5 个模拟传感器。

图 7.19 展示了树莓派、16 位的 ADS1115 模块和 LM35 温度传感器的电路图。树莓派使用 I2C 与 ADS1115 通信，ADS1115 通过通道 A0 读取 LM35 温度传感器的值。

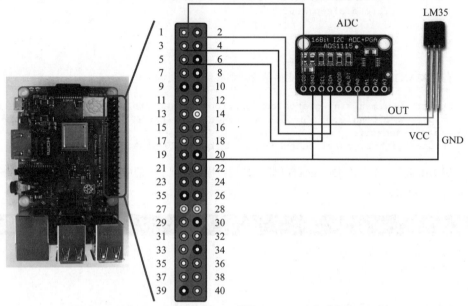

图 7.19　树莓派、16 位的 ADS1115 模块和 LM35 温度传感器的电路图

图 7.20 展示了树莓派、16 位的 ADS1115 模块和光敏电阻(LDR)传感器的电路图。

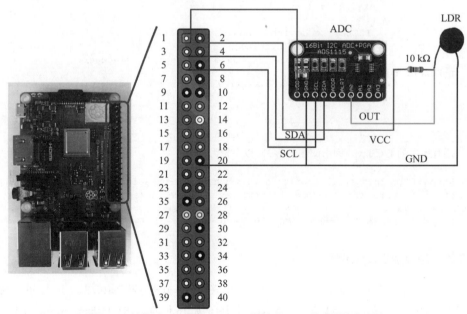

图 7.20　树莓派、16 位的 ADS1115 模块和 LDR 传感器的电路图

在 **Pi4J** 库中，有一个名为 ADS1115GpioExample.java 的示例程序，用于从 ADS1115 模块读取四个通道的值。以下是完整的代码：

```
/*
 * #%L
 * **********************************************************************
 * ORGANIZATION  :  Pi4J
 * PROJECT       :  Pi4J :: Java Examples
 * FILENAME      :  ADS1115GpioExample.java
 *
 * This file is part of the Pi4J project. More information about
 * this project can be found here:  http://www.pi4j.com/
 * **********************************************************************
 * %%
 * Copyright (C) 2012 - 2018 Pi4J
 * %%
 * This program is free software: you can redistribute it and/or modify
 * it under the terms of the GNU Lesser General Public License as
 * published by the Free Software Foundation, either version 3 of the
 * License, or (at your option) any later version.
 *
 * This program is distributed in the hope that it will be useful,
 * but WITHOUT ANY WARRANTY; without even the implied warranty of
 * MERCHANTABILITY or FITNESS FOR A PARTICULAR PURPOSE. See the
 * GNU General Lesser Public License for more details.
 *
 * You should have received a copy of the GNU General Lesser Public
 * License along with this program. If not, see
 * <http://www.gnu.org/licenses/lgpl-3.0.html>.
 * #L%
 */


import java.io.IOException;
import java.text.DecimalFormat;

import com.pi4j.gpio.extension.ads.ADS1115GpioProvider;
import com.pi4j.gpio.extension.ads.ADS1115Pin;
import com.pi4j.gpio.extension.ads.ADS1x15GpioProvider.ProgrammableGainAmplifierValue;
import com.pi4j.io.gpio.GpioController;
import com.pi4j.io.gpio.GpioFactory;
import com.pi4j.io.gpio.GpioPinAnalogInput;
import com.pi4j.io.gpio.event.GpioPinAnalogValueChangeEvent;
import com.pi4j.io.gpio.event.GpioPinListenerAnalog;
import com.pi4j.io.i2c.I2CBus;
import com.pi4j.io.i2c.I2CFactory.UnsupportedBusNumberException;

/**
 * <p>
 * This example code demonstrates how to use the ADS1115 Pi4J GPIO
 * interface for analog input pins.
 * </p>
 *
 * @author Robert Savage
```

```java
*/
public class ADS1115GpioExample {


    public static void main(String args[]) throws InterruptedException,
UnsupportedBusNumberException, IOException {

        System.out.println("<--Pi4J--> ADS1115 GPIO Example ... started.");

        // number formatters
        final DecimalFormat df = new DecimalFormat("#.##");
        final DecimalFormat pdf = new DecimalFormat("###.#");

        // create gpio controller
        final GpioController gpio = GpioFactory.getInstance();

        // create custom ADS1115 GPIO provider
        final ADS1115GpioProvider gpioProvider = new
ADS1115GpioProvider(I2CBus.BUS_1, ADS1115GpioProvider.ADS1115_ADDRESS_0x48);

        // provision gpio analog input pins from ADS1115
        GpioPinAnalogInput myInputs[] = {
            gpio.provisionAnalogInputPin(gpioProvider, ADS1115Pin.
                INPUT_A0, "MyAnalogInput-A0"),
            gpio.provisionAnalogInputPin(gpioProvider, ADS1115Pin.
                INPUT_A1, "MyAnalogInput-A1"),
            gpio.provisionAnalogInputPin(gpioProvider, ADS1115Pin.
                INPUT_A2, "MyAnalogInput-A2"),
            gpio.provisionAnalogInputPin(gpioProvider, ADS1115Pin.
                INPUT_A3, "MyAnalogInput-A3"),
        };

        // ATTENTION !!
        // It is important to set the PGA (Programmable Gain Amplifier)
        // for all analog input pins.
        // (You can optionally set each input to a different value)
        // You measured input voltage should never exceed this value!
        //
        // In my testing, I am using a Sharp IR Distance Sensor
        // (GP2Y0A21YK0F) whose voltage never exceeds 3.3 VDC
        // (http://www.adafruit.com/products/164)
        //
        // PGA value PGA_4_096V is a 1:1 scaled input,
        // so the output values are in direct proportion to the detected
        // voltage on the input pins
        gpioProvider.setProgrammableGainAmplifier(
ProgrammableGainAmplifierValue.PGA_4_096V, ADS1115Pin.ALL);


        // Define a threshold value for each pin for analog value change
        // events to be raised.
        // It is important to set this threshold high enough so that you don't
        // overwhelm your program with change events for insignificant changes
        gpioProvider.setEventThreshold(500, ADS1115Pin.ALL);
```

```
        // Define the monitoring thread refresh interval (in milliseconds)
        // This governs the rate at which the monitoring thread will
        // read input values from the ADC chip
        // a value less than 50 ms is not permitted
        gpioProvider.setMonitorInterval(100);

        // create analog pin value change listener
        GpioPinListenerAnalog listener = new GpioPinListenerAnalog()
        {
            @Override
            public void handleGpioPinAnalogValueChangeEvent(
GpioPinAnalogValueChangeEvent event)
            {
                // RAW value
                double value = event.getValue();

                // percentage
                double percent = ((value * 100) / ADS1115GpioProvider.
                    ADS1115_RANGE_MAX_VALUE);

                // approximate voltage ( *scaled based on PGA setting )
                    double voltage = gpioProvider.getProgrammableGainAmplifier(
                    event.getPin()).getVoltage() * (percent/100);

                // display output
                System.out.println(" (" + event.getPin().getName() +") : 
                    VOLTS=" + df.format(voltage) + " | PERCENT=" + pdf.format(percent) +
                    "% | RAW=" + value + " ");
            }
        };

        myInputs[0].addListener(listener);
        myInputs[1].addListener(listener);
        myInputs[2].addListener(listener);
        myInputs[3].addListener(listener);

        // keep program running for 10 minutes
        Thread.sleep(600000);

        // stop all GPIO activity/threads by shutting down the GPIO controller
        // this method will forcefully shutdown all GPIO monitoring
        // threads and scheduled tasks
        gpio.shutdown();

        System.out.println("Exiting ADS1115GpioExample");
    }
}
```

7.6.8　Java 数字传感器示例

树莓派可以直接连接到数字传感器。图 7.21 展示了带有 DHT11 数字温度和湿度传感器的树莓派的电路图。DHT11 传感器的数据引脚被连接到树莓派引脚 40。

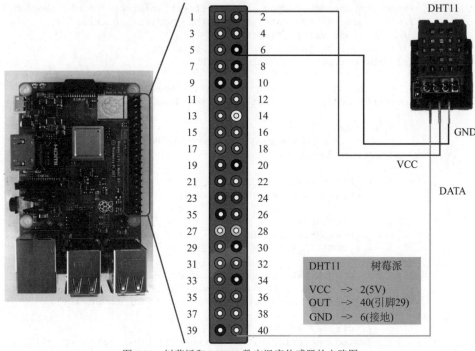

图 7.21　树莓派和 DHT11 数字温度传感器的电路图

例 7.11 展示的 Java 程序可以从 DHT11 传感器读取温度和湿度值，并进行奇偶校验以验证接收到的数据。

例 7.11　从 DHT11 传感器读取数据的 Java 程序

```java
//Example 7.11
import com.pi4j.wiringpi.Gpio;
import com.pi4j.wiringpi.GpioUtil;

public class DHT11Example1 {
    private static final int MAXTIMINGS = 85;
    private final int[]      dht11_dat = { 0, 0, 0, 0, 0 };
    private int pin = 29; //the default sensor data pin GPIO 29.
    public DHT11Example1(int pin) {

        // setup wiringPi
        if (Gpio.wiringPiSetup() == -1) {
            System.out.println(" ==>> GPIO SETUP FAILED");
            return;
        }

        GpioUtil.export(3, GpioUtil.DIRECTION_OUT);
        this.pin = pin;
    }
    public String readTemp() {
        String result = "";
```

```java
        int laststate = Gpio.HIGH;
        int j = 0;
        dht11_dat[0] = dht11_dat[1] = dht11_dat[2] = dht11_dat[3] = dht11_
           dat[4] = 0;

        Gpio.pinMode(pin, Gpio.OUTPUT);
        Gpio.digitalWrite(pin, Gpio.LOW);
        Gpio.delay(18);

        Gpio.digitalWrite(pin, Gpio.HIGH);
        Gpio.pinMode(pin, Gpio.INPUT);

        for (int i = 0; i < MAXTIMINGS; i++) {
           int counter = 0;
           while (Gpio.digitalRead(pin) == laststate) {
              counter++;
              Gpio.delayMicroseconds(1);
              if (counter == 255) {
                 break;
              }
           }

           laststate = Gpio.digitalRead(pin);

           if (counter == 255) {
              break;
           }

           /* ignore first 3 transitions */
           if (i >= 4 && i % 2 == 0) {
              /* shove each bit into the storage bytes */
              dht11_dat[j / 8] <<= 1;
              if (counter > 16) {
                 dht11_dat[j / 8] |= 1;
              }
              j++;
           }
        }
        // check we read 40 bits (8bit x 5 ) + verify checksum in the last
        // byte
        if (j >= 40 && checkParity()) {
           float h = (float) ((dht11_dat[0] << 8) + dht11_dat[1]) / 10;
           if (h > 100) {
              h = dht11_dat[0]; // for DHT11
           }
           float c = (float) (((dht11_dat[2] & 0x7F) << 8) + dht11_dat[3]) / 10;
           if (c > 125) {
              c = dht11_dat[2]; // for DHT11
           }
           if ((dht11_dat[2] & 0x80) != 0) {
              c = -c;
           }
           result = "Humidity = " + h + "% Temperature = " + c + "C";
        } else {
```

```
            result = "";
        }
        return result;
    }

    private boolean checkParity() {
        return dht11_dat[4] == (dht11_dat[0] + dht11_dat[1] + dht11_dat[2]
                        + dht11_dat[3] & 0xFF);
    }

    public static void main(String ars[]) throws Exception {
        int DHTpin =29;
        DHT11Example1 dht = new DHT11Example1(DHTpin );

        for (int i = 0; i < 10; i++) {
            System.out.println(dht.readTemp());
            Thread.sleep(1000);
        }
    }
}
```

为了编译和执行上述 Java 程序，请输入以下命令(或将它们放入 Linux shell 脚本):

```
$ javac -classpath ".:pi4j-core.jar" DHT11Example1.java
$ sudo java -classpath ".:pi4j-core.jar" DHT11Example1
```

图 7.22 给出了 DHT11Example1.java 程序的执行结果，该程序可以从 DHT11 传感器读取温度和湿度值。

有关树莓派和 DHT11 传感器的更多信息，请参阅以下资源:

http://www.circuitbasics.com/how-to-set-up-the-dht11-humidity-sensor-on-the-raspberry-pi/
https://tutorials-raspberrypi.com/raspberry-pi-measure-humidity-temperature-dht11-dht22/

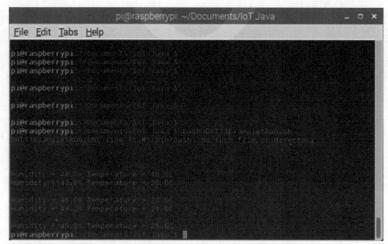

图 7.22　DHT11Example1.java 程序的执行结果

7.6.9 Java MQTT 示例

消息队列遥测传输(MQTT)是一种 IoT 协议，这种 IoT 协议允许 IoT 设备通过代理发布关于某个主题的消息，然后由代理将消息转发给所有订阅者。在本例中，我们将在树莓派上创建一个 Java MQTT 程序。我们使用 iot.eclipse.org 作为 MQTT 代理，将一些温度和湿度信息发布到名为 PX Temperature and Humidity 的主题，然后使用计算机(或其他移动设备)订阅该主题并使用 IBM WMQTT A92 程序接收消息。图 7.23 给出了概念图。

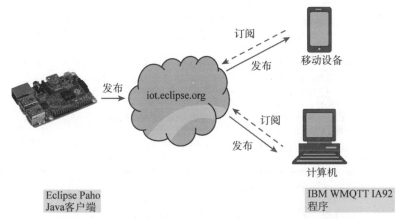

图 7.23 用于发布和查看 MQTT 消息的 Java MQTT 示例的概念图

为了完成这个示例，我们需要下载如下两个软件。

- Eclipse Paho Java 客户端(在树莓派上)：https://www.eclipse.org/paho/clients/java/。
- IBM WMQTT IA92 程序(在计算机上)：https://github.com/mqtt/mqtt.github.io/wiki/ia92。

Eclipse Paho Java 客户端有不同的版本可以选择。在这里，我们选择使用 MQTT V3 1.0.2 版，你可以从以下链接直接下载：

```
https://repo.eclipse.org/content/repositories/paho/org/eclipse/paho/org.eclipse.paho.client.mqttv3/1.0.2/org.eclipse.paho.client.mqttv3-1.0.2.jar
```

对于 IBM WMQTT IA92 程序，只需要访问 GitHub 网站并按照说明进行下载并安装即可。最简单的方法是下载为压缩文件，然后解压缩到计算机上。

例 7.12 展示了一个简单的 Java MQTT 程序，该程序从 Eclipse Paho Java 客户端网站上的 Java 示例 MqttPublishSample 修改而来。该程序使用 iot.eclipse.org:1883 作为代理，主题为"温度和湿度"，消息内容为 T=30C and RH=40%，QoS 为 2 级，这意味着正好执行一次。

例 7.12 一个简单的 Java MQTT 程序

```
//Example 7.12 Modified from MqttPublishSample program in
//https://www.eclipse.org/paho/clients/java/
import org.eclipse.paho.client.mqttv3.MqttClient;
import org.eclipse.paho.client.mqttv3.MqttConnectOptions;
```

```java
import org.eclipse.paho.client.mqttv3.MqttException;
import org.eclipse.paho.client.mqttv3.MqttMessage;
import org.eclipse.paho.client.mqttv3.persist.MemoryPersistence;

public class MqttExample {
    public static void main(String[] args) {
        String topic       = "PX Temperature and Humidity";
        String content     = "T=30C and RH=40%";
        int qos            = 2;
        String broker      = "tcp://iot.eclipse.org:1883";
        String clientId    = "JavaMQTTExmple";
        MemoryPersistence persistence = new MemoryPersistence();

        try {
            MqttClient sampleClient = new MqttClient(broker, clientId,
                persistence);
            MqttConnectOptions connOpts = new MqttConnectOptions();
            connOpts.setCleanSession(true);
            System.out.println("Connecting to broker: "+broker);
            sampleClient.connect(connOpts);
            System.out.println("Connected");
            System.out.println("Publishing message: "+content);
            MqttMessage message = new MqttMessage(content.getBytes());
            message.setQos(qos);
            sampleClient.publish(topic, message);
            System.out.println("Message published");
            sampleClient.disconnect();
            System.out.println("Disconnected");
            System.exit(0);
        } catch(MqttException me) {
            me.printStackTrace();
        }
    }
}
```

将MQTT JAR文件和Java程序放在同一文件夹中,然后使用以下命令编译并执行程序(在这里,将MQTT JAR文件包括在类路径中以进行编译和执行非常重要):

```
$ javac -classpath ".:org.eclipse.paho.client.mqttv3-1.0.2.jar" MqttExample.java
$ sudo java -classpath ".:org.eclipse.paho.client.mqttv3-1.0.2.jar" MqttExample
```

图7.24给出了MQTTExample.java程序的编译、执行和输出结果。

要查看MQTT消息,请进入IBM的名为ia92的WMQTT IA92解压缩文件夹,然后在J2SE子目录中找到wmqttSample.jar文件,双击这个文件。如果程序没有运行,那么可以在Windows命令提示符窗口中执行 java –jar wmqttSample.jar 命令。确保首先在端口1883上连接到iot.eclipse.org,并订阅"温度和湿度"主题。这样,每当Java程序从树莓派发送输出时,就可以接收到温度和湿度信息,如图7.25所示。

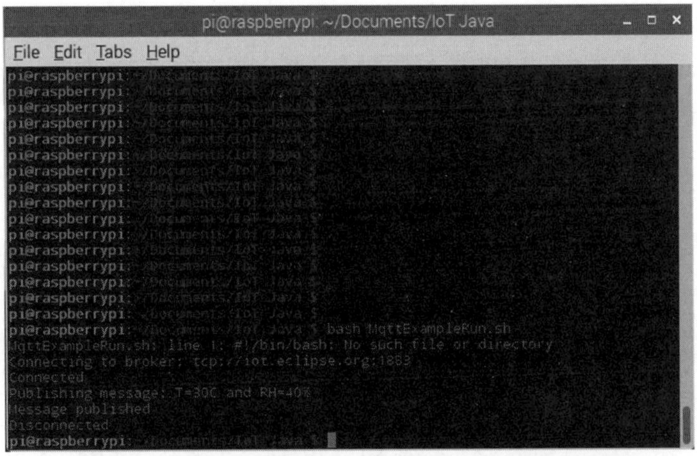

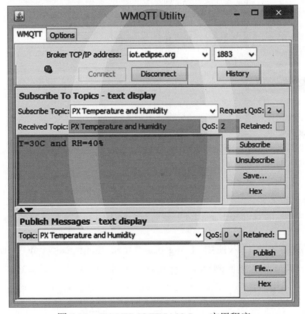

图 7.24　MQTTExample.java 程序的编译、执行和输出结果

图 7.25　IBM WMQTT IA92 Java 实用程序

7.6.10　Java REST 示例

代表性状态传输(REST)是另一种流行的 IoT 通信协议，类似于 MQTT。但是与 MQTT 不同，REST 是基于 Web 的。因此，只需要使用 Web 浏览器就能查看消息，无须再使用其他软件。在本例中，你将看到如何在树莓派上创建一个 Java REST 程序，用于将消息发送到 REST 服务器 Thingspeak，并使用 Web 浏览器在计算机上查看消息。

为了完成这个示例，需要提前做两件事：

- 下载 Unirest Java 库，链接为 http://unirest.io/java.html。

- 在 Thingspeak 网站上注册，网址为 https://thingspeak.com/。

Unirest Java 库有许多下载和使用方法。这里选择从以下链接下载 Unirest Java JAR 1.4.9 文件，其中包含所有依赖项：

https://jar-download.com/artifacts/com.mashape.unirest/unirest-java/1.4.9/source-code

将下载的 jar_files.zip 文件解压缩到一个文件夹，例如/home/pi/Downloads/unirest-1.4.9/，里面应该包含 unirest-java-1.4.9.jar 文件和其他 8 个 JAR 文件，如图 7.26 所示。

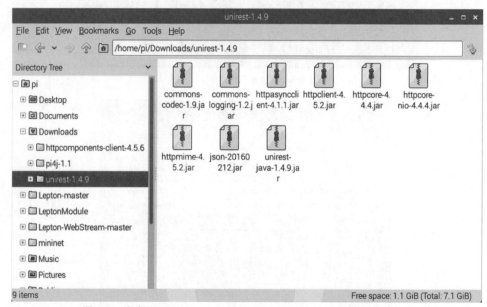

图 7.26　名为 unirest-java-1.4.9.jar 的 Unirest Java JAR 文件和所有依赖项

对于 REST 服务器，可以使用 Thingspeak(https://thingspeak.com/)，Thingspeak 是具有 MATLAB 分析功能的开放式物联网平台。如果还没有账户，那么需要先注册。注册后，就可以登录并创建频道。在本例中，笔者创建了一个名为 PX Sensor 的通道，通道 ID 为 599274。在图 7.27 展示的 Thingspeak 网页中，不同的选项卡显示了不同的视图。Private View 和 Public View 选项卡分别展示了 Web 网页如何查找私人用户和公共用户。Channel Settings 选项卡展示了频道统计信息和字段。对于许多传感器值来说，可能会有多个字段。这里只使用一个字段，名为 Temperature。API Keys 选项卡展示了用于从 Thingspeak REST 服务器写入和读取的安全密钥。Data Import/Export 选项卡展示了如何使用 API 请求从 REST 服务器更新和检索数据。

例 7.13 展示了如何使用 REST 协议将不会说话的温度传感器数据发送到 Thinkspeak REST 服务器网站。

第 7 章 面向物联网应用的 Java 编程 225

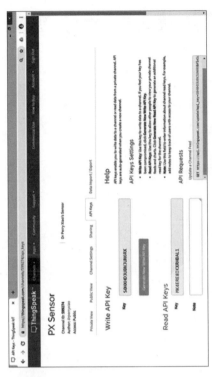
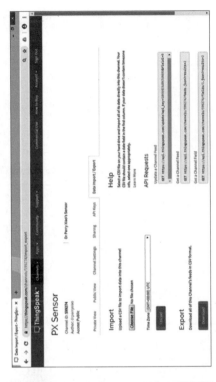
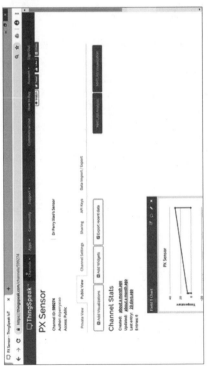
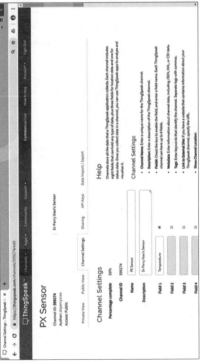

图 7.27 Thinkspeak 网页以及各种不同的选项卡

例 7.13　使用 REST 协议发送不会说话的温度传感器数据

```java
//Example 7.13
import java.io.InputStream;
import com.mashape.unirest.http.*;
import com.mashape.unirest.http.async.Callback;
import com.mashape.unirest.http.exceptions.UnirestException;


public class RESTCall implements Callback<JsonNode>{

    public void sendDataOverRest(double temp) {

        Unirest.post("https://api.thingspeak.com/update.json")
            .header("accept", "application/json")
            .field("api_key", "S0HXHD3UBNJUN6RX")
            .field("field1",temp)
            .asJsonAsync(this);
        }

        @Override
        public void cancelled() {
            System.out.println("The request has been cancelled");
        }

        @Override
        public void completed(HttpResponse<JsonNode> response) {
            int code = response.getStatus();

                JsonNode body =response.getBody();
                InputStream rawBody = response.getRawBody();

                    System.out.println(code);
                    System.out.println(body);
                    System.out.println(rawBody);
        }

        @Override
        public void failed(UnirestException arg0) {
            System.out.println("The request has failed");
        }

        public static void main(String[] args) throws InterruptedException {
            RESTCall http = new RESTCall();

            double temp=30.0;
            http.sendDataOverRest(temp);
        }
}
```

输入以下命令以编译并执行程序，确保在类路径中包含 Unirest Java 库文件以及所有依赖。

```
$ javac -classpath ".:/home/pi/Downloads/unirest-1.4.9/*" RESTCall.java
$ sudo java -classpath ".:/home/pi/Downloads/unirest-1.4.9/*" RESTCall
```

打开 Web 浏览器，从 Private View 或 Public View 选项卡登录到 Thinkspeak 账户，你应该能看到从 Java REST 程序发送过来的温度值。

有关 Java 和树莓派的更多信息，请参阅以下资源：

```
http://www.oracle.com/technetwork/articles/java/raspberrypi-1704896.html#Java%20
http://www.oracle.com/technetwork/articles/java/cruz-gpio-2295970.html
http://www.oracle.com/webfolder/technetwork/tutorials/obe/java/RaspberryPi
_GPIO/RaspberryPi_GPIO.html
https://iot.eclipse.org/java/tutorial/
http://agilerule.blogspot.com/2016/06/java-raspberry-pi-pi4j-pir-motion-
sensor.html
http://www.robo4j.io/2017/05/be-ready-and-prepare-raspberry-pi-for.html
```

7.7　Oracle Java ME 嵌入式客户端

除 Java Standard Edition 外，还有 Java Micro Edition 和 Java ME(https://www.oracle.com/technetwork/java/embedded/javame/index.html)。Java ME 为那些在物联网中的嵌入式设备和移动设备上运行的应用(例如微控制器、智能仪表/传感器、网关、移动电话、电视和打印机)提供了强大且灵活的环境。嵌入式 Java ME 是一种利用了核心 Java ME 技术的 Java 运行时环境，可以在许多嵌入式设备上运行，例如 Freescale FRDM K64F、STM32429I-EVAL(Cortex-M4 / RTX)、STM 32F746GDISCOVERY(Cortex-M7 / RTX)、英特尔 Galileo Gen 2 和树莓派(ARM 11 / Linux)。

有关更多信息，请参阅以下资源：

```
https://www.oracle.com/technetwork/java/embedded/javame/embedded-client/
overview/meembeddedclientgetstarted-2177401.html
https://docs.oracle.com/javame/8.1/get-started-freescale-k64/install.htm
http://thomasweldon.com/tpw/courses/eegr6114/javambed/dspJavaMbedNetbeans.html
```

7.8　适用于 Java 的物联网平台

除了在嵌入式系统中运行 Java 之外，还可以使用 Java 与各种物联网平台进行通信。

7.8.1　Eclipse Open IoT Stack

Eclipse Open IoT Stack 是一组开源技术，能使 Java 开发人员更轻松地构建 IoT 解决方案。该物联网平台的重点是使开发人员能够连接和管理作为物联网解决方案一部分的设备、传感器和执行器。Eclipse Open IoT Stack 提供对许多流行的物联网标准的支持，例如 MQTT、CoAP、轻量级 M2M(LWM2M)以及一组用于构建物联网网关的服务。所有信息都可以从 Eclipse Open IoT Stack 网站(https://iot.eclipse.org/java/)获得。Eclipse Open IoT Stack 简化了物联网解决方案的开发，例如构建智能温室(https://iot.Eclipse.org/java/tutorial/)和构建智能家居(https://www.Eclipse.org/smarthome/gettingstarted.html)。

7.8.2 IBM Watson IoT

IBM Watson IoT 是另一个流行的物联网平台(https://internetofthings.ibmcloud.com/#/)，图 7.28 展示了 IBM Watson IoT 的演示网站(http://discover-iot.eu-gb.mybluemix.net/#/play)，可以将任何嵌入式系统连接到 IBM 物联网云并显示传感器，无须注册。

这里有几个有趣的网站可供你参考，例如 IBM Watson IoT Java 客户端库网站(https://github.com/ibm-watson-iot/iot-java)、IBM Watson IoT 的树莓派诀窍(https://developer.ibm.com/recipes/tutorials/?s=Raspberry)、IBM Watson IoT 教程以及使用树莓派摄像头和 Watson Visual Recognition 确定图像中是否存在感兴趣的对象(https://developer.ibm.com/recipes/tutorials/use-a-raspberry-pi-camera-and-watson-visual-recognition-to-determine-if-object-of-interest-is-in-the-image/)。

图 7.28　IBM Watson IoT 的演示网站

7.8.3　AWS IoT

Amazon Web Services (AWS) 提供了可靠、可扩展且廉价的云计算服务。参与是免费的，只需要为使用付费。你可以在适用于 Java 的 AWS SDK 网站(https://aws.amazon.com/sdk-for-java/)上找到所有文档，并且可以在 GitHub 网站(https://GitHub.com/o-can/aws-java-iot-example)上找到适用于 Java 的 AWS IoT SDK 示例程序。

有关适用于 Java 的 AWS IoT SDK 的更多信息，请参阅以下资源：

```
https://docs.aws.amazon.com/iot/latest/developerguide/iot-sdks.html#iotjava-sdk
https://aws.amazon.com/blogs/iot/introducing-aws-iot-device-sdksfor-java-and-python/
```

7.8.4 Microsoft Azure IoT

Microsoft Azure 是一种开放的、灵活的企业级云计算平台，可提供基础设施即服务(Infrastructure as a Service，IaaS)、平台即服务(Platform as a Service，PaaS)和软件即服务(Software as a Service，SaaS)，并且支持许多不同的编程语言，包括 Java。IaaS 允许用户启动通用的 Microsoft Windows 和 Linux 虚拟机。PaaS 允许开发人员轻松地发布和管理网站。SaaS 允许用户通过 Internet 连接并使用基于云的软件，例如电子邮件、日历和办公工具(如 Microsoft Office 365)。Microsoft Azure IoT 让用户无须编写代码即可连接、监视和控制 IoT 设备，对实时数据流进行分析，存储物联网数据以及自动化数据访问和跨云使用数据。

使用 Microsoft Azure Java SDK 可以开发 Java 应用以访问大量的 Microsoft Azure 云服务，例如存储服务、媒体服务、队列服务、服务总线队列和 SQL 数据库。对于 Microsoft Azure 来说，启动 Java 的最佳位置是 Java 开发人员中心(https://azure.microsoft.com/en-us/develop/java/)，在那里可以免费获得 200 美元的赠金和 12 个月的免费服务。

在面向 Java 开发人员的 Microsoft Azure Java 文档中心(https://docs.microsoft.com/zh-cn/java/azure/)可以找到入门指南以及一些更深入的信息。Microsoft Azure SDK 是开源软件，因此可以随意修改。可从 GitHub 网站(https://github.com/Azure/azure-sdk-for-java)获取完整的源代码。

7.9 小结

本章介绍了 IoT(物联网)的概念、解释了 IoT 的工作原理和 IoT 技术，还为不同的物联网平台(如 Eclipse Open IoT Stack、IBM Watson IoT、AWS IoT 和 Microsoft Azure IoT)引入了带有树莓派的 Java 物联网应用。

7.10 本章复习题

1. 什么是物联网？
2. 物联网是如何工作的？
3. 什么是潜在的物联网应用？
4. 什么是工业 4.0？
5. 什么是物联网通信协议？
6. 什么是物联网平台？举几个例子。
7. 什么是树莓派？市面上还有其他类似的嵌入式计算机吗？
8. 树莓派的 GPIO 引脚是什么？
9. 什么是 PWM？树莓派的默认 PWM 引脚是哪些？

10. 什么是I2C？树莓派的默认I2C引脚是哪些？
11. 什么是MQTT？
12. 什么是REST？与HTTP进行比较。
13. 什么是IaaS、PaaS和SaaS？
14. 比较不同的物联网平台并列出各自的特点。

第8章

面向人工智能应用的Java编程

"生活就像骑自行车,只有不断前进,才能保持平衡。"

——Albert Einstein

8.1　什么是人工智能

8.2　神经网络

8.3　机器学习

8.4　深度学习

8.5　Java AI 库

8.6　神经网络方面的 Java 示例

8.7　机器学习方面的 Java 示例

8.8　深度学习方面的 Java 示例

8.9　适用于 Java 的 TensorFlow

8.10　AI 资源

8.11　小结

8.12　本章复习题

8.1　什么是人工智能

当前,人工智能(Artificial Intelligence,AI)是另一个热门话题。人工智能到底是什么?人工智能将如何影响我们的生活?人工智能会让我们都失业吗?

人工智能是计算机科学的众多领域之一,研究的是如何创造机器来做智能的事情,例如学习、规划、解决问题、预测、人脸和语音识别等。人工智能的起源可以追溯到 20 世纪 50 年代，当时英国计算机科学家艾伦·图灵(Alan Turing)提出了模仿游戏测试以观察计算机是否能与人类无差别地思考和做出行为。这就是著名的图灵测试。到目前为止，还没有一台计算机通过图灵测试。

人工智能作为一门研究学科在 1956 年举办的一场研讨会上建立。"人工智能"这个术语是由斯坦福大学的传奇计算机科学家 John McCarthy 创造的，他也是人工智能研究方面最具影响力的创始人和领导者之一。

人工智能可分为狭义人工智能、通用人工智能和超级人工智能。

狭义人工智能又称弱人工智能，是执行单一任务的智能。狭义人工智能的例子包括天气预报、购买建议、销售预测、计算机视觉、自然语言处理、语音识别、下棋和谷歌翻译等。狭义人工智能是我们迄今已取得的成就。狭隘人工智能也是我们目前所处的阶段，通用人工智能才是我们的目标。

通用人工智能又称强人工智能，是能够处理更复杂、更一般任务的智能。通用人工智能具有认知能力，能够理解环境。通用人工智能能够像人类一样观察、思考、分析、学习、发明并拥有感觉。根据著名的未来学家、谷歌工程总监 Ray Kurzweil 的说法，到 2029 年，人工智能将通过图灵测试；到 2045 年，技术奇点将出现。奇点是指人工智能开始超越人类的时间，图 8.1 阐释了这一概念。

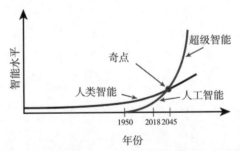

图 8.1 与人类智能相比，人工智能的技术奇点和时间线

超级人工智能又称超智能，是奇点之后的人工智能。对于超级人工智能会发生什么，人们有不同的看法。一些人表达了担忧和恐惧。例如，SpaceX 创始人兼特斯拉公司首席执行官埃隆·马斯克(Elon Musk)曾将 AI 称为最大的生存威胁。英国理论物理学家斯蒂芬·霍金(Stephen Hawking)警告说，人工智能终有一天会终结人类。许多其他人，如微软创始人比尔·盖茨(Bill Gates)和 Facebook 创始人马克·扎克伯格(Mark Zuckorberg)，都认为人工智能能使人类受益。就像许多早期的技术革命一样，人工智能虽然会代替人类的一些工作，但也将创造更多新的工作。不管你是否准备好，也不管喜不喜欢，人工智能都会到来。因此，我们需要做好准备，确保从人工智能中获益。

8.1.1 人工智能的研究历史

人工智能的研究历史大致可以分为三个阶段：神经网络(20 世纪 50 年代至 20 世纪 70 年代)、机器学习(20 世纪 80 年代至 21 世纪的前十年)和深度学习(当前)，如图 8.2 所示(https://www.sas.com/cn_gb/insights/analytics/what-is-artificial-intelligence.html)。

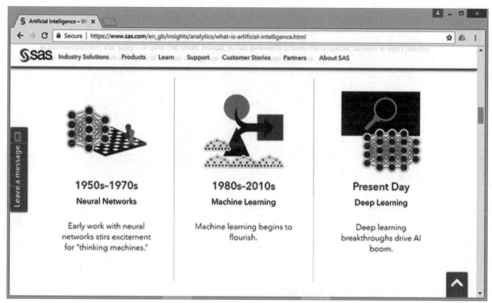

图 8.2 人工智能的研究历史

神经网络 神经网络是基于人类的生物神经网络开发的。神经网络通常由三个不同的层组成：输入层、隐藏层和输出层。一旦使用大量给定数据训练了神经网络，就可以将其用于预测未知数据的输出。从 20 世纪 50 年代到 70 年代，神经网络引起广泛关注，并激发了人们的热情和乐观情绪。

机器学习 机器学习是一套用于自动数据分析的数学算法。机器学习在 20 世纪 80 年代到 21 世纪的前十年开始蓬勃发展，产生了一些十分流行的算法，例如支持向量机(Support Vector Machine，SVM)、K 均值聚类、线性回归和朴素贝叶斯等。

深度学习 深度学习使用了具有多个隐藏层的神经网络。自 2010 年以来，随着现有计算能力、特定图形处理单元(Graphics Processing Unit，GPU)和改进算法的增加，深度学习开始登上舞台。如今，随着海量的标记数据集的不断增加和 GPU 计算能力的不断提高，深度学习已在许多应用领域显示出巨大的潜力。

图 8.3 展示了人工智能、机器学习和深度学习的时间线以及它们之间的差异(https://blogs.nvidia.com/blog/2016/07/29/whats-difference-artificial-intelligence-machine-learning-deep-learning-ai/)。

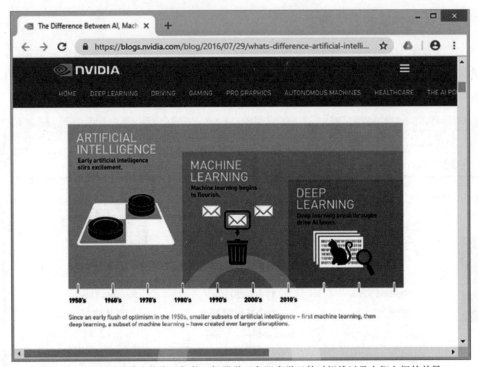

图 8.3 Nvidia 网站给出的人工智能、机器学习和深度学习的时间线以及它们之间的差异

8.1.2 云人工智能与边缘人工智能

许多人工智能应用需要大量的训练数据集和强大的计算能力。为此，在云上运行人工智能是有益的。谷歌、IBM、微软、亚马逊、阿里巴巴和百度等许多技术巨头都提供基于云的人工智能服务。使用基于云的人工智能服务的客户有很多优势，比如不需要购买昂贵的硬件，而只需要为使用的服务付费，并且无须担心软件的安装、配置、故障排除和升级等问题。基于云的人工智能的缺点在于延迟、带宽要求和安全性。因为需要将数据发送到云端并获取结果，所以基于云的人工智能会有延迟。向云端发送大量数据也需要大量带宽。最后，如果云服务遭到黑客攻击，数据或信息可能就会丢失或被窃取。对于许多物联网应用或其他实时应用，延迟、带宽和安全性可能成为问题。这可能是边缘人工智能发挥作用的地方。

边缘人工智能是指在边缘设备(例如微控制器、智能手机或其他设备)上运行的人工智能软件，又称为设备上的人工智能。边缘人工智能的优势在于实施性好、运行稳定可靠。语音识别、人脸识别、目标检测、无人驾驶汽车等应用都可以使用边缘人工智能。云人工智能和边缘人工智能将在很大程度上相互补充。

下面是一些边缘人工智能设备和应用的例子。

- OpenMV：单片机上的计算机视觉(http://docs.openmv.io/)。
- JeVois：支持 TensorFlow 的摄像头模块(http://jevois.org/)。
- Google Edge TPU：TensorFlow 处理单元(https://cloud.google.com/edge-tpu/)。

- Movidius：英特尔的视觉处理单元(https://www.movidius.com/)。
- Nvidia JETSON：基于 GPU 的人工智能(https://www.nvidia.com/en-us/autonomous-machines/embedded-systems-dev-kits-modules/)。
- UP AI Edge：由英特尔的 CPU、GPU、VPU 和 FPGA 支持的人工智能板(https://up-shop.org/25-up-ai-edge)。
- Ultra96：Xilinx FPGA(https://www.96boards.ai/products/ultra96/)。
- TensorFlow Lite：适用于移动和嵌入式设备的轻量级 TensorFlow(https://www.tensorflow.org/lite/)。
- uTensor：基于 Mbed 和 TensorFlow 构建的极为轻量级的机器学习推理框架(https://github.com/uTensor/uTensor)。
- Qualcomm Neural Processing SDK for AI：高通的人工智能 SDK(https://developer.qualcomm.com/software/qualcomm-neural-processing-sdk)。
- 华为 NPU：神经网络处理单元(https://developer.huawei.com/consumer/en/service/HiAI.html)。

8.2 神经网络

神经网络(Neural Network，NN)又称为人工神经网络(Artificial Neural Network，ANN)，是用来解决问题的数学算法。人工神经网络的概念最早是在 1943 年由美国的 Warren S.McCulloch(神经科学家)和 Walter Pitts(逻辑学家)提出的，灵感来自构成人脑的生物神经网络。生物神经网络由大量相互连接的神经元组成。人脑通常有大约 1000 亿个神经元。每个神经元由三个主要部分组成：树突、细胞体和轴突。

树突是从周围神经元接收输入信号的树状结构；细胞体用于处理输入信号；轴突用于连接另一个神经元的树突，并通过突触进行接触。突触允许神经元将电或化学信号传递到另一个神经元，并且不同突触的连接强度是不同的。神经元将对所有输入进行相加，然后通过轴突向下一个神经元发射输出信号。信号可以是兴奋性的或抑制性的，这就意味着可以取决于某些条件增加或减小发射强度。图 8.4 展示了不同类型神经元的典型结构(https://en.wikipedia.org/wiki/Neuron)。

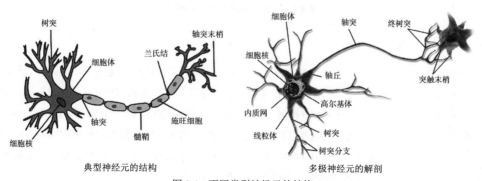

图 8.4 不同类型神经元的结构

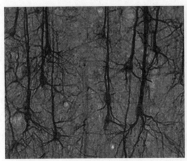

大脑皮层内 SMI32 染色的锥体神经元　　　高尔基染色的海马神经元

图 8.4(续)

图8.5 展示的神经网络教程解释了如何使用数学函数基于真实的神经元创建人工神经元(https://leonardoaraujosantos.gitbooks.io/artificial-inteligence/content/neural_networks.html)。

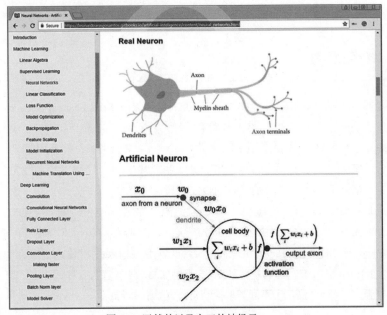

图 8.5　天然的以及人工的神经元

8.2.1　感知器

与生物神经网络一样,人工神经网络也由相互连接的神经元组成,称为感知器。感知器是神经网络中最基本的元件。感知器算法由美国心理学家弗兰克·罗森布拉特(Frank Rosenblatt)于 1957 年在美国康奈尔航空实验室提出。图 8.6 展示了感知器的结构,分为输入(树突)、主体和输出(轴突)。每个输入的权重反映了突触的连接强度。感知器根据输入的权重和偏差将所有输入相加,然后将结果输入激活函数中,激活函数将决定感知器的输出。感知器是典型的前馈网络,节点之间的连接不形成循环,各层之间没有反馈。

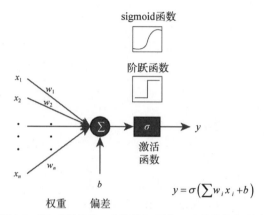

图 8.6 感知器的结构，分为输入(树突)、主体和输出(轴突)

如果 $x_1, x_2, \cdots, x_n$ 是感知器的输入，$\omega_1, \omega_2, \cdots, \omega_n$ 是输入的相应权重，n 是输入的总数，b 是相应的偏差，那么输入的加权和可以如下计算：

$$z = \sum_{i=1}^{n} \omega_i x_i + b$$

这里，$\omega_i x_i$ 表示 ω_i 乘以 x_i，i 表示每组数据的第 i 项，$\sum$ 表示计算第 $1\sim n$ 项的 $\omega_i x_i$ 的和。然后将计算结果输入激活函数以生成感知器的输出。

$$y = \delta(z) = \delta\left(\sum_{i=1}^{n} \omega_i x_i + b\right)$$

激活函数有很多。最简单且最常用的激活函数 $\delta(z)$ 是阶跃函数，输出为 0 或 1。

$$\delta(z) = \begin{cases} 1 & \text{如果 } z > 0 \\ 0 & \text{其他} \end{cases}$$

与 sigmoid 函数类似，另一种选择是使用双曲正切函数 tanh。

$$\delta(z) = \tanh(z) = \frac{e^z - e^{-z}}{e^z + e^{-z}}$$

接下来，需要训练感知器。为此，我们需要一组具有给定输入和期望输出的数据样本。训练是通过不断调整 ω 和 b 的值来完成的，直到用给定的输入 x 得到所需的输出 y。这需要通过多次迭代才能完成。以下伪代码显示了逻辑：

```
生成随机权重和偏差
对于每次迭代
    对于每组样本
        //计算输出
        y' = δ(∑ωᵢxᵢ + b)
        //计算误差
        ε = y - y'
```

```
        //计算调整量(梯度)
        delta=learnrate*x*ε
        //更新权重
        w=w+delta
        //更新偏差
        b=b+delta
直到样本完成

如果误差足够小或者次数达到限制
    停止
否则
    继续
迭代结束
```

这就是所谓的反向传播算法,关键是计算梯度来调整权重和偏差。一旦训练,感知器就应该能够为任何看不见的数据产生输出。感知器已被成功地应用于许多应用,如逻辑运算中的与、或、非、异或。但是,单个感知器或感知器层还不足以解决复杂的问题,因此人们研究出了多层感知器。8.6 节将提供用于单个感知器和多层感知器的 Java 示例程序。

有关感知器的更多信息,请参阅以下资源:

```
https://appliedgo.net/perceptron/
http://neuralnetworksanddeeplearning.com/chap1.html
https://github.com/mnielsen/neural-networks-and-deep-learning
https://towardsdatascience.com/what-the-hell-is-perceptron-626217814f53
https://natureofcode.com/book/chapter-10-neural-networks/
```

8.2.2 多层感知器与反向传播/前馈神经网络

常规的神经网络由多层感知器(MultiLayered Perceptron,MLP)组成,这些感知器通常具有三层:输入层、隐藏层和输出层。每一层可以有多个感知器。图 8.7 展示了一个传统的神经网络,它在输入层有四个感知器(表示有四个输入),在隐藏层有三个感知器,在输出层有两个感知器。

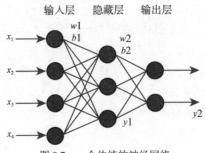

图 8.7 一个传统的神经网络

如果 x_1、x_2、x_3、x_4 是输入层的输入,$\omega1_1$、$\omega1_2$、$\cdots$、$\omega1_{12}$ 是输入层输出的相应权重,$b1_1$、$b1_2$、$b1_3$ 是隐藏层相应的偏差,那么隐藏层的输出 $y1_1$、$y1_2$、$y1_3$ 可以如下计算:

$$y1_j = \sum_{i=1}^{4} \omega 1_{((i-1) \cdot 3+1)} x_i + b1_j$$

然后，如果 $\omega 2_1$、$\omega 2_2$、$\cdots$、$\omega 2_6$ 是隐藏层输出的相应权重，$b2_1$、$b2_2$ 是输出层的相应偏差，输出层的输出 $y2_1$、$y2_2$ 可以如下计算：

$$y2_j = \sum_{i=1}^{3} \omega 2_{((i-1) \cdot 3+1)} y1_i + b2_j$$

通过使用 sigmoid 激活函数，可以如下计算神经网络的最终输出：

$$y = \delta(y2) = \frac{1}{1-e^{y2}}$$

同样，也可以使用反向传播算法训练网络。你需要一组具有给定输入和期望输出的数据样本。可通过连续调整 ω 和 b 的值来完成训练，直到对于给定的输入 x，可以获得所需的输出 y。以下伪代码显示了逻辑：

```
生成随机权重和偏差
对于每次迭代
   对于每组样本
     //计算隐藏层的输出
     y1=∑w1*x+b1
     //计算隐藏层的输出
     y2=δ(∑w2*y1+b2)
     //计算输出层误差
     delta2=y-y2
     //计算隐藏层误差
     delta1=delta2*w2
     //更新隐藏层的权重
     w1=w1+learnrate*x*delta1*y1(1-y1)
     //更新隐藏层的权重
     w2=learnrate*x*delta2*y2(1-y2)
     //更新隐藏层的偏差
     b1=b1+delta1
     //更新输出层的偏差
     b2=b2+delta2
   直到样本完成
   如果误差足够小或者次数达到限制
     停止
   否则
     继续
迭代结束
```

有关神经网络的更多信息，请参见以下资源：

```
https://www.nnwj.de/
https://www.cse.unsw.edu.au/~cs9417ml/MLP2/
https://kunuk.wordpress.com/2010/10/11/neural-network-backpropagationwith
-java/
https://www.doc.ic.ac.uk/~nd/surprise_96/journal/vol4/cs11/report.html
http://diffsharp.github.io/DiffSharp/0.6.3/examples-neuralnetworks.html
http://www.theprojectspot.com/tutorial-post/introduction-to-artificial-
neural-networks-part-1/7
```

http://www.theprojectspot.com/tutorial-post/introduction-to-artificial-neural-networks-part-2-learning/8
https://machinelearningmastery.com/neural-networks-crash-course/

8.3 机器学习

机器学习(Machine Learning，ML)是一类数学算法，可以使软件在预测给定数据集的结果方面更加准确。"机器学习"这个术语是在 1959 年由 Arthur Samuel(美国计算机和人工智能先驱)在 IBM 提出的。机器学习可分为监督学习、无监督学习、半监督学习和强化学习四种。

监督学习　在监督学习中，使用标记数据训练算法。机器学习算法使用给定的输入计算输出，并对计算的输出与期望的输出进行比较，然后相应地调整算法。一个很好的例子是使用支持向量机根据虹膜数据的萼片长度、萼片宽度、花瓣长度和花瓣宽度对虹膜类型进行分类。其他的例子包括语音识别、手写识别、模式识别、垃圾邮件检测和光学字符识别。

无监督学习　在无监督学习中，使用的是无标记数据。机器学习算法将研究数据的结构，并将它们分成具有最接近特征的组。K 均值聚类算法是无监督学习算法的典型例子。无监督学习的应用示例包括根据客户的购买行为进行分组、将某些客户与某些类型的产品相关联，等等。

半监督学习　在半监督学习中，标记数据和无标记数据都会使用。当在训练过程中因标记的成本太高而无法进行完全标记时，或者当并不是所有数据都可以标记时，半监督学习特别适用。半监督学习的例子包括语音分析和网页内容分析。

强化学习　在强化学习中，机器学习算法通过反复试验来学习找出哪个动作可以产生最大的回报，这通常是在没有现有训练数据的情况下完成的。强化学习通常用于机器人技术、游戏和导航。

以下是一些常用的机器学习算法：

- 线性回归算法
- 逻辑回归算法
- 线性判别分析算法
- 分类回归树算法
- 朴素贝叶斯算法
- K 均值聚类算法
- 学习矢量量化算法
- 支持向量机算法
- 袋装和随机森林算法
- Boosting 和 AdaBoost 算法

图 8.8 展示了 SAS 的机器学习网站(https://www.sas.com/en_gb/insights/analytics/machine-learning.html)。

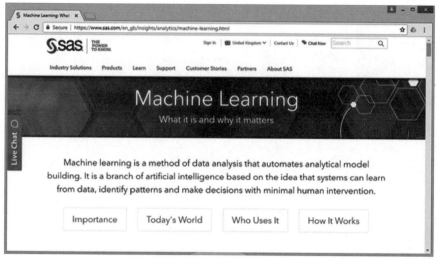

图 8.8　SAS 的机器学习网站

有关机器学习的更多信息，请参阅以下资源：

    https://www.toptal.com/machine-learning/machine-learning-theory-an-introductory
    -primer
    https://www.kaggle.com/kanncaa1/machine-learning-tutorial-for-beginners
    https://www.digitalocean.com/community/tutorials/an-introduction-to-machine
    -learning

8.4　深度学习

传统的神经网络只有三层：输入层、隐藏层和输出层。隐藏层只有一个，这是因为神经网络的训练是使用一种名为梯度下降(gradient descent)的算法来完成的。梯度下降算法用于查找函数的最小值，首先从初始值开始，然后在当前点采取与函数的梯度负值成比例的步骤，直到达到最小值(梯度接近零)为止。随着隐藏层数目的增加，训练也变得缓慢且困难，这称为消失梯度(vanishing gradient)问题。

2009 年，斯坦福大学人工智能教授 Fei-Fei Li 推出了免费数据库 ImageNet，其中包含 1400 多万张带标签的图片，目的是利用大数据改进机器学习。2010 年，ImageNet 大规模视觉识别挑战赛(ILSVRC)正式启动。在 2010 年的年度挑战赛中，主办方鼓励参赛者使用 ImageNet 训练算法并提交他们的预测。突破出现在 2012 年，当时由多伦多大学的 Alex Krizhevsky、Ilya Sutskever 和 Geoff Hinton 开发的卷积神经网络 AlexNet，能将现有的预测错误率减半至 15.3%，领先亚军 10.8 个百分点。AlexNet 有几个关键功能。首先，AlexNet 有八层，其中前五层是卷积层，后三层是完全连接层，如图 8.9 所示(http://www.mdpi.com/2072-4292/9/8/848)。

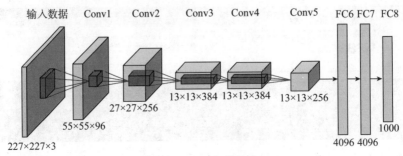

图 8.9 AlexNet 的架构

卷积层将卷积运算应用于输入，减少了参数的数量，从而使网络更深入且参数更少。首先，在完全连接层，每一层的每个神经元都连接到另一层的每个神经元。这种八层架构共有 6000 万个参数。其次，AlexNet 使用图形处理单元(Graphics Processing Unit，GPU)对模型进行训练。GPU 在本质上是并行浮点计算器，相比传统的中央处理器(CPU)快得多。使用 GPU 意味着可以训练更大的模型，从而降低错误率。最后，他们使用了非饱和修正线性单元(Rectified Linear Activation Unit，ReLU)激活函数，与其他激活函数(如 tanh 和 sigmoid 函数)相比，这减少了过度拟合并提高了训练性能。今天，AlexNet 已对深度学习(尤其是机器视觉)产生了重大影响。截至 2018 年，AlexNet 被引用超过 25 000 次。图 8.10 展示了神经网络中激活函数的一些示例(https://towardsdatascience.com/activation-functions-neural-networks-1cbd9f8d91d6)。

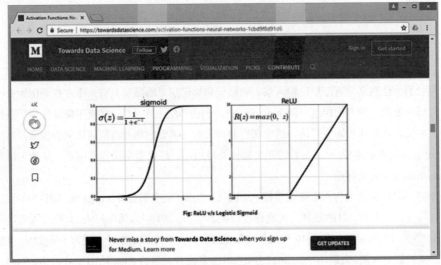

图 8.10 神经网络中的激活函数

2014 年，另一位令人印象深刻的赢家是 ILSVRC 挑战赛中的 GoogLeNet。GoogLeNet 的错误率降到了惊人的 6.67%。GoogLeNet 采用的是 22 层的卷积神经网络，它将参数的数量从 6000 万(AlexNet)减少到了 400 万。图 8.11 展示了卷积神经网络的工作原理(https://indoml.com/2018/03/07/student-notes-convolutional-neural-networks-cnn-introduction/)。卷积神经网络对于以图

像数据为输入的任何类型的预测问题都是有效的。

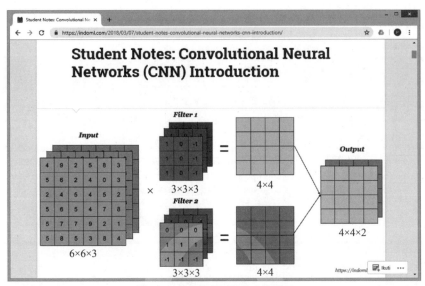

图 8.11　卷积神经网络的工作原理

最后也是最重要的一点是，2015 年的 ILSVRC 挑战赛的获胜者是由何开明等人开发的残差神经网络(ResNet)。ResNet 来自微软，错误率降到了 3.57%。ResNet 使用了一种带有"跳过连接"的新结构，并且拥有大量的批处理规范化功能，这使得他们能够训练 152 层的神经网络，同时仍然具有较低的复杂度。

深度学习网络的另一种类型是递归神经网络(Recurrent Neural Network，RNN)，RNN 被设计用于处理序列预测问题。序列预测问题的示例包括一对多、多对一和多对多问题。一对多问题是将观察结果作为输入映射到多个输出，多对一问题是将多个输入的序列映射到单个输出，多对多问题是将多个输入的序列映射到多个输出。可以将 RNN 用于文本数据、语音数据和时间序列数据等。

图 8.12 展示了 SAS 的深度学习信息网站(https://www.sas.com/en_gb/insights/analytics/deep-learning.html)。图 8.13 展示了有关 Python 深度学习的 Keras 教程网站(https://www.datacamp.com/community/tutorials/deep-learning-python)。

有关深度学习的更多信息，请参阅以下资源：

```
https://papers.nips.cc/paper/4824-imagenet-classification-with-deep
-convolutional-neural-networks.pdf
    https://machinelearningmastery.com/crash-course-convolutional-neural-networks/
    https://machinelearningmastery.com/crash-course-recurrent-neural-networks
-deep-learning/
```

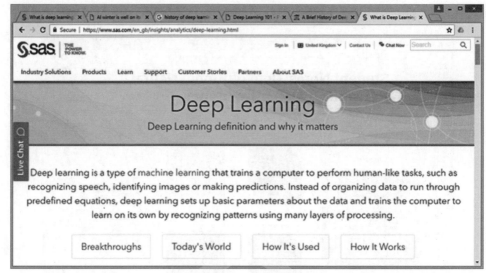

图 8.12　SAS 的深度学习信息网站

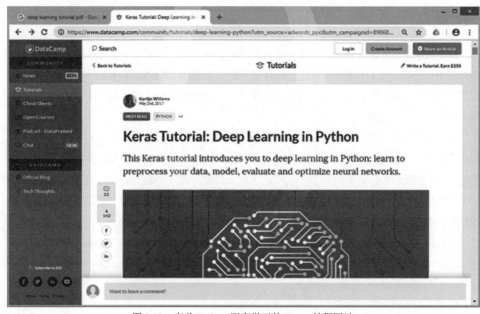

图 8.13　有关 Python 深度学习的 Keras 教程网站

8.5　Java AI 库

下面分类列出了一些十分有用的 Java AI 库。

专家系统

Apache Jena：免费、开源的 Java 语义 Web 框架(http://jena.apache.org/)。

神经网络

Neurph：轻量级、灵活的 Java 神经网络框架(http://neuroph.sourceforge.net/index.html)。

Deeplearning4j：深度学习编程库,能够广泛支持深度学习算法(https://deeplearning4j.org/index.html)。

自然语言处理

Apache OpenNLP：处理自然语言文本的机器学习库(https://opennlp.apache.org/)。

Stanford CoreNLP：用 Java 编写的人类语言技术工具(https://stanfordnlp.github.io/CoreNLP/)。

机器学习

Java ML：用 Java 编写的机器学习算法(http://java-ml.sourceforge.net/)。

Weka 3：Java 中的数据挖掘软件,用于数据挖掘任务的机器学习算法的集合(https://www.cs.waikato.ac.nz/ml/weka/)。

SMILE：用于大数据处理的快速通用机器学习引擎(http://haifengl.github.io/smile/)和 https://github.com/haifengl/smile)。

计算机视觉

JavaCV：OpenCV 的 Java 接口(https://github.com/bytedeco/javacv)。

OpenCV Java：使用 OpenCV 的 Java 应用,详细信息可参阅以下资源。

```
https://opencv-java-tutorials.readthedocs.io/en/latest/
https://opencv-java-tutorials.readthedocs.io/en/latest/01-installing-open
cv-for-java.html
https://docs.opencv.org/2.4/doc/tutorials/introduction/java_eclipse/java
_eclipse.html#java-eclipse
https://opencv.org/opencv-java-api.html
https://www.tutorialspoint.com/java_dip/introduction_to_opencv.htm
https://www.behance.net/gallery/9972461/Java-OpenCV-Webcam
```

其他资源

```
https://www.baeldung.com/java-ai
https://skymind.ai/wiki/java-ai
```

8.6 神经网络方面的 Java 示例

现在让我们看看神经网络方面的一些 Java 示例。

8.6.1 Java 感知器示例

例 8.1(分为例 8.1A 和例 8.1B)展示了一个简单的 Java 感知器(单神经元)应用,它有两个 Java 文件：Neuron1.java 和 Neuron1Demo.java。Neuron1.java 文件用于定义输入和输出,输出只是所有输入的加权和。Neuron1Demo.java 是使用 Neuron1 类创建感知器的示例程序。图 8.14 给出了 Neuron1Demo.java 程序的编译、执行和输出结果。

例 8.1A Neuron1.java 文件中的代码

```java
//Example 8.1A Java Perceptron (single neuron) example
public class Neuron1 {
    final double x[], w[];
    Neuron1(double x[], double w[]) {
        this.x = x;
        this.w = w;
    }
    public double Output() {
        double sum = 0.0;
        for(int i=0;i<x.length;i++)
        {
            sum += w[i]*x[i];
        }
        return Math.tanh(sum);
    }
}
```

例 8.1B Neuron1Demo.java 文件中的代码

```java
//Example 8.1B Java Perceptron (single neuron) demo example
import java.util.Random;

public class Neuron1Demo{
    public static void main(String[] args) {
        double x[] = {1.4, -0.33};
        double w[] = {new Random().nextDouble(), new Random().nextDouble()};
        Neuron1 n = new Neuron1(x, w);

        System.out.println(x[0]);
        System.out.println(x[1]);
        System.out.println(n.Output());
    }
}
```

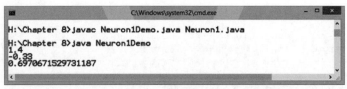

图 8.14 Neuron1Demo.java 程序的编译、执行和输出结果

例 8.2(分为例 8.2A 和例 8.2B)展示了另一个 Java 感知器应用,它也包含两个 Java 文件:Neuron2.java 和 Neuron2Demo.java。Neuron2.java 文件用于定义输入和输出,输出只是所有输入的加权和,使用这个文件中定义的 Train()方法,可以训练感知器,使其表现为逻辑 AND 函数。Neuron2Demo.java 使用 Neuron2 类创建感知器的示例程序。图 8.15 给出了 Neuron2Demo.java 程序的编译、执行和输出结果。

例 8.2A Neuron2.java 文件中的代码

```
//Example 8.2A Java Perceptron (single neuron) example 2 with training
```

```java
import java.util.Random;
public class Neuron2 {
    double w[];          //weights
    double threshold;    //threshold
    public int Output(double x[]) {
        double sum = 0.0;
        for(int i=0;i<x.length;i++)
        {
            sum += w[i]*x[i];
        }
        if(sum>threshold)
            return 1;
        else
            return 0;
    }
    public void Train(double[][] x, int[] y, double threshold, double learnrate, int epoch)
    {
        this.threshold = threshold;
        int N = x[0].length;
        w = new double[N];
        Random r = new Random();

        //initialize weights
        for(int i=0;i<N;i++)
        {
            w[i] = r.nextDouble();
        }
        //do the training
        for(int i=0;i<epoch;i++)
        {
            int totalError = 0;
            for(int j =0;j<y.length;j++)
            {
                //calculate the output and error
                int output = Output(x[j]);
                int error = y[j] - output;
                totalError +=error;
                //update the weights
                for(int k = 0;k<N;k++)
                {
                    double delta = learnrate * x[j][k] * error;
                    w[k] += delta;
                }
            }
            if(totalError == 0)
                break;
        }
    }
}
```

例 8.2B　Neuron2Demo.java 文件中的代码

```java
//Example 8.2B Java Perceptron (single neuron) example 2 with training

public class Neuron2Demo{
```

```
    public static void main(String[] args) {
        double x[][] = {{0,0},{0,1},{1,0},{1,1}};
        int y[] = {0,1,1,1};                       //Logical AND
        Neuron2 n = new Neuron2();
        n.Train(x, y,0.2, 0.5, 1000);

        System.out.println(n.Output(new double[]{0,0}));
        System.out.println(n.Output(new double[]{0,1}));
        System.out.println(n.Output(new double[]{1,0}));
        System.out.println(n.Output(new double[]{1,1}));
    }
}
```

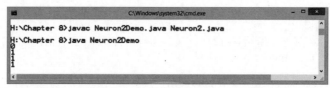

图 8.15　Neuron2Demo.java 程序的编译、执行和输出结果

8.6.2　Java 神经网络反向传播示例

例 8.3 展示的 Java 反向传播神经网络示例改编自以下代码示例：

https://supundharmarathne.wordpress.com/2012/11/23/a-simple-backpropagation-example-of-neural-network/

这个简单的 Java 反向传播神经网络包含一个输入层、一个隐藏层和一个输出层。输入层有四个神经元，隐藏层有三个神经元，输出层有两个神经元。图 8.16 给出了经过 10 次迭代后 BackpropagationDemo1.java 程序的编译、执行和输出结果。图 8.17 给出了经过 1000 次迭代后 BackpropagationDemo1.java 程序的编译、执行和输出结果。

例 8.3　一个 Java 反向传播神经网络示例

```
//Example 8.3 BackpropagationDemo1.java Java backpropagation example
//Modified from https://supundharmarathne.wordpress.com/
//2012/11/23/a-simple-backpropagation-example-of-neural-network/

public class BackpropagationDemo1 {

    //Simple NN with 1 input layer, 1 hidden layer, and 1 output layer
    static int nInputs =4, nHidden=3, nOutput=2;

    static double[][] input = {{0,1,1,0}, {1,0,0,1}, {1,1,0,0}};
    static int[][] target = {{0,1}, {1,0}, {1,1}};

    //initialize input layer weights w1 and hidden layer weights w2
    static double[] w1 = new double [nInputs*nHidden];
    static double[] w2 = new double [nHidden*nOutput];

    //initialize hidden layer output y1 and output layer output y2
    static double[] y1 = new double [nHidden];
```

```java
static double[] y2 = new double [nOutput];

//initialize hidden layer errors delta1 and output layer errors delta2
static double[] delta1 = new double [nHidden];
static double[] delta2 = new double [nOutput];

//initialize hidden layer bias b1 and output layer bias b2
static double[] b1 = new double [nHidden];
static double[] b2 = new double [nOutput];

//learning rate
static double learningRate=0.4;
static int count = 0;
static int maxCount = 1000;
static boolean loop = true;
public static void main(String[] args) {

    generateWR() ;
    while(loop){
        for(int i=0;i<input.length;i++){
            calculateY(input[i]);
            calculateDelta(i);
            calculateNewWeights(i);
            calculateNewBias();
            count++;
            System.out.println(y2[0] + ", "+ y2[1]);
        }
        System.out.println("==============================");
        if(count>=maxCount){
            loop = false;
        }
    }

}
static private void generateWR() {
    //Generate random w1 and w2
    for(int i=0; i<nInputs*nHidden; i++) {
        w1[i] = Math.abs(Math.random() - 0.5);
    }
    for(int i=0; i<nHidden*nOutput; i++) {
        w2[i] = Math.abs(Math.random() - 0.5);
    }
    //Generate random b1 and b2
    for(int i=0; i<nHidden; i++) {
        b1[i] = Math.abs(Math.random() - 0.5);
    }
    for(int i=0; i<nOutput; i++) {
        b2[i] = Math.abs(Math.random() - 0.5);
    }
}
static void calculateY(double x[]){
    //Calculate the hidden layer output y1
    for(int i=0; i<nHidden; i++){
        y1[i]=0;
        for(int j=0; j<nInputs; j++){
            y1[i] += (x[j]*w1[i+j*nHidden]);
```

```
            }
            y1[i] = sigmoid(y1[i]+b1[i]);
        }
        //Calculate the output layer output y2
        for(int i=0; i<nOutput; i++) {
            y2[i]=0;
            for(int j=0; j<nHidden; j++){
                y2[i] += (y1[j]*w2[i+j*nOutput]);
            }
            y2[i] = sigmoid(y2[i]+b2[i]);
        }
    }

    static void calculateDelta(int j){
        for(int i=0; i<nOutput; i++) {
            delta2[i] = target[j][i] - y2[i];
        }
        for(int i=0; i<nHidden; i++) {
            delta1[i]=0;
            for(int k=0; k<nOutput; k++){
                delta1[i] += (delta2[k]*w2[i*nOutput+k]);
            }
        }
    }

    private static void calculateNewWeights(int j){
        for(int i=0; i<nInputs; i++) {
            for(int k=0; k<nHidden; k++){
                int n= i*nHidden+k;
                w1[n] += (learningRate*delta1[k]*input[j]
                        [i]*y1[k]*(1-y1[k]));
            }
        }

        for(int i=0; i<nHidden; i++) {
            for(int k=0; k<nOutput; k++){
                int n= i*nOutput+k;
                w2[n] += (learningRate*delta2[k]*y1[i]*y2[k]*(1-y2[k]));
            }
        }
    }
    private static void calculateNewBias(){
        for(int i=0; i<nHidden; i++) {
            b1[i] +=delta1[i];
        }
        for(int i=0; i<nOutput; i++) {
            b2[i] += delta2[i];
        }
    }
    static double sigmoid(double exponent){
        return (1.0/(1+Math.pow(Math.E,(-1)*exponent)));
    }
}
```

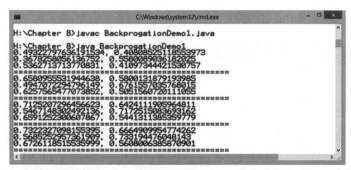

图 8.16　经过 10 次迭代后，BackpropagationDemo1.java 程序的编译、执行和输出结果

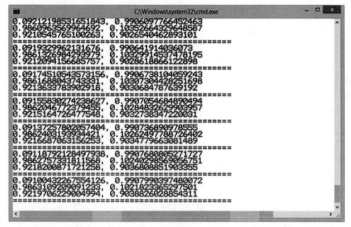

图 8.17　经过 1000 次迭代后，BackpropagationDemo1.java 程序的编译、执行和输出结果

8.7　机器学习方面的 Java 示例

有几个基于 Java 的库可用于机器学习。其中一个常用的库是 Waikato Environment for Knowledge Analysis (Weka)，它是新西兰 Waikato 大学开发的用于数据挖掘任务的机器学习算法的集合，里面包含了一些用于数据准备、分类、回归、聚类、关联规则挖掘和可视化的工具。

要想使用 Weka，首先需要下载 Weka 库。

http://www.cs.waikato.ac.nz/ml/weka/snapshots/weka_snapshots.html

下载文件 stable-3-8.zip 并将其解压缩，找到 JAR 文件 weka.jar 和数据文件 iris.arff。

然后使用 IntelliJ IDEA 创建一个新的 Java 项目，并将 weka.jar 和 iris.arff 文件添加到 IntelliJ IDEA 项目中。创建一个空的名为 WekaTest.java 的 Java 类，并将例 8.4 所示的源代码复制到 WekaTest.java 文件中。这只是简单的 Weka 分类演示，可基于以下链接提供的示例代码进行修改：

https://www.programcreek.com/2013/01/a-simple-machine-learning-examplein-java/

这个示例的目的是从文件中读取 Iris 分类数据，将它们分成训练集和测试集，并运行 J48

决策树分类器，最后打印出结果。

例 8.4　Weka 测试程序

```java
//Example 8.4 modified from
//https://www.programcreek.com/2013/01/a-simple-machine-learning-example-
//in-java/

import java.io.BufferedReader;
import java.io.FileNotFoundException;
import java.io.FileReader;
import weka.classifiers.Classifier;
import weka.classifiers.Evaluation;
import weka.classifiers.evaluation.NominalPrediction;

import weka.classifiers.trees.J48;
import weka.core.FastVector;
import weka.core.Instances;

public class WekaTest {
    public static BufferedReader readDataFile(String filename) {
        BufferedReader inputReader = null;

        try {
            inputReader = new BufferedReader(new FileReader(filename));
        } catch (FileNotFoundException ex) {
            System.err.println("File not found: " + filename);
        }

        return inputReader;
    }

    public static Evaluation classify(Classifier model,
Instances trainingSet, Instances testingSet) throws Exception {
        Evaluation evaluation = new Evaluation(trainingSet);

        model.buildClassifier(trainingSet);
        evaluation.evaluateModel(model, testingSet);

        return evaluation;
    }

    public static double calculateAccuracy(FastVector predictions) {
        double correct = 0;
        for (int i = 0; i < predictions.size(); i++) {
                NominalPrediction np = (NominalPrediction) predictions.
                   elementAt(i);
            if (np.predicted() == np.actual()) {
                correct++;
            }
        }

        return 100 * correct / predictions.size();
    }

    public static Instances[][] crossValidationSplit(Instances data, int numberOfFolds) {
```

```java
        Instances[][] split = new Instances[2][numberOfFolds];

        for (int i = 0; i < numberOfFolds; i++) {
            split[0][i] = data.trainCV(numberOfFolds, i);
            split[1][i] = data.testCV(numberOfFolds, i);
        }

        return split;
    }

    public static void main(String[] args) throws Exception {
        BufferedReader datafile = readDataFile("iris.arff");

        Instances data = new Instances(datafile);
        data.setClassIndex(data.numAttributes() - 1);

        // Do 10-split cross validation
        Instances[][] split = crossValidationSplit(data, 10);

        // Separate split into training and testing arrays
        Instances[] trainingSplits = split[0];
        Instances[] testingSplits = split[1];

        // Use a classifier
        Classifier models = new J48(); // a decision tree

        // Collect every group of predictions for current model in a FastVector
        FastVector predictions = new FastVector();

        // For each training-testing split pair, train and test the classifier
        for (int i = 0; i < trainingSplits.length; i++) {
            Evaluation validation = classify(models, trainingSplits[i],
                testingSplits[i]);
            predictions.appendElements(validation.predictions());
        }

        // Calculate overall accuracy of current classifier on all splits
        double accuracy = calculateAccuracy(predictions);

        // Print current classifier's name and accuracy in a complicated,
        // but nice-looking way.
        System.out.println("Accuracy of " + models.getClass().getSimpleName()
            + ": "
            + String.format("%.2f%%", accuracy)
            + "\n---------------------------------");
    }
}
```

图 8.18 展示了 iris.arff 文件的内容，图 8.19 展示了 Eclipse 项目 WekaTest 的内容及其输出。有关 Weka 的更多信息，请参阅以下资源：

  https://www.cs.waikato.ac.nz/ml/weka/
  http://www.cs.umb.edu/~ding/history/480_697_spring_2013/homework/WekaJava-APITutorial.pdf
  http://www.cs.ru.nl/P.Lucas/teaching/DM/weka.pdf

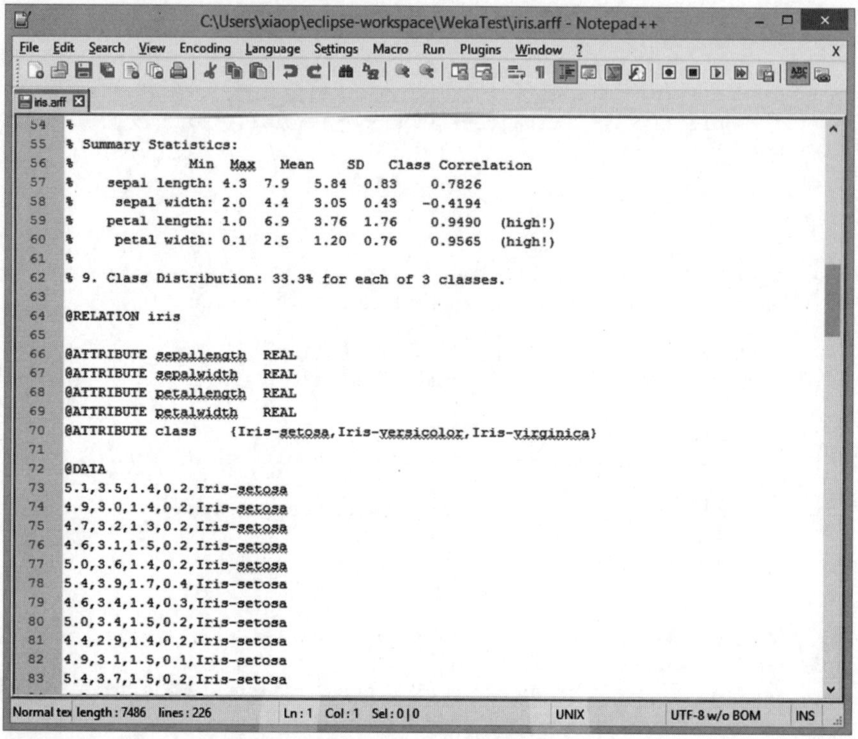

图 8.18 iris.arff 文件的内容

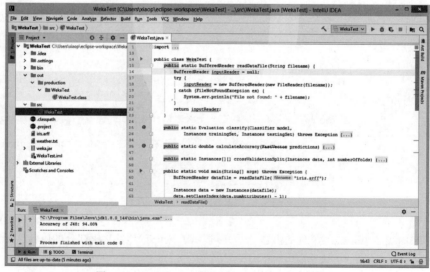

图 8.19 IntelliJ IDEA 项目 WekaTest 的内容及其输出

另一个流行的支持 Java 编程语言的机器学习库是支持向量机库(LIBSVM)，如图 8.20 所示 (https://www.csie.ntu.edu.tw/~cjlin/LIBSVM/)。LIBSVM 不仅支持向量分类和分布估计，还支持多类分类。

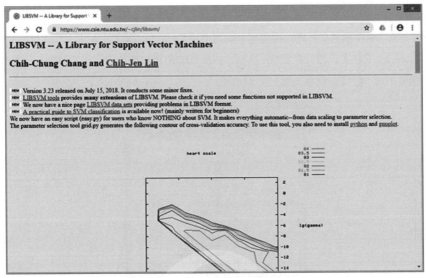

图 8.20　支持向量机库 LIBSVM

8.8　深度学习方面的 Java 示例

深度学习是当前另一个十分热门的研究主题。使用 Java 进行深度学习的最佳方法是使用 Deeplearning4J(简称 DL4J)库，图 8.21 给出了下载页面(https://deeplearning4j.org/docs/latest/deeplearning4-jquickstart)。也可从如下 GitHub 网站以压缩文件的形式下载整个 DL4J 库：

https://github.com/deeplearning4j/dl4j-examples

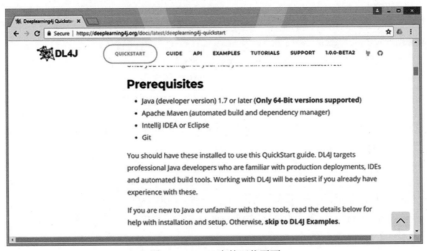

图 8.21　DL4J 库的下载页面

解压缩下载的文件后，找到名为 dl4j-examples 的子文件夹，其中有许多深度学习方面的示例应用。

为了运行 DL4J 示例,需要再次使用 IntelliJ IDEA。在 IntelliJ IDEA 中,打开一个项目,选择 dl4j-examples 文件夹,然后单击 OK 按钮,如图 8.22 所示。打开项目后,如图 8.23 所示,可以运行现有示例、修改示例或创建自己的程序。

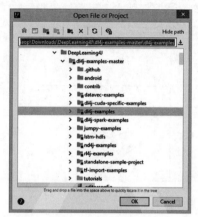

图 8.22　在 IntelliJ IDEA 中打开一个项目,然后选择 dl4j-examples 文件夹

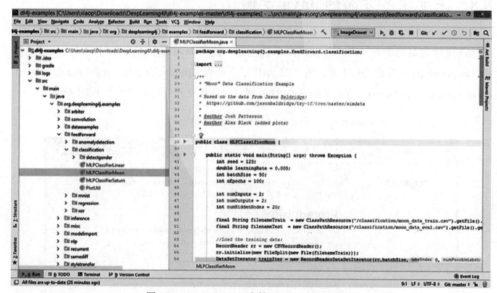

图 8.23　IntelliJ IDEA 中的 dl4j-examples 项目

图 8.24 给出了 XorExample.java 程序的运行结果。XorExample.java 程序使用简单的多层前馈神经网络实现了 XOR 函数。它有两个输入神经元,一个带有四个隐藏神经元的隐藏层,另外还有两个输出神经元。

图 8.25 给出了 MLPClassifierLinear.java 程序的运行结果。MLPClassifierLinear.java 程序使用多层感知器神经网络作为线性分类器来分离两组数据。

第 8 章 面向人工智能应用的 Java 编程

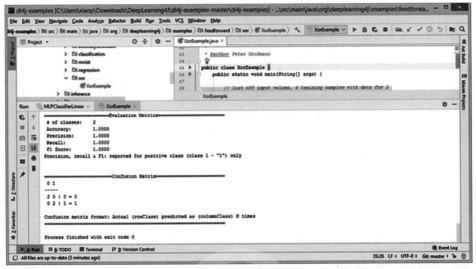

图 8.24 XorExample.java 程序的运行结果

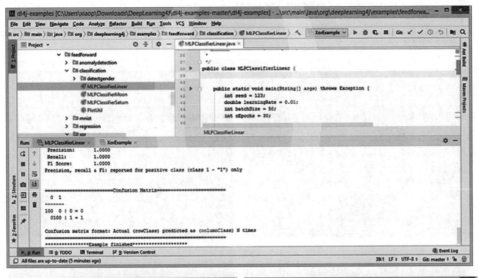

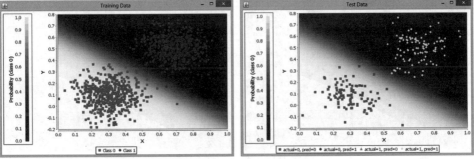

图 8.25 MLPClassifierLinear.java 程序的运行结果

图 8.26 给出了 ImageDrawer.java 程序的运行结果。这个程序使用深度学习神经网络逐像素地重新绘制给定的图像：首先得到非常粗略的目标图像，然后继续微调，直至得到的图像效果令人满意，这通常需要耗费长达几个小时的时间。

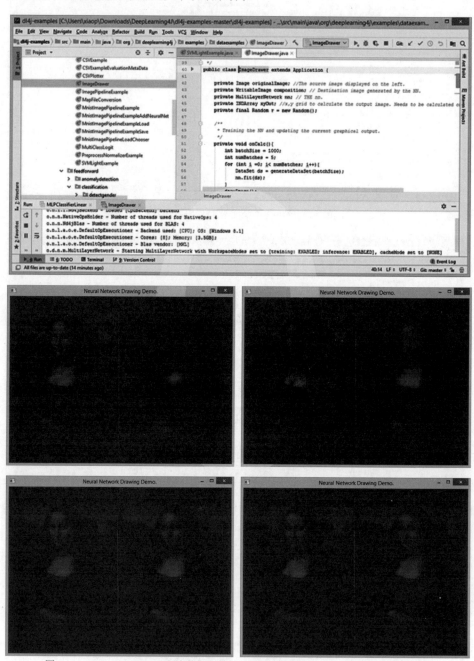

图 8.26　ImageDrawer.java 程序的运行结果，这里对比显示了大约 5 分钟、30 分钟、6 小时和 7 小时后的目标图像以及重新绘制的图像

以下 Google 论文介绍了如何使用递归神经网络生成图像：

https://arxiv.org/pdf/1502.04623.pdf

以下是关于深度学习以及如何使用 Deeplearning4J 库的有趣且免费的短期课程：

http://www.whatisdeeplearning.com/course/

也可以参考以下链接，使用 Deeplearning4J 库在 Android 设备上创建和训练神经网络：

https://deeplearning4j.org/docs/latest/deeplearning4j-android

8.9　适用于 Java 的 TensorFlow

TensorFlow 是 Google 专为机器学习而开发的开源软件库。TensorFlow 可以在多种不同的操作系统上运行，例如 Ubuntu Linux、Windows、macOS 甚至 Raspbian。

TensorFlow 默认使用的编程语言是 Python，但是 TensorFlow 提供了用于 Java 程序的 API，如图 8.27 所示(https://www.tensorflow.org/install/install_java)。

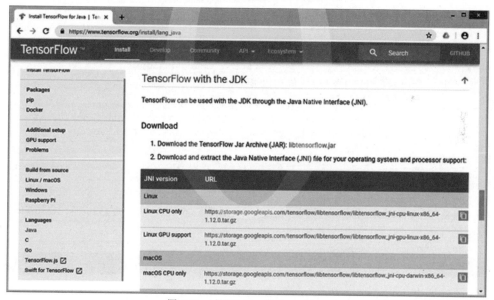

图 8.27　适用于 Java 的 TensorFlow

为了使用适用于 Java 的 TensorFlow，需要从网上下载两个文件。

1) 从以下网址下载名为 libtensorflow.jar 的 TensorFlow JAR 文件：https://storage.googleapis.com/tensorflow/libtensorflow/libtensorflow-1.12.0.jar。

2) 下载并解压缩用于支持操作系统和处理器的 Java 本地接口(Java Native Interface，JNI)文件。在这里，我们下载了用于 Windows CPU 的 JNI 文件，并从下载的压缩文件中提取出名为

tensorflow_jni.dll 的文件。

接下来，将 libtensorflow.jar 和 tensorflow_jni.dll 文件都放入你的 Java 程序文件夹(在这里是 H:\Chapter 8)中，然后创建一个名为 HelloTensorFlow.java 的文件。你可以从 TensorFlow for Java 网站上获取这个文件的内容，如下所示：

```java
//Example code from https://www.tensorflow.org/install/lang_java

import org.tensorflow.Graph;
import org.tensorflow.Session;
import org.tensorflow.Tensor;
import org.tensorflow.TensorFlow;

public class HelloTensorFlow {
  public static void main(String[] args) throws Exception {
    try (Graph g = new Graph()) {
      final String value = "Hello from " + TensorFlow.version();

      // Construct the computation graph with a single operation, a constant
      // named "MyConst" with a value "value".
      try (Tensor t = Tensor.create(value.getBytes("UTF-8"))) {
        // The Java API doesn't yet include convenience functions for
        // adding operations.
        g.opBuilder("Const", "MyConst").setAttr("dtype", t.dataType()).
                   setAttr("value", t).build();
      }

      // Execute the "MyConst" operation in a Session.
      try (Session s = new Session(g);
          // Generally, there may be multiple output tensors,
          // all of them must be closed to prevent resource leaks.
          Tensor output = s.runner().fetch("MyConst").run().get(0)) {
            System.out.println(new String(output.bytesValue(), "UTF-8"));
      }
    }
  }
}
```

例 8.5 是例 8.4 的简化版本，例 8.5 仅仅打印 Hello from xxx，其中的 xxx 表示 TensorFlow 的版本。

例 8.5　TensorFlow Java 示例程序

```java
//Example 8.5 TensorFlow Java Example
import org.tensorflow.TensorFlow;

public class TensorFlowExample {
    public static void main(String[] args) {
        final String value = "Hello from " + TensorFlow.version();
        System.out.println(value);
    }
}
```

为了编译并执行上述程序，需要运行以下命令：

```
javac -cp libtensorflow-1.12.0.jar HelloTensorFlow.java
java -cp libtensorflow-1.12.0.jar;. -Djava.library.path=. HelloTensorFlow
```

-cp libtensorflow-1.12.0.jar 命令在类路径中包含了 libtensorflow.jar 文件，-Djava.library.path=. 命令指定在何处可以找到 tensorflow_ujni.dll 文件。在本例中，.表示当前文件夹。图 8.28 给出了 HelloTensorFlow.java 程序的编译、执行和输出结果。TensorFlow 的版本是 1.12.0。

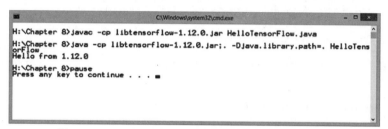

图 8.28　HelloTensorFlow.java 程序的编译、执行和输出结果

网络上有许多教程和 TensorFlow Java 示例程序可供你参考。图 8.29(https://sites.google.com/view/tensorflow-example-java-api)展示了 Google TensorFlow Java API 示例站点，该站点使用 YOLO 模型(https://pjreddie.com/darknet/yolo/)进行物体检测，例如检测图片中的猫。

图 8.29　Google TensorFlow Java API 示例站点

以下 GitHub 站点为你提供了简单的说明性教程，演示了如何使用 Java 学习 TensorFlow；另外还提供了一个 Hello TensorFlow 示例(与前面的示例相同)和一个 LabelImage 示例，后者使用 TensorFlow 模型文件 tensorflow_inception_graph.pb 进行图像分类。

https://github.com/loretoparisi/tensorflow-java

以下链接为你提供了在 Web 浏览器中使用 Java 和 JavaScript 学习 TensorFlow 的入门指导：

https://dzone.com/articles/getting-started-with-tensorflowusing-java-javascr

以下 GitHub 站点为你提供了一些有趣的 TensorFlow Java 示例，包括 hello-world、图像分类、情感分析、音频分类器、audio-recommender 和音频搜索引擎。

https://github.com/chen0040/java-tensorflow-samples

以下是 TensorFlow Java API 文档站点：

https://www.tensorflow.org/api_docs/java/reference/org/tensorflow/package-summary

更多 TensorFlow Java 示例可从以下站点获得：

https://github.com/tensorflow/models/tree/master/samples/languages/java
https://github.com/szaza/tensorflow-example-java

8.10　AI 资源

本节提供一些有趣的 AI 资源，包括书籍和教程。
以下链接提供了一些有关人工智能和深度学习的好书：

https://leonardoaraujosantos.gitbooks.io/artificial-inteligence/content/

以下链接给出了 AI 领域的首屈一指的科学家和最有影响力的人物之一——Andrew Ng 的最新大作 *Machine Learning Yearning*：

https://www.deeplearning.ai/copy-of-machine-learning-yearning

以上神经网络和深度学习站点提供了免费的在线书籍：

http://neuralnetworksanddeeplearning.com/

以下链接演示了一些实际操作，所有示例代码都可以从 GitHub 站点获得：

https://github.com/mnielsen/neural-networks-and-deep-learning

以下链接向你介绍了一些与众不同的机器学习算法，并以 C++、Scala、Java 和 Python 示例程序进行说明。

http://www.kareemalkaseer.com/books/ml

以下链接给出了 *The Nature of Code* 一书的第 10 章，其中包含许多短代码示例。

https://natureofcode.com/book/chapter-10-neural-networks/#

以下链接给出了斯坦福大学的深度学习教程，非常受欢迎。

http://deeplearning.stanford.edu/wiki/index.php/UFLDL_Tutorial

以下链接给出了 Andrew Ng 主讲的另一十分流行的机器学习课程。如果想学习 AI，这是必学课程之一。

https://www.coursera.org/learn/machine-learning

8.11 小结

本章首先介绍了人工智能(AI)的概念，然后解释并说明了 AI 应用的一些 Java 示例。你学习了 AI 的类型——狭义 AI、通用 AI 和超级 AI，还学习了 AI 开发的各个阶段，了解到了神经网络始于 20 世纪 50 年代，机器学习始于 20 世纪 80 年代，深度学习始于 21 世纪的第一个十年。对于神经网络，本章介绍了感知器(也就是单个神经元)以及多层感知器、反向传播神经网络和前馈神经网络的概念。本章还介绍了机器学习的不同类型——监督学习、无监督学习、半监督学习和强化学习，并且介绍了深度学习的历史以及流行的深度学习网络类型，如 AlexNet、GoogLeNet 和 ResNet。最后，本章介绍了一系列用于神经网络、机器学习和深度学习的 Java 示例，并探讨了适用于 Java 的 TensorFlow 机器学习库。

8.12 本章复习题

1. 什么是人工智能？
2. 人工智能的三种类型是什么？
3. 人工智能的三个发展阶段是什么？
4. 什么是神经网络？
5. 什么是感知器？
6. 什么是多层感知器(MultiLayer Perceptron，MLP)？
7. 什么是前馈神经网络？
8. 什么是机器学习？机器学习有哪些不同类型？
9. 什么是深度学习？
10. 对卷积神经网络(Convolutional Neural Network，CNN)进行一些研究。CNN 的主要特征是什么？
11. 对递归神经网络(Recurrent Neural Network，RNN)进行一些研究。RNN 的主要功能是什么？

第 9 章

面向网络安全应用的Java编程

"安全与其说取决于拥有多少，不如说取决于能够获得多少。"
——Joseph Wood Krutch

9.1 什么是网络安全

9.2 什么是加密

9.3 哈希函数和消息摘要

9.4 数字签名

9.5 数字证书

9.6 案例研究 1：安全电子邮件

9.7 案例研究 2：安全网络

9.8 Java 私钥加密示例

9.9 Java 公钥加密示例

9.10 Java 数字签名/消息摘要示例

9.11 Java 数字证书示例

9.12 其他 Java 示例

9.13 小结

9.14 本章复习题

9.1 什么是网络安全

网络安全与来自互联网的威胁有关。随着我们的生活越来越依赖互联网，网络安全也变得越来越重要。比如，发生于 2017 年的 WannaCry 病毒攻击影响了 150 个国家/地区的二十多万台计算机。WannaCry 病毒以较旧的 Windows 操作系统为目标，锁定受影响的计算机并要求以比特币支付赎金。感染了这种计算机病毒的医院不得不取消手术，工厂也不得不停止生产。再比如，2015 年，英国电信公司 TalkTalk 被北爱尔兰的一名 15 岁中学生入侵，成千上万客户的在线详细信息被盗。TalkTalk 表示，这次黑客攻击使公司损失 4200 万英镑，并失去 9.8 万名宽带用户。

综上，在数字技术时代，网络安全至关重要。本章将首先介绍网络安全中使用的关键术语和技术，然后给出一些用于网络安全应用的 Java 编程示例。

网络安全解决了以下关键问题。

机密性 机密性是指发送方和接收方之间传输的信息只能由相应各方访问。

身份验证 身份验证是指确定已获取信息的来源准确无误。

完整性 完整性是指访问的信息必须是原始的、完好无损的。

可用性 可用性意味着始终可以访问被授权访问的信息。

改善网络安全的常用方法是加密。

9.2 什么是加密

加密是最古老的安全技术之一。这一技术至少可以追溯到罗马时代，当时是以 Julius Caesar 的名字命名的密码加密系统(现在称为凯撒密码)。在这种简单的移位密码中，文本中的字母向下移动固定数量的位置，从而变得不可读。

例如，如果将每个字母向左移动三个位置，那么以下明文

```
attack London
```

将变为

```
xqqxzh ilkalk
```

虽然原始的明文是不可读的，但是这种加密很容易解密。现代加密技术使用了更加复杂和高级的数学算法，这些算法更加安全。现代加密技术可以分为私钥加密(也称为对称密钥加密)和公钥加密(也称为非对称密钥加密)。旨在研究加密和安全通信的技术领域则被称为密码学。

9.2.1 私钥加密

如图 9.1 所示，在私钥加密中，发送者和接收者使用相同的密钥来加密和解密消息。可以将私钥想象成一个随机数序列。由于双方使用相同的密钥，因此也称为对称密钥加密。

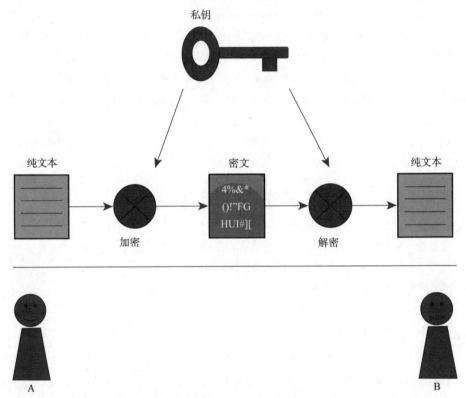

图 9.1　私钥加密：发件方 A 使用密钥将纯文本加密为密文，接收方 B 使用相同的
　　　　　密钥将密文解密回纯文本

最受欢迎的私钥加密算法是 AES(Advanced Encryption Standard，高级加密标准)，也称为 Rijndael 算法，由比利时密码学家 Vincent Rijmen 和 Joan Daemen 开发。在 AES 中，纯文本被分为 128 位的块。可通过密钥扩展算法将密钥(可以是 128、192 或 256 位)扩展为几个循环密钥。然后，对组合的纯文本和相应的密钥进行几轮数学运算，例如添加循环密钥(AddRoundKey)、替换字节(SubBytes)、移动行(ShiftRows)和混合列(MixColumns)。轮数取决于密钥大小：128 位密钥为 10 轮，192 位密钥为 12 轮，而 256 位密钥为 14 轮。密文将显示在末尾。这种方案的优点在于，加密和解密执行的是相同的操作，但顺序相反，如图 9.2 所示。

迄今为止，AES 算法相比其他任何算法(包括 DES(Data Encryption Standard，数据加密标准)、三重 DES、Twofish 等算法)都更安全，使用更广泛。如果想蛮力破解 AES 256 位加密(即搜索 2^{256} 种可能性)，那么即使使用 50 台每秒可检查 10^{18} 个 AES 密钥的超级计算机，估计也将花费 3×10^{51} 年。

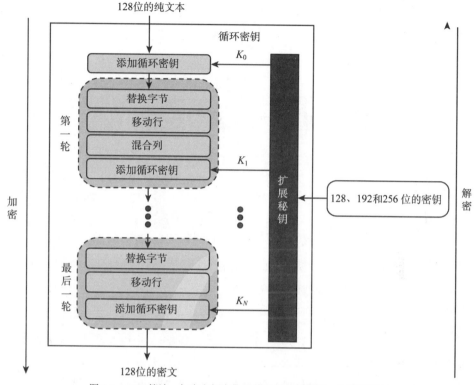

图 9.2 AES 算法：加密和解密执行的是相同的操作，但顺序相反

私钥加密的优点是很安全，可用于大文本加密。缺点是发送方和接收方都需要使用相同的密钥，因此密钥分发始终是个问题。

有关 AES 的更多信息，请参阅以下资源：

https://parsiya.net/blog/2015-01-06-tales-from-the-crypto---leaking-aes-keys/
https://www.tutorialspoint.com/cryptography/advanced_encryption_standard.htm
https://iis-people.ee.ethz.ch/~kgf/acacia/c3.html#tth_sec3.2

9.2.2 公钥加密

与私钥加密相反，使用公钥加密，发送方和接收方便可以使用不同的密钥来加密和解密消息，如图 9.3 所示。首先，每个人都需要生成一对自己的密钥——一个私钥和一个公钥。他们将保留私钥给自己，并将公钥提供给其他所有人。然后，如果 A 只想发送秘密消息给 B，那么 A 将使用 B 的公钥来加密消息，而 B 将使用自己的私钥来解密消息。在这种情况下，只有 B 可以解密消息。由于发送方和接收方使用不同的密钥，因此也称为非对称密钥加密。这种加密方法的优点在于不存在密钥分发问题。其他人永远无法从公钥中弄清对方的私钥是什么。

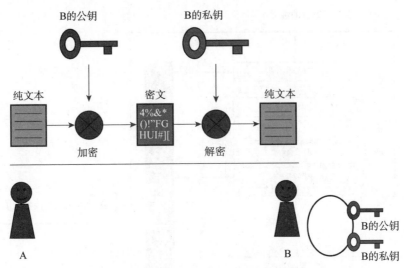

图 9.3 公钥加密：发件人 A 使用 B 的公钥来加密消息，而接收方 B 使用自己的私钥来解密消息

公钥加密也可以用于身份验证。例如，为了证明消息来自 A，发送方 A 可以使用自己的私钥对消息进行加密，如果 B 可以使用 A 的公钥对消息进行解密，接收方 B 就可以确认消息的确来自 A，如图 9.4 所示。最受欢迎的公钥加密算法是 Rivest-Shamir-Adelman(RSA)，RSA 算法是由 Ron Rivest、Adi Shamir 和 Leonard Adelman 于 1977 年设计出来的。RSA 算法涉及三个步骤：生成密钥、加密和解密。

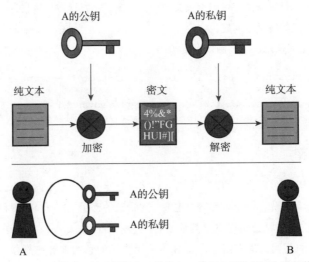

图 9.4 用于身份验证的公钥加密：发件人 A 使用自己的私钥加密邮件，而接收方 B 使用 A 的公钥解密邮件

生成密钥

选择两个不同的质数 p 和 q。

计算 $n = p \times q$。

计算 $z = (p-1) \times (q-1)$。

选择整数 e，其中 gcd(z, e) = 1，1 < e < z。

选择整数 d，以便 $(e \times d - 1)$ 可以被 z 整除。

公钥：(n, e)。

私钥：(n, d)。

在这里，gcd 表示最大公约数，而 mod 是模数运算符，用于计算两个数字相除后的余数。公钥是(n, e)，私钥是(n, d)。

加密

纯文本： $M (< n)$

密文： $C = M^e \bmod n$

解密

密文： C

密文： $M = C^d \bmod n$

下面用一个例子来说明 RSA 算法是如何工作的。为了创建公钥和私钥，你需要选择两个不同的质数 p=11 和 q=19。然后计算 $n = p \times q = 11 \times 19 = 209$，并计算 $z = (p-1) \times (q-1) = 10 \times 18 = 180$，选择 e=13，以使 z 和 e 的最大公约数为 1，最后选择 d=97，得到 97×13-1=1260=180×7。因此，公钥为(209, 13)，私钥为(209, 97)。

假设想要发送数字 5(字母也可以视为数字)以加密纯文本 M(M = 5)。使用公共密钥，可以使密文 $C = M^e \bmod n$ = 513 mod 209 = 169。这是将要发送的密文。为了解密消息，需要计算 $M = C^d \bmod n$ = 16997 mod 209 = 5。

你也可以使用私钥对消息进行加密，然后使用公钥对消息进行解密。

公钥加密的优点是易于管理密钥。每个人都可以发布自己的公钥，同时将私钥保留。在此过程中不存在安全问题。公钥加密的缺点是计算量大，因为需要计算数字的幂。公钥加密不适用于大文本加密。

因此，实际上公钥加密和私钥加密经常一起使用。发送方首先使用公钥加密来建立连接并共享临时会话私钥。临时会话密钥是随机生成的，并且仅对当前会话有效。然后，双方将使用共享的临时会话私钥来加密后续消息，如图 9.5 所示。

使用图 9.6 所示的在线 RSA 密钥生成器(http://travistidwell.com/jsencrypt/demo/)可以生成一对大小不同的私钥和公钥，提供纯文本消息并转换为密文。

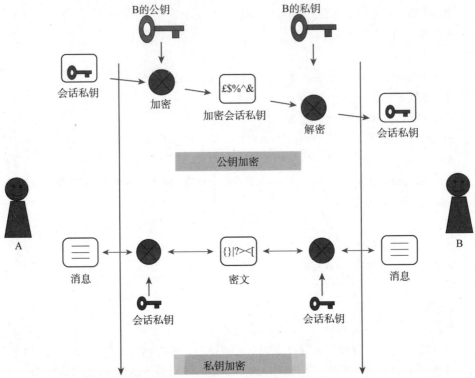

图 9.5　发送方 A 首先通过公钥加密与接收方 B 共享临时会话私钥，然后发送方 A 和接收方 B 可以通过私钥加密来使用共享的临时私钥加密所有后继消息

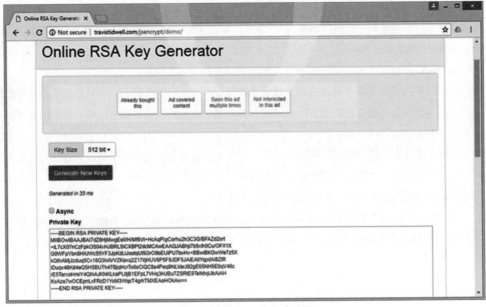

图 9.6　在线 RSA 密钥生成器

9.3 哈希函数和消息摘要

哈希函数是一种可以将大型的长文本消息转换为固定长度的短消息(称为消息摘要)的函数,如图 9.7 所示。哈希函数和消息摘要很重要,并且在密码学中得到了广泛应用。最简单的哈希函数是校验和计算函数,该函数首先将长消息分成相同固定长度的许多小块,然后将所有小块加在一起。哈希函数是多对一函数;其中一项重要的功能是,当长文本消息发生变化时,消息摘要也将发生变化。有了哈希函数和消息摘要,我们可以介绍数字签名和数字证书的概念了。

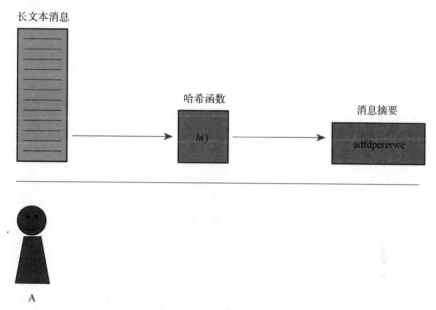

图 9.7 使用哈希函数将长文本消息转换为消息摘要

最常用的哈希函数之一是由美国国家安全局(NSA)设计的安全哈希算法(SHA)。SHA 有三个版本:SHA-1,原始版本;SHA-2,第二个版本,也是当前正在使用的版本;SHA-3,计划使用的最新版本。SHA-2 支持长度为 224、256、384 或 512 位的哈希值,最常用的是 SHA-256 和 SHA-512。例如,最受欢迎的加密货币比特币使用 SHA-256 来验证交易和计算工作量。第 10 章将介绍有关加密货币和哈希函数的更多详细信息。图 9.8 展示了一个有趣的在线 SHA-256 哈希生成器,可以在其中输入文本,它将生成相应的哈希值(https://passwordsgenerator.net/sha256-hash-generator/)。另一个使用广泛的哈希函数是消息摘要算法(MD5)。MD5 最初被设计为加密哈希函数,但是由于存在许多漏洞,D5 通常用作校验和以验证数据的完整性。MD5 能产生 128 位的哈希值。当从 Internet 下载文件时,许多网站还提供了相应的 MD5 消息摘要,用于检查下载文件的完整性。

图 9.8　一个有趣的在线 SHA256 哈希生成器

9.4　数字签名

当人们书写纸质信函时，通常会在底部签名以证明信函已写好。那么，如何签名数字信息呢？ 答案是使用数字签名，如图 9.9 所示。当发送方 A 想要向接收方 B 发送长文本消息时，发送方 A 使用哈希函数从长文本消息中生成消息摘要，然后使用私钥对消息摘要进行加密以创建数字签名，最后将合并的长文本消息和数字签名发送到接收方 B。

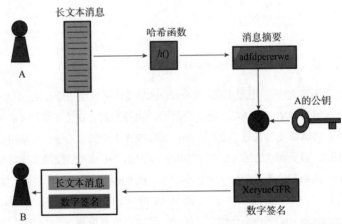

图 9.9　当发送人 A 要发送长文本消息时，A 使用哈希函数生成消息摘要，然后使用私钥对消息摘要进行加密以创建数字签名，最后将合并的长文本消息和数字签名发送到接收方 B

接收方 B 得到合并的消息后，首先将长文本消息和数字签名分开，然后使用相同的哈希函数生成消息摘要，并使用 A 的公钥从数字签名中解密消息摘要。最后，接收方 B 对计算出的消息摘要与得到的消息摘要进行比较。如果两者匹配，那么说明消息是真实的，否则消息就是假冒的，如图 9.10 所示。

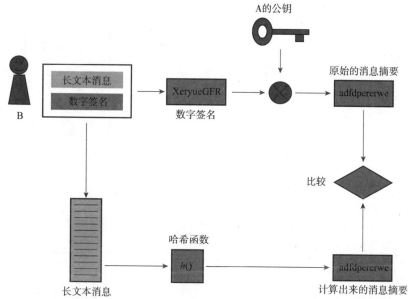

图 9.10 接收方 B 得到合并的消息后，首先将长文本消息和数字签名分开，然后使用相同的哈希函数生成消息摘要，并使用 A 的公钥从数字签名中解密消息摘要。最后，接收方 B 对计算出的消息摘要与得到的消息摘要进行比较

9.5 数字证书

当收到某人的公钥时，怎么判断其真实性？答案是使用数字证书。1988 年，由国际电信联盟(International Telecommunications Union，ITU)开发的 ITU X.509 是最常用的数字认证标准。目前，ITU X.509 认证已在许多 Internet 协议中得到应用，包括安全的超文本传输协议(HTTPS)，HTTPS 是超文本传输协议(HTTP)的安全版本。HTTP 用于标准的万维网(WWW)服务，而 HTTPS 用于需要安全通信的在线银行、在线购物等。

图 9.11 展示了 ITU X.509 数字证书的工作方式。在这种情况下，需要发送方 A 和接收方 B 都可以信任的第三方，称为证书颁发机构(Certificate Authority，CA)。如果发送方 A 需要签名的数字证书，那么发送方 A 需要通过证书签名请求(Certificate Signing Request，CSR)执行如下操作：

(1) 发送方 A 生成一对密钥(私钥和公钥)。

(2) 发送方 A 将自己的身份信息(姓名、组织、地址等)和公钥发送给 CA。发送方 A 还使用自己的私钥对消息进行签名。

(3) CA 验证发送方 A 发送的信息和公钥，然后使用 CA 的私钥生成数字签名。

(4) CA 将发送方 A 的原始消息与自己的签名结合起来，创建数字证书，然后将其发送给接收方 B。

(5) 接收方 B 收到数字证书后，按照之前图 9.9 描述的步骤，使用 CA 的公钥获取发送方 A 的身份信息和公钥。

图 9.12 给出了 ITU X.509 数字证书的内容。

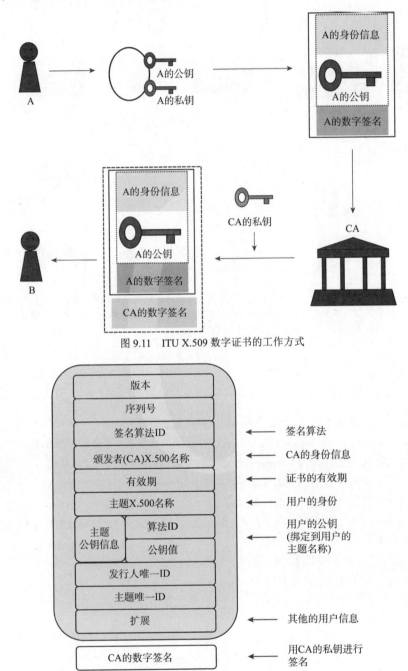

图 9.11 ITU X.509 数字证书的工作方式

图 9.12 ITU X.509 数字证书的内容

9.6 案例研究1：安全电子邮件

本节和接下来的 9.7 节将分别介绍一个案例，这两个案例演示了如何在现实生活中使用加密技术。本节介绍的案例是安全电子邮件。传统的电子邮件以明文形式发送和接收消息，但是，对于需要安全地发送和接收消息的商业通信来说，这是商家不希望看到的。"良好隐私密码法"(Pretty Good Privacy，PGP)——由 Phil Zimmermann 于 1991 年开发的一种加密程序——可以用来保护电子邮件。PGP 主要用于加密、解密和签名电子邮件或文件。PGP 遵循用于加密和解密数据的 OpenPGP 标准(RFC 4880)。在 PGP 中，发件人使用随机生成的密钥对电子邮件(或数据)进行加密，并使用收件人的公钥和 RSA 算法对随机生成的密钥进行加密。然后，发件人将两者都发送给收件人。发件人使用自己的私钥和 RSA 算法解密随机生成的密钥，然后使用随机生成的密钥解密电子邮件(或数据)。图 9.13 展示了 PGP 的工作方式(https://en.wikipedia.org/wiki/Pretty_Good_Privacy)。

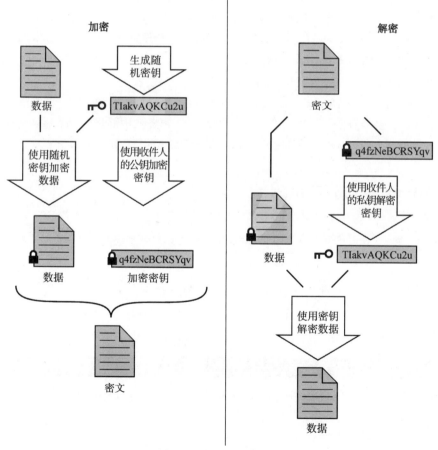

图 9.13　PGP 的工作方式

9.7 案例研究 2：安全网络

本书介绍的案例是安全网络。像电子邮件一样，基于 HTTP 的传统网络服务也使用明文发送和接收消息。但是，人们对于诸如在线银行和在线购物的商业应用也是不希望使用明文传输信息的，这就需要用到传输层安全性(Transport Layer Security，TLS)和安全套接字层(Secure Sockets Layer，SSL)。安全性高的网络往往使用基于 TLS 或 SSL 的 HTTPS 来提供安全的网络通信，步骤如下：

(1) Web 客户端将 ClientHello 消息发送到 Web 服务器，消息内容包括客户端支持的最高 TLS 协议版本、随机数以及首选加密算法的列表。

(2) Web 服务器使用 ServerHello 消息进行答复，消息内容包括从客户端提供的选项中选择的 TLS 协议版本和加密算法，另外还包括随机数和 Web 服务器的数字证书。数字证书包含由 CA 认证的 Web 服务器的公钥。

(3) Web 服务器发送 ServerHelloDone 消息，指示初始握手后得出的结论。

(4) Web 客户端拥有受委托 CA 的列表和所有 CA 的公钥。首先，Web 客户端会验证 Web 服务器的数字证书并生成对称的会话密钥。然后，Web 客户端使用 Web 服务器的公钥对密钥进行加密，最后将加密的对称会话密钥发送到 Web 服务器。

(5) Web 服务器使用私钥解密对称的会话密钥。

(6) Web 客户端发送另一条消息，指示所有后续消息都将使用对称的会话密钥进行加密。

(7) Web 服务器回复一条消息，以确认后续所有消息都将使用相同的会话密钥进行加密。

(8) Web 客户端和 Web 服务器现在可以安全地使用加密的消息进行通信。

9.8 Java 私钥加密示例

例 9.1 演示了使用私钥(对称密钥)的加密和解密过程。例 9.1 所示的 Java 程序基于如下网站上的示例：

http://jexp.ru/index.php/Java_Tutorial/Security/Key_Generator

首先，这个 Java 程序基于 AES 算法生成一个私钥，然后对明文消息 Hello World 进行加密和解密，此外还能显示私钥信息。图 9.14 给出了程序的编译、执行和输出结果。

例 9.1　Java 私钥加密

```
//Example 9.1 Private Key Encryption
//Modified from
//http://jexp.ru/index.php/Java_Tutorial/Security/Key_Generator
import java.security.*;
import javax.crypto.*;
```

```java
public class PrivateKeyDemo1 {
   static String algorithm = "AES";
   static Key key ;
   static Cipher cipher;
   public static void main(String[] args) throws Exception {
     key = KeyGenerator.getInstance(algorithm).generateKey();
     cipher = Cipher.getInstance(algorithm);
     String text="Hello World";
     byte[] encryptionBytes = encrypt(text);
     System.out.println("Original Text: " + text);
     System.out.println("Key: " + key.toString());
     System.out.println("Encrypted Text: " + encrypt(text));
     System.out.println("Decrypted Text: " + decrypt(encryptionBytes));
   }
   private static byte[] encrypt(String input) throws Exception {
     cipher.init(Cipher.ENCRYPT_MODE, key);
     byte[] inputBytes = input.getBytes();
     return cipher.doFinal(inputBytes);
   }
   private static String decrypt(byte[] encryptionBytes) throws Exception {
     cipher.init(Cipher.DECRYPT_MODE, key);
     byte[] recoveredBytes = cipher.doFinal(encryptionBytes);
     String recovered = new String(recoveredBytes);
     return recovered;
   }
}
```

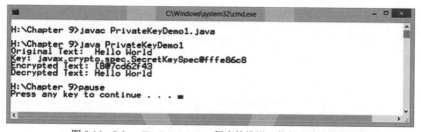

图 9.14　PrivateKeyDemo1.java 程序的编译、执行和输出结果

9.9　Java 公钥加密示例

例 9.2 演示了使用非对称密钥的加密和解密过程。例 9.2 所示的 Java 程序基于如下网站上的示例：

http://www.javacirecep.com/java-security/java-rsa-encryption-decryption-example/

首先，这个 Java 程序基于 RSA 算法生成一对公钥/私钥，然后使用公钥加密明文消息 Hello World，再使用私钥解密消息，此外还可以显示公钥/私钥信息。图 9.15 给出了程序的编译、执行和输出结果。

例 9.2　Java 公钥加密

```java
//Example 9.2 Java Public Key Encryption
//Modified from http://www.javacirecep.com/java-security/
//java-rsa-encryption-decryption-example/

import java.security.*;
import javax.crypto.*;
import java.util.*;

public class PublicKeyDemo1 {
    private static KeyPair keyPair;
    private static String algorithm = "RSA"; //DSA DH etc
    public static void main(String[] args) throws Exception{
        KeyPairGenerator keyPairGenerator = KeyPairGenerator.getInstance(algorithm);
        keyPairGenerator.initialize(1024);
        keyPair = keyPairGenerator.generateKeyPair();
          final Cipher cipher = Cipher.getInstance(algorithm);
        final String plaintext = "Hello World";

        System.out.println("Public Key = " + keyPair.getPublic().
            toString());
        System.out.println("Private Key = " + keyPair.getPrivate().
            toString());

        // ENCRYPT using the PUBLIC key
        cipher.init(Cipher.ENCRYPT_MODE, keyPair.getPublic());
        byte[] encryptedBytes = cipher.doFinal(plaintext.getBytes());
        String ciphertext = new String(Base64.getEncoder().
            encode(encryptedBytes));
        System.out.println("encrypted (ciphertext) = " + ciphertext);

        // DECRYPT using the PRIVATE key
        cipher.init(Cipher.DECRYPT_MODE, keyPair.getPrivate());
        byte[] ciphertextBytes = Base64.getDecoder().decode(ciphertext.
            getBytes());
        byte[] decryptedBytes = cipher.doFinal(ciphertextBytes);
        String decryptedString = new String(decryptedBytes);
        System.out.println("decrypted (plaintext) = "+ decryptedString);
    }
}
```

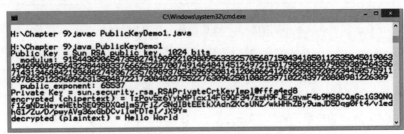

图 9.15　PublicKeyDemo1.java 程序的编译、执行和输出结果

有关将 Java 用于非对称加密的更多示例，请参阅以下资源：

https://www.mkyong.com/java/java-asymmetric-cryptography-example/
https://www.devglan.com/java8/rsa-encryption-decryption-java
https://javadigest.wordpress.com/2012/08/26/rsa-encryption-example/

9.10 Java 数字签名/消息摘要示例

例 9.3 展示了如何使用 Java 中的 SHA-256 哈希函数创建消息摘要。图 9.16 给出了程序的编译、执行和输出结果。

例 9.3　Java 消息摘要 1

```java
//Example 9.3 Java Message Digest

import java.security.MessageDigest;
public class MessageDigestDemo1 {
  public static void main(String[] args) throws Exception {
    String stringToEncrypt="Hello World";
    MessageDigest messageDigest = MessageDigest.getInstance("SHA-256");
    messageDigest.update(stringToEncrypt.getBytes());
    String encryptedString = new String(messageDigest.digest());
    System.out.println("Original Text: " + stringToEncrypt);
    System.out.println("Message Digest: " + encryptedString);
  }
}
```

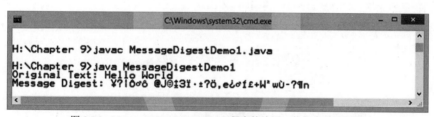

图 9.16　MessageDigestDemo1.java 程序的编译、执行和输出结果

例 9.4 是例 9.3 的另一个版本，例 9.4 展示了如何使用 bytesToHex()方法将消息摘要转换为十六进制值。图 9.17 给出了程序的编译、执行和输出结果。

例 9.4　Java 消息摘要 2

```java
//Example 9.4 Java Message Digest 2

import java.security.MessageDigest;
public class MessageDigestDemo2 {
  public static void main(String[] args) throws Exception {
    String stringToEncrypt="Hello World";
    MessageDigest messageDigest = MessageDigest.getInstance("SHA-256");
    byte[] encodedhash = messageDigest.digest(stringToEncrypt.getBytes());
    String encryptedString = bytesToHex(encodedhash);
    System.out.println("Original Text: " + stringToEncrypt);
    System.out.println("Message Digest: " + encryptedString);
```

```
        }
        private static String bytesToHex(byte[] hash) {
            StringBuffer hexString = new StringBuffer();
            for (int i = 0; i < hash.length; i++) {
                String hex = Integer.toHexString(0xff & hash[i]);
                if(hex.length() == 1) hexString.append('0');
                   hexString.append(hex);
            }
            return hexString.toString();
        }
}
```

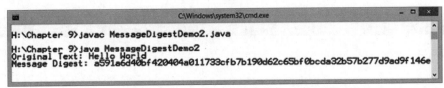

图 9.17　MessageDigestDemo2.java 程序的编译、执行和输出结果

例 9.5 用来演示数字签名，其中的 Java 程序基于如下网站上的示例：

http://tutorials.jenkov.com/java-cryptography/signature.html

首先，这个 Java 程序会生成一对公钥/私钥，并使用私钥基于 SHA256 with DSA 算法创建数字签名，然后使用数字签名对明文消息 Hello World 进行签名，最后使用公钥创建另一个数字签名以验证原始签名，详细过程类似于图 9.18。

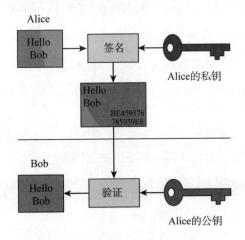

图 9.18　使用数字签名验证消息的流程

例 9.5　Java 数字签名

```
//Example 9.5 Java Digital Signature
//created based on examples from
//http://tutorials.jenkov.com/java-cryptography/signature.html
import java.security.*;
```

```java
public class DigitalSignatureDemo1 {
  public static void main(String[] args) throws Exception {
    String m = "Hello World";
    Signature signature = Signature.getInstance("SHA256WithDSA");
    SecureRandom secureRandom = new SecureRandom();
    KeyPairGenerator keyPairGenerator = KeyPairGenerator.getInstance("DSA");
    KeyPair keyPair = keyPairGenerator.generateKeyPair();

    //initialize the digital signature
    signature.initSign(keyPair.getPrivate(), secureRandom);
    byte[] data = m.getBytes("UTF-8");
    signature.update(data);
    byte[] digitalSignature = signature.sign();
    System.out.println("Create Digital Signature: " + digitalSignature.
        toString());

    Signature signature2 = Signature.getInstance("SHA256WithDSA");
    signature2.initVerify(keyPair.getPublic());
    byte[] data2 = m.getBytes("UTF-8");
    signature2.update(data2);
    boolean verified = signature2.verify(digitalSignature);
    System.out.println("signature verifies: " + verified);
  }
}
```

图 9.19 给出了这个程序的编译、执行和输出结果。

图 9.19 DigitalSignatureDemo1.java 程序的编译、执行和输出结果

有关数字签名的更多示例，详见下面的网址：

https://www.mkyong.com/java/java-digital-signatures-example/

图 9.20 展示了来自 Tools4noobs(https://www.tools4noobs.com/online_tools/encrypt/)的说明性在线私钥加密工具，该工具分为两个页面：一个页面用于加密消息，另一个页面用于解密消息。可以选择密钥值和消息内容，然后选择用于加密和解密的算法。

图 9.21 为美国佐治亚大学计算机科学系的公钥加密演示截图(http://cobweb.cs.uga.edu/~dme/csci6300/Encryption/Crypto.html)。

图 9.22 为 Travis Tidwell 的在线 RSA 密钥生成器演示截图,可以在其中指定密钥大小为 512、1024、2048 或 4096 位(http://travistidwell.com/jsencrypt/demo/)。

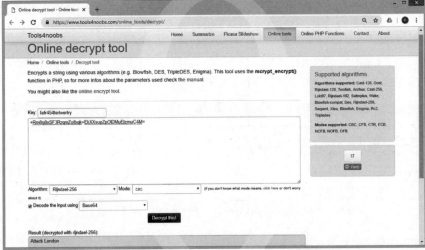

图 9.20 来自 Tools4noobs 的说明性在线私钥加密工具，上方的页面用于加密消息，下方的页面用于解密消息

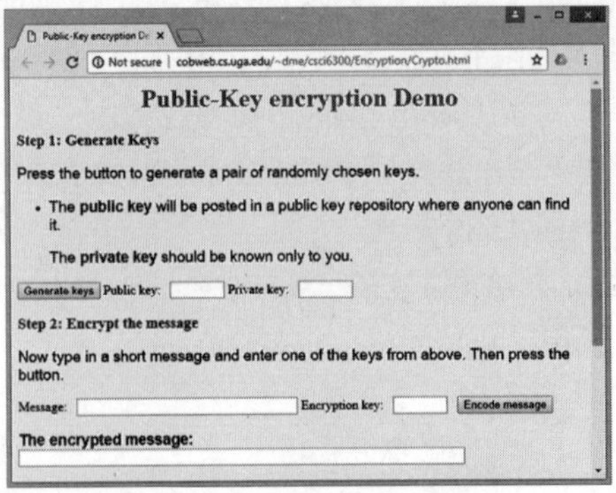

图 9.21 公钥加密演示截图

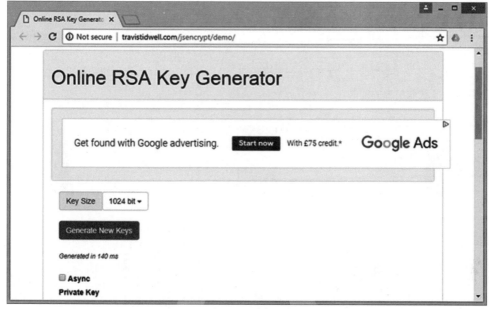

图 9.22　在线 RSA 密钥生成器演示截图

图 9.23 为基于 JavaScript 的在线哈希函数演示截图，这里实现了 FIPS PUB 180-4 和 FIPS PUB 202 中定义的整个 SHA 哈希系列(SHA-1、SHA-224、SHA3-224、SHA-256、SHA3-256、SHA-384、SHA3-384、SHA-512、SHA3-512、SHAKE128 和 SHAKE256)，详见 https://caligatio.github.io/jsSHA/。

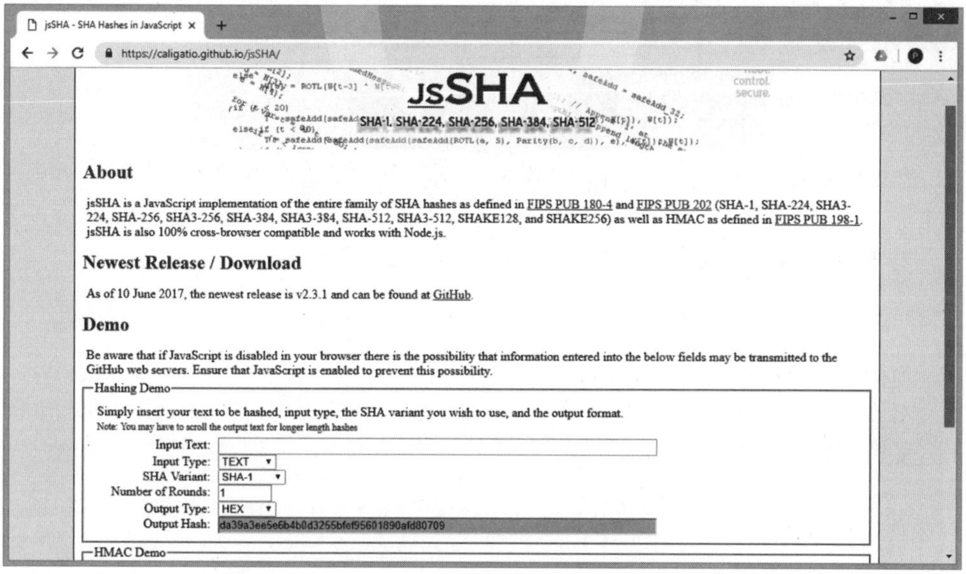

图 9.23　基于 JavaScript 的在线 SHA 哈希函数演示截图

9.11　Java 数字证书示例

图 9.24 展示了数字证书方面的一个典型示例。用户将自己的详细信息和公钥提供给证书颁发机构,并获取相应的证书。

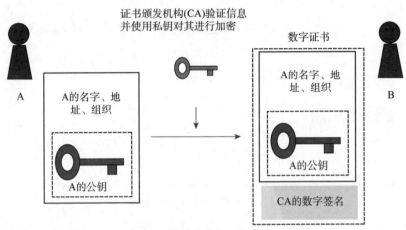

图 9.24　用户的数字证书及公钥

在 Java 中,可以使用 SUN 认证机构提供的 keytool 命令创建数字证书。请执行以下步骤:

(1) 通过在一行中输入以下命令来创建自签名的服务器数字证书。我们使用 RSA 生成公钥,服务器别名为 LSBU,存储密码为 storepassword,存储类型为 PKCS12,Java 密钥的存储文件名为 keystore.pfx。PKCS12 文件的文件扩展名是.pfx 或.p12。

```
keytool -genkey -alias LSBU -keyalg RSA -storepass storepassword -storetype PKCS12 -keystore keystore.pfx
```

按 Enter 键时,keytool 会提示输入服务器名称、组织、位置、州和国家/地区代码,参见图 9.25。

(2) 使用以下命令验证生成的密钥库文件。图 9.26 给出了输出结果。

```
keytool -list -v -keystore keystore.pfx -storetype pkcs12
```

(3) 将 keystore.pfx 文件中生成的自签名服务器证书导出到 server.cer 文件中,如图 9.27 所示。

(4) 将服务器证书添加到信任库文件 cacerts.pfx 中。出现提示时,输入 yes,然后按 Enter 键,如图 9.28 所示。

```
keytool -import -v -trustcacerts -alias LSBU -file server.cer
 -keystore cacerts.pfx -storepass storepassword -storetype pkcs12
```

```
H:\Chapter 9>keytool -genkey -alias LSBU -keyalg RSA -storepass storepassword -storetype PKCS12 -keystore keystore.pfx
What is your first and last name?
  [Unknown]:  Perry Xiao
What is the name of your organizational unit?
  [Unknown]:  School of Engineering
What is the name of your organization?
  [Unknown]:  London South Bank University
What is the name of your City or Locality?
  [Unknown]:  London
What is the name of your State or Province?
  [Unknown]:  London
What is the two-letter country code for this unit?
  [Unknown]:  UK
Is CN=Perry Xiao, OU=School of Engineering, O=London South Bank University, L=London, ST=London, C=UK correct?
  [no]:  yes

H:\Chapter 9>
```

图 9.25　使用 keytool 创建自签名的数字证书

```
H:\Chapter 9>keytool -list -v -keystore keystore.pfx -storetype pkcs12
Enter keystore password:
Keystore type: PKCS12
Keystore provider: SUN

Your keystore contains 1 entry

Alias name: lsbu
Creation date: 9 Apr 2019
Entry type: PrivateKeyEntry
Certificate chain length: 1
Certificate[1]:
Owner: CN=Perry Xiao, OU=School of Engineering, O=London South Bank University, L=London, ST=London, C=UK
Issuer: CN=Perry Xiao, OU=School of Engineering, O=London South Bank University, L=London, ST=London, C=UK
Serial number: 75067fd6
Valid from: Tue Apr 09 08:10:44 CST 2019 until: Mon Jul 08 08:10:44 CST 2019
Certificate fingerprints:
         SHA1: 9F:75:65:87:1D:71:87:49:D3:10:58:88:E7:C0:03:53:BA:37:BD:10
         SHA256: 8D:13:49:5F:F0:2A:82:FA:B2:67:5A:20:7C:F4:CA:C7:DA:E5:E8:AE:72:32:C0:87:DA:38:F9:AD:26:BD:41:0B
Signature algorithm name: SHA256withRSA
Subject Public Key Algorithm: 2048-bit RSA key
Version: 3

Extensions:

#1: ObjectId: 2.5.29.14 Criticality=false
SubjectKeyIdentifier [
KeyIdentifier [
0000: 0D 9A 2B BD A4 11 31 4D   E4 11 9D 6A C6 35 99 2C  ..+...1M...j.5.,
0010: 83 C3 BD 82                                        ....
]
]
```

图 9.26　使用 keytool 验证数字证书存储文件

```
H:\Chapter 9>keytool -export -alias LSBU -storepass storepassword -file server.cer -keystore keystore.pfx -storetype pkcs12
Certificate stored in file <server.cer>

H:\Chapter 9>
```

图 9.27　使用 keytool 将 keystore.pfx 文件中生成的自签名服务器证书导出到 server.cer 文件中

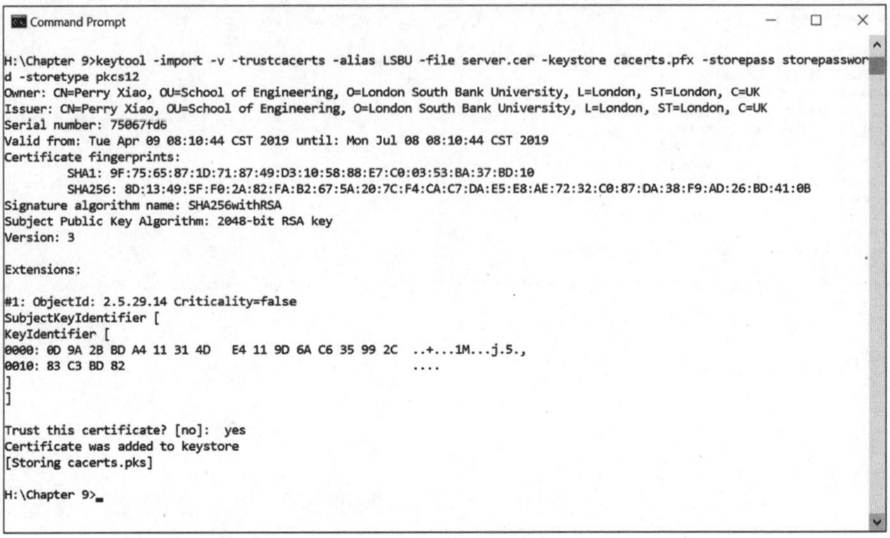

图 9.28　使用 keytool 将服务器证书添加到信任库文件 cacerts.pfx 中

例 9.6 展示的 Java 程序用于从 keystore.jks 文件中读取信息。图 9.29 给出了程序的部分输出结果。

例 9.6　Java 密钥库

```java
//Example 9.6 Java Key Store
import java.io.FileInputStream;
import java.security.*;
import java.security.cert.Certificate;

public class PKCS12Example {
  public static void main(String[] argv) throws Exception {
    String storefile ="keystore.pfx";
    String alias = "LSBU";
    String storepass ="storepassword";

    FileInputStream is = new FileInputStream(storefile);
    KeyStore keystore = KeyStore.getInstance("PKCS12");
    keystore.load(is, storepass.toCharArray());

    Key key = keystore.getKey(alias, storepass.toCharArray());
    if (key instanceof PrivateKey) {
      // Get certificate of public key
      Certificate cert = keystore.getCertificate(alias);
      System.out.println(cert.toString());
      // Get public key
      PublicKey publicKey = cert.getPublicKey();
      System.out.println(publicKey.toString());
      // Return a key pair
      KeyPair kp = new KeyPair(publicKey, (PrivateKey) key);
      System.out.println(kp.toString());
    }
  }
}
```

```
Command Prompt

H:\Chapter 9>javac PKCS12Example.java

H:\Chapter 9>java PKCS12Example
[
[
  Version: V3
  Subject: CN=Perry Xiao, OU=School of Engineering, O=London South Bank University, L=London, ST=London, C=UK
  Signature Algorithm: SHA256withRSA, OID = 1.2.840.113549.1.1.11

  Key:  Sun RSA public key, 2048 bits
  modulus: 21784245738210549611025619537078407056017077257886662295959724838066175485107207154428480842915872838460615161
1851152760274358472139305424847225050292468651941131114686928590928801845354621795996225264935606493736019509451101554627
4907616183134845445240273073774056319220977376686810806604267836219490057869576811814010088196192892716938793283961944238
2040509207471574076555841854377392088420792314218908088322891162232882139673848756288857625416460230783761221392983126
12349219871511246775769033158361883167427687735318802536483628655686015015277689735107857297275247586203734543450031353188
88587186328199904086878473
  public exponent: 65537
  Validity: [From: Tue Apr 09 08:10:44 CST 2019,
               To: Mon Jul 08 08:10:44 CST 2019]
  Issuer: CN=Perry Xiao, OU=School of Engineering, O=London South Bank University, L=London, ST=London, C=UK
  SerialNumber: [    75067fd6]

Certificate Extensions: 1
[1]: ObjectId: 2.5.29.14 Criticality=false
SubjectKeyIdentifier [
KeyIdentifier [
0000: 0D 9A 2B BD A4 11 31 4D   E4 11 9D 6A C6 35 99 2C  ..+...1M...j.5.,
0010: 83 C3 BD 82                                        ....
]
]

]
  Algorithm: [SHA256withRSA]
  Signature:
0000: 19 7E 91 26 21 33 9C 31   D0 E6 78 A1 B2 A3 CA 85  ...&!3.1..x.....
```

```
Command Prompt
0010: 83 C3 BD 82                                        ....
]
]

]
  Algorithm: [SHA256withRSA]
  Signature:
0000: 19 7E 91 26 21 33 9C 31   D0 E6 78 A1 B2 A3 CA 85  ...&!3.1..x.....
0010: C6 0E 3B 03 94 30 D6 85   BD 0B 79 E4 54 3F 6D B0  ..;..0....y.T?m.
0020: CF 03 4F 15 C0 15 A0 73   94 06 BA A4 D3 1C 96 7D  ..O....s........
0030: CF 42 1B 29 0E BD CC EB   45 64 2D 44 76 DC FF F0  .B.)....Ed-Dv...
0040: A3 3A BD 47 6A B9 54 4D   3C 88 C2 3F D9 B0 7E A3  .:.Gj.TM<..?....
0050: 29 79 F0 A8 59 AB 27 BB   13 E3 1D 36 DD B1 8E 81  )y..Y.'....6....
0060: B8 8D 69 BC 21 79 96 1D   42 15 F3 CE A4 46 A8 29  ..i.!y..B....F.)
0070: 8E 40 64 C2 B0 70 40 55   D0 96 C8 A9 AF 14 8D CC  .@d..p@U........
0080: 3E 97 60 34 9A FD E5 59   CC 65 E2 22 FF FE 36 B8  >.`4...Y.e."..6.
0090: C9 34 3F CF ED 21 B4 DE   56 4A 39 C5 9E A0 09 B5  .4?..!..VJ9.....
00A0: 69 EB A6 7F ED 61 0E E1   B2 ED 1C BE D3 9C E9 BC  i....a..........
00B0: 94 63 75 93 47 B1 06 CD   3D A2 15 FB F8 D7 69 E9  .cu.G...=.....i.
00C0: 85 20 AB 7F 66 D4 07 8D   DF AE 4C 34 28 EE AE 37  . ..f.....L4(..7
00D0: 3F BA A4 3C 1D 8E AB 0C   62 E2 4A A4 16 DD E2 7B  ?..<....b.J.....
00E0: 56 18 1D A0 6D 19 25 28   20 80 B8 F4 3C 35 90 34  V...m.%( ...<5.4
00F0: 8C 2C D6 D3 C4 55 04 96   DC 1D C7 49 BD 62 32 1D  .,...U.....I.b2.

]
Sun RSA public key, 2048 bits
  modulus: 21784245738210549611025619537078407056017077257886662295959724838066175485107207154428480842915872838460615161
1851152760274358472139305424847225050292468651941131114686928590928801845354621795996225264935606493736019509451101554627
4907616183134845445240273073774056319220977376686810806604267836219490057869576811814010088196192892716938793283961944238
2040509207471574076555841854377392088420792314218908088322891162232882139673848756288857625416460230783761221392983126
12349219871511246775769033158361883167427687735318802536483628655686015015277689735107857297275247586203734543450031353188
88587186328199904086878473
  public exponent: 65537
java.security.KeyPair@1f28c152

H:\Chapter 9>
```

图 9.29 PKCS12Example.java 程序的部分输出结果

有关使用 keytool 进行数字认证的更多详细信息，可参见以下 Oracle 文章：

https://docs.oracle.com/cd/E19798-01/821-1841/gjrgy/

图 9.30 展示了如何在 Google Chrome 浏览器中查看网站的证书。只需要单击网站 URL 旁边的锁图标，然后选择 Certificate(Valid)即可。

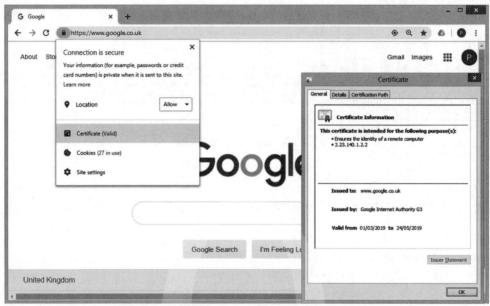

图 9.30　在 Google Chrome 浏览器中查看网站的证书

图 9.31 展示了如何在 Google Chrome 浏览器中查看受信任 CA 的列表。只需要在浏览器的右上角，选择 Settings | Advanced | Privacy And Security | Manage Certificates 即可。

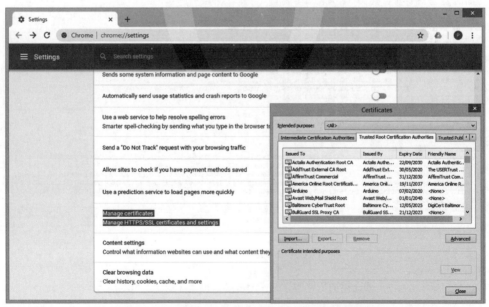

图 9.31　在 Google Chrome 浏览器中查看受信任 CA 的列表

9.12 其他 Java 示例

以下是 David Hook 撰写的 *Beginning Cryptography with Java* 一书的支持网站。该书通过简单的 Java 代码对加密技术做了有趣的介绍，适合初学者使用。

www.wrox.com/WileyCDA/WroxTitle/productCd-0764596330,descCd-DOWNLOAD.html

该书对应的 GitHub 网站如下，可以从中下载所有 Java 示例代码：

https://github.com/boeboe/be.boeboe.spongycastle

通过使用 Bouncy Castle Java Cryptography API，以下网站提供了使用 Java 进行加密的一些示例：

https://www.bouncycastle.org/java.html

以下网站提供了 Java2S 的综合性 Java 安全教程：

www.java2s.com/Tutorial/Java/0490_Security/Catalog0490_Security.htm

9.13 小结

本章介绍了网络安全的基础知识，解释了加密、哈希函数、消息摘要、数字签名和数字证书等关键概念。本章还提供了一些 Java 示例，分别用于私钥加密、公钥加密、数字签名、消息摘要和数字证书。最后，本章列举了一些有趣的 Java 安全编程资源。

9.14 本章复习题

1. 什么是网络安全？
2. 在新闻报道中找到两个最新的网络安全漏洞事件。
3. 什么是加密？
4. 什么是私钥加密(或对称密钥加密)？
5. 什么是公钥加密(或非对称密钥加密)？
6. 什么是哈希函数？
7. 什么是消息摘要？
8. 什么是数字签名？
9. 什么是数字证书？
10. 什么是 PGP？
11. TTL 和 SSL 有什么区别？
12. HTTP 和 HTTPS 有什么区别？

第10章

面向区块链应用的Java编程

"你可能一时骗过一些人,也可能在一段时间里骗过所有人,但你不可能永远骗过所有人。"

—— Abraham Lincoln

10.1 什么是区块链

10.2 如何验证区块链

10.3 如何挖掘区块

10.4 区块链的工作方式

10.5 区块链的应用

10.6 关于区块链的一些问题

10.7 Java 区块链示例

10.8 Java 区块链交易示例

10.9 Java BitcoinJ 示例

10.10 Java Web3j 示例

10.11 Java EthereumJ 示例

10.12 Java Ethereum 智能合约示例

10.13 更进一步:选择区块链平台

10.14 小结

10.15 本章复习题

10.1 什么是区块链

区块链是当前最热门的话题之一。什么是区块链？它与比特币有什么关系？区块链是一项在 2008 年由神秘人中本聪发明的技术，中本聪的真实身份目前仍然不得而知(人们普遍认为中本聪是居住在日本的一个人或一群人)。区块链技术的最初目的是充当加密货币比特币的分布式、数字化、公开化的交易账本，以解决双重支付问题。在数字货币中，当同一数字代币可以被多次使用时，就会出现双重支付。使用公开的交易账本，当第二次试图使用相同的数字代币时，交易将得不到确认。如今，区块链在其他许多领域显示出了潜力，如智能合约、智能物业、保险、音乐、医疗保健、制造、供应链、艺术、物联网等。区块链是另一种可能会彻底改变人们生活和工作方式的数字技术。

简而言之，如图 10.1 所示，区块链基本上是一条包含区块的链，每个区块都包含一些信息(数据)。对于数字货币领域的区块链，数据可以是转账金额、发送者和接收者的身份、日期和时间等。每个区块还包含索引、前一区块的哈希值、时间戳、随机数以及区块自己的哈希值。每个区块可通过哈希值连接到前一区块，因此所有区块都像菊花链一样连接在一起，只是这里还获得了区块链的名称。区块链中的第一个区块名为创世区块，没有连接前一区块，因此前一区块的哈希值为 NULL。索引是每个区块的唯一编号。第一个区块的索引为 0，第二个区块的索引为 1，第三个区块的索引为 2，依此类推。时间戳是区块的创建日期和时间，随机数是 32 位(4 字节)的整数，用于控制区块的哈希计算结果。

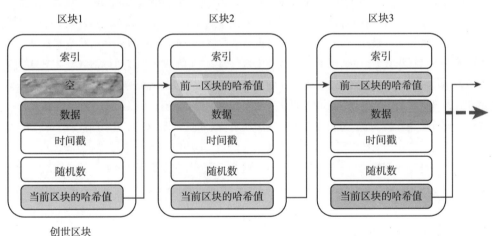

图 10.1 区块链示意图

每个区块都使用自己的索引、前一区块的哈希值、数据、时间戳和随机数以填充诸如 SHA-256 的哈希函数，从而创建自己的哈希值，如图 10.2 所示。如第 9 章所述，哈希函数是一种数学函数，可以用来将任何大小的数据映射到固定大小的数据。与加密不同，哈希是不可逆的。如果有人得到当前块的哈希值，他们就不可能在区块中找出信息。

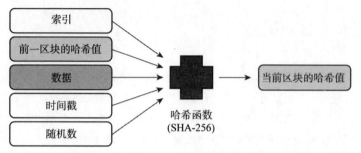

图 10.2　每个区块使用自己的索引、前一区块的哈希值、数据、时间戳和随机数填充哈希函数(如 SHA-256)以计算当前区块的哈希值

10.2　如何验证区块链

区块链是一系列信息的记录,例如一系列交易。区块链在形成之后,必须进行验证。在验证区块链时,必须验证区块链中的每一个区块,可从创世区块开始。

验证区块并不困难,只需要执行两个步骤。首先,需要测试前一区块的哈希值是否与当前区块的哈希值相同(对第一个区块的验证可以跳过此步骤)。其次,需要用区块的索引、前一区块的哈希值、数据、时间戳和随机数填充哈希函数(例如 SHA-256),以计算当前区块的哈希值,从而查看新计算的哈希值与区块中原始的哈希值是否相同。如果这两个步骤能执行成功,就说明区块有效。如果区块链中的所有区块都有效,就说明整个区块链是有效的。

现在,如果尝试在创建区块链之后更改图 10.1 中区块 2 的数据,会发生什么?当更改数据时,区块将变得无效,因为新的哈希值将与现有哈希值不匹配。可以使用新的哈希值替换现有的哈希值以使区块有效,但是这样区块 3 将无效,因为之前的哈希值不再匹配区块 2 中新的哈希值。当然,也可以使用区块 2 中新的哈希值替换区块 3 中前一区块的哈希值,但是之后需要重新计算区块 3 的哈希值以使区块 3 再次有效。然后,你必须对区块 4、5、6 等重复执行这个步骤,直到区块链的末端。这就是在区块链中使用哈希的好处。一旦创建了区块链,就无法再更改。当你试图从区块链中删除区块或将新的区块插入其中时,也会发生同样的情况。

对区块链进行更改的唯一方法是在更改区块之后重新计算所有区块的哈希值,但是区块链技术提供了一种共识机制来防止这种情况的发生,这种机制被称为工作量证明。

10.3　如何挖掘区块

区块链技术中的工作量证明系统意味着当计算哈希时,哈希需要以一定数量的 0 开始。0 越多意味着难度越大,0 越少意味着难度越小。区块链使用难度来控制区块的创建时间。如果难度值为 5,那么意味着哈希值必须以 5 个 0 开头。因为区块中的所有其他信息(例如索引、前一区块的哈希值、数据和时间戳)都是固定的,所以唯一可以更改的就是随机数。通过尝试不同的随机数,反复计算哈希,直到获得以所需数量的 0 开始的哈希值,这称为区块挖掘。显然,难度值越高,挖掘区块所需的时间就越长。平均而言,挖掘一个区块大约需要 10 分钟。如果区块链较长,那么挖掘所有区块的时间将更长。这还只是一条区块。如果有数百万用户,那就需要更改数百万条区块链。工作量证明系统几乎不可能更改区块链中的信息。这就是区块链与其他技术有所不同的原因,也是区块链在许多应用中有潜在价值的原因。

10.4 区块链的工作方式

为了帮助你了解区块链是如何工作的,让我们以汇款为例。传统方式下,如果 A 要向 B 汇款,A 需要通过银行进行这项操作,因为双方都信任银行,参见图 10.3 的左图。这是一种集中式汇款,在这种方式下,银行就像一本总账。这种方式存在几个问题。首先,银行会对这项服务收费。特别是,如果要汇一大笔钱,费用可能非常高。其次,交易需要时间,尤其是国际交易。最后,如果银行出现问题,就像 2008 年金融危机时那样,A 和 B 都会遭殃。

相反,区块链使用的是分布式汇款。在这种方式下,如果 A 要向 B 汇款,A 只需要在现有的区块链中添加一个新的区块,参见图 10.3 的右图。新的区块将包含发送方的名称、接收方的名称以及要汇的金额之类的信息。然后,将区块链复制给网络中的每个用户,称为 peer。peer 形成点对点网络,所有 peer 一起表现为分布式账本的一种形式。这带来如下好处:免费、快速且透明。

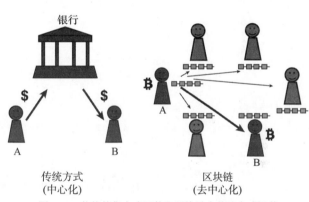

图 10.3 传统的集中式汇款和区块链中的分布式汇款

与传统的集中式汇款不同的是,A 实际上并没有向 B 发送任何东西——没有账户,也没有余额。所有用户只共享一条区块链,然后计算谁发送了什么,谁接收了什么,最后的余额是多少。因此,数字货币就是存在于一系列交易记录中的数字。

如果 B 想要通过改变区块链中的信息来进行欺诈,那么 B 不仅需要更改自己的区块链,还需要更改网络中所有用户的区块链,由于存在工作量证明系统,这实际上是不可能实现的。

下面总结区块链技术的主要特点和优点。

去中心化 这是区块链技术的核心概念,也是核心价值所在。没有使用集中的账本来记录和确认交易;相反,网络中的用户共同扮演着分布式账本的角色。因此,服务是免费的,而且速度很快,交易时间可能只有几小时,而不是几天。

不可篡改 一旦创建了区块,就无法更改。既不能在区块链中添加或删除区块,也不能更改区块中的任何信息。唯一可以做的就是在区块链的末尾添加新的区块。这种不可篡改性是区块链的另一主要优势,从而为人们带来了安全性和可靠性。

透明性 在点对点网络中,因为所有的交易记录都是与所有用户共享的,所以一切都是透明的,没有黑盒子。

高可用性 由于交易记录在点对点网络中与所有用户共享,因此不存在单点故障问题,数据高度可用。图 10.4 展示了一个虚拟的区块链演示网站(https://blockchaindemo.io/),你可以看到区块的内容,将新的区块添加到区块链中,并添加更多的 peer。

图 10.4　一个虚拟的区块链演示网站

10.5　区块链的应用

区块链技术有许多应用,下面介绍一些示例。

10.5.1　比特币

第一个也是最重要的应用是数字货币,也称为加密货币,因为数字货币是通过加密来进行保护的。有许多数字货币。最受欢迎的是比特币(BTC),符号为₿。比特币由中本聪发明,并于 2009 年作为开源软件发布。比特币可以转移和兑换成真实的货币、产品和服务。世界上有数百万比特币用户,每个月产生数百万笔交易。一枚比特币的价格已经从几美分上涨到上万美元。交易时,不需要购买一枚完整的比特币。可以买卖的最小比特币单位为"聪"——一枚比特币的 1 亿分之一。与可以无限供应的真实货币不同,比特币的供应有限。比特币是通过名为比特币挖矿的过程产生的,如 10.3 节所述。比特币挖矿基本上是一个验证最近交易、向区块链中添加新的区块并获得奖励的过程。通过比特币挖矿,新的比特币被创造出来,但每小时只能创建少量比特币,并受底层算法严格控制。新的比特币被不断创造,直至达到 2100 万枚的上限。这种有限的供应使比特币成为一种潜在资产:随着需求的增加,价值也会增加。

图 10.5 展示了比特币区块的内部结构。每个区块都有头和主体。头部分包含前一区块的头部分的哈希值、时间戳、目标难度、随机数和 Merkle 根。主体部分包含所有交易。每个区块包含约 500 笔交易。每笔交易都将计算自己的哈希值,然后将所有的哈希值一起计算得到最终的哈希值。最终的哈希值存储在 Merkle 根或 Tx_Root 中,如图 10.5 所示。

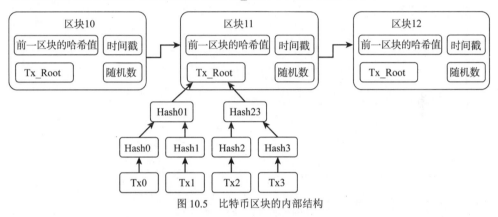

图 10.5　比特币区块的内部结构

图 10.6 展示了三层的比特币服务架构。底层是点对点网络,这是用来将所有节点或用户连接在一起的地方。中间层是去中心化的账本,中间层用来实现区块链技术。顶层是应用层,用来实现诸如比特币钱包的应用。

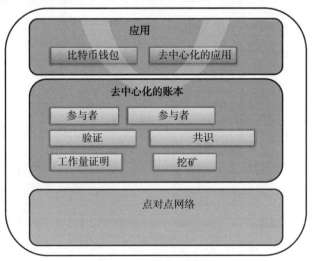

图 10.6　三层的比特币服务架构

有几个网站可以供你研究比特币区块链的结构。图 10.7 展示了来自 Block Explorer

(https://www.blockchain.com/explorer)的示例，你可以看到比特币主链中的总块数(称为高度)在当时为 539 602，最近的区块创建于大约 6 分钟之前(称为区块链的年龄)，其中包含 652 笔交易，发送的比特币数量为 1686.25，区块大小为 313.35 KB。

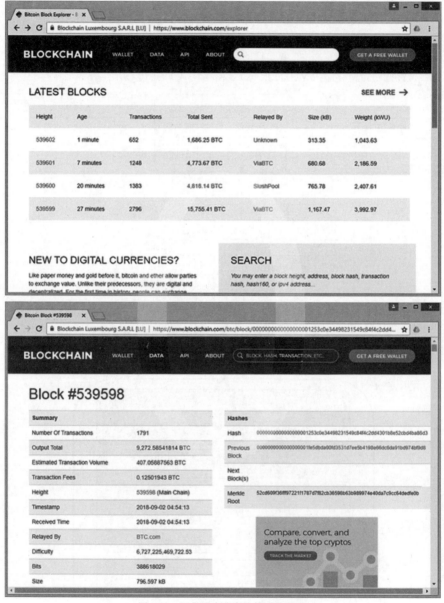

图 10.7 可供用来研究比特币的网站

通过访问如下链接，可以查看最新的比特币交易和比特币区块的结构：

https://blockexplorer.com/

另一个有趣的网站是 Learn Me a Bitcoin，通过该网站可以学习比特币方面的术语，还可以浏览比特币区块的详细信息：

```
http://learnmeabitcoin.com/guide/
```

但是，随着比特币的普及，出现了一些可扩展性问题：比特币使用 1 MB 大小的区块，每秒只能进行 7 笔交易，每笔交易需要 10 分钟的处理时间。这导致一种新的加密货币的出现：比特币现金(Bitcoin Cash，BCH)，BCH 是比特币的另一版本，因为比特币和 BCH 基于相同的技术，而 BCH 正作为新的分支进行发展。使用 8 MB 大小的区块，难度级别可调，并且交易只需要两分钟时间。

如果有兴趣尝试使用比特币或其他加密货币，例如 BCH、以太币、Ripple、Litecoin、Peercoin 或 Dogecoin，可下载并安装以下主流的加密货币数字钱包。

Bitcoin Core 由中本聪为比特币创建，又称为中本聪客户端(https://bitcoin.org/en/bitcoin-core/)。

Ethereum 一种开源的、公共的、基于区块链的分布式计算平台，提供了一种名为 Ether 的加密货币(https://www.ethereum.org/)。

Ripple 一种基于共享的公共 XRP 总账的实时支付系统，已被联合信贷银行(UniCredit)、瑞银(UBS)和桑坦德(Santander)等银行普遍采购。

Bitcoin Cash 比特币的另一版本，但使用起来更快、更便宜，区块大小为 8 MB(https://www.bitcoincash.org/)。

Electrum 轻量级的比特币客户端，同时具有硬件钱包和软件钱包的功能。硬件钱包允许用户在硬件设备(例如 USB 记忆棒)上存储比特币信息，例如私钥。与软件钱包不同，硬件钱包可以物理保护，并且不受病毒的侵害，病毒往往从软件钱包中窃取信息(https://electrum.org/)。

Coinbase 最知名、最值得信赖的数字货币买卖和管理应用之一(https://www.coinbase.com/)。

Blockchain Luxembourg 具有易于使用的精美用户界面以及许多用于加密货币的有用功能(https://www.blockchain.com/)。

图 10.8 展示了来自 bitcoin.org 网站(https://bitcoin.org/en/choose-your-wallet)的更多加密货币钱包。图 10.9 展示了 Live Coin Watch 网站(https://www.livecoinwatch.com/)，可以在上面找到每种加密货币的更多详细信息，例如价格、市值、交易量和趋势。根据 Live Coin Watch 网站上的数据以及它们的市值，最受欢迎的三种加密货币是 BTC(比特币)、XRP(瑞波币)和 ETH(Ethereum)。BTC 约为 700 亿美元，XRP 约为 150 亿美元，ETH 约为 120 亿美元，远远领先于其他加密货币。

想要了解更多信息，请阅读中本聪的白皮书，并访问维基百科和 Bitcoin.org 页面。

```
https://bitcoin.org/bitcoin.pdf
https://en.bitcoin.it/wiki/Main_Page
https://bitcoin.org/en/
```

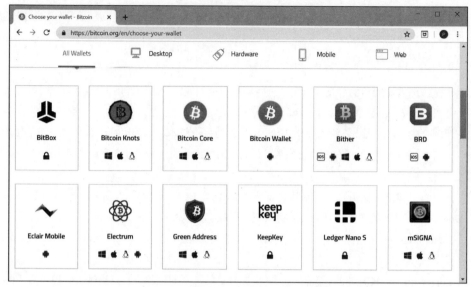

图 10.8　来自 Bitcoin.org 的加密货币钱包

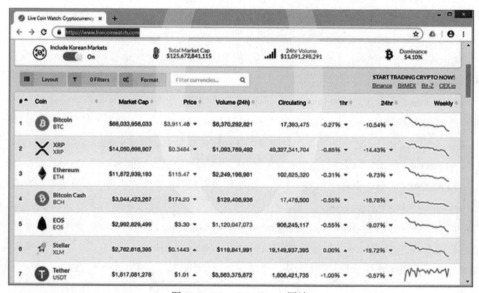

图 10.9　Live Coin Watch 网站

10.5.2　智能合约

智能合约是自动执行的合同，旨在执行协议中的条款。智能合约用于在特定条件下控制各方之间数字货币或资产的转移。使用智能合约，无需中间人即可交换金钱、财产、股份或任何有价物品。智能合约不仅定义了所有规则和所有处罚，而且会自动执行这些义务。

例如，购买房屋通常涉及买方、卖方和多个第三方，例如房地产经纪人和律师。有了智能合约和数字货币，就可以在买卖双方之间完成交易。一旦满足所有条件，智能合约在本质上就是软件程序，它将使用数字货币自动完成交易。具体过程如下：

(1) 买卖双方就以数字货币衡量的房屋价格达成一致，将合同写入区块链。合同包括买方和卖方的所有信息、条款和条件。因为合同在公共总账上，所以不能更改。

(2) 当满足所有条件并发生触发事件(如日子到期)时，合同将按照约定的条款执行。

(3) 卖方获得数字货币形式的付款，买方获得房屋，包括所有法律文件，例如土地契约。所有清算和结算都是自动化的，所有权无可争议。

图 10.10 展示了智能合约的运作，它改编自 BlockGeeks 网站。

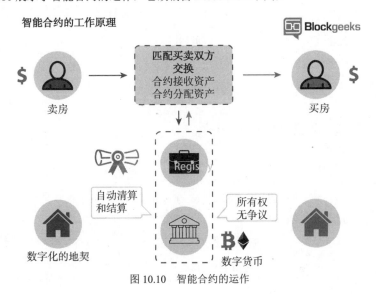

图 10.10　智能合约的运作

10.5.3　医疗

个人健康数据也可以存储在受加密保护的区块链中，只允许特定用户(例如医生和保险提供商)访问。医生将拥有患者完整的病史，因此可以提供更好的诊断。病历将自动与保险公司共享，以支持索赔。区块链也可用于验证药物的真实性，防止假冒药物和医疗器械，以及提高临床试验数据的质量和可靠性。

10.5.4　制造业和供应链

区块链技术还可以用于制造业和供应链，以记录任何交换、协议/合同、跟踪和付款。由于每笔交易都记录在一个区块中，并且记录在分布于用户之间的多个账本的副本中，因此这种记录方法是高度透明的，也是高度安全的，非常有效并且可扩展。你将能够查看产品或组件的完整记录，包括来源和去向，这叫作可追溯性。区块链技术可以提高供应链的效率和透明度。例如，跨国零售商沃尔玛(Walmart)已与 IBM 合作，利用 Hyperledger Fabric 区块链追踪从供应商到货架的主要食品。这项技术还被用于追踪艺术品、古董、珠宝和其他贵重物品，以进行身份验证和所有权证明，并打击仿冒行为。

10.5.5 物联网

有了物联网(IoT)，数十亿台设备将相互连接。安全地存储物联网系统产生的海量数据是人们将要面临的一项巨大挑战。借助区块链技术的分布式账本，物联网数据将能够以一种不可信任的方式进行分布式存储，并且可以更好地进行组织和分析，从而产生有价值的见解。区块链非常适合于公共和私营的物联网系统，因为它具有识别、验证和数据传输能力。区块链可以在双方和设备之间建立信任，通过消除中间人来降低成本，还可以通过缩短结算时间来加速交易。

10.5.6 政务

选举是政府最重要的民主程序之一。借助区块链，选举可以更加公开，成本更低并且更不容易出现欺诈。许多政府信息和数据也可以存储在区块链中，这将使访问变得更容易，并使政府行为更加透明并减少腐败。

10.6 关于区块链的一些问题

就像其他技术一样，区块链也存在一些局限性。

安全性 因为区块链依赖用户来存储记录，所以如果大多数用户决定更改记录，那么对记录所做的更改将被接受，这被称为51%攻击。中本聪在比特币问世时就强调了这一安全漏洞。为了将这种51%攻击的可能性降到最低，网络必须大于一定的大小，以使任何人或团体都不可能拥有超过51%的控制权。黑客攻击也是数字货币面临的一种常见威胁。

复杂性 区块链基于庞大、复杂的技术，涉及复杂的数学和大量的软件编程。区块链的复杂性使得任何人都很难理解它的工作原理以及带来的好处，从而阻碍了应用。

缺乏监管 因为区块链技术太新了，所以几乎没有法规。缺乏法规会造成危险的环境，在这种环境中，欺诈、诈骗和市场操纵司空见惯。

量子准备 区块链建立在加密的基础上，对于当今的计算机而言是安全的。但是随着量子计算等技术的出现，计算机的运行速度可能比现在快1亿倍，这将对当今使用的所有形式的加密构成严重威胁。为了面对这一挑战，需要开发一种抗量子密码系统。

可伸缩性 随着网络的增长，由于存在复杂性、加密性和分布式特性，区块链中的交易处理可能会变得非常缓慢和烦琐。例如，像比特币和Ethereum这么流行的区块链平台平均每秒能处理7到15笔交易，而Visa目前平均每秒能处理5000到8000笔交易。所以，人们还需要进行更多的研究以提高可伸缩性。

10.7 Java 区块链示例

让我们来看一些Java区块链示例。例10.1(分为例10.1A、例10.1B、例10.1C和例10.1D)是一个简单的Java区块链演示程序，它由两个Java文件Block.java和BlockChainMain.java组成。Block.java(参见例 10.1A)的任务是创建单个区块。每个区块中有 6 个属性和 3 个方法。这 6 个属性与前面所讲的属性相同：index、timestamp、currentHash、previousHash、data、nonce。这 3 个方法分别是 calculateHash()、mineBlock()和 toString()。calculateHash()用于计算区块的当前哈希值，mineBlock()用于根据指定的难度(前导 0 的数量)挖掘区块，toString()用于显示区块的信息。为了创建区块，需要提供有关 index、previousHash 和 data 的信息。

例 10.1A　Block.java 文件的内容

```java
//Example 10.1A Block.java
import java.security.*;
import java.util.*;

public class Block {
    public int index;
    public long timestamp;
    public String currentHash;
    public String previousHash;
    public String data;
    public int nonce;

    public Block(int index, String previousHash, String data) {
        this.index = index;
        this.timestamp = System.currentTimeMillis();
        this.previousHash = previousHash;
        this.data = data;
        nonce = 0;
        currentHash = calculateHash();
    }
    public String calculateHash(){
        try {

            String input = index + timestamp + previousHash + data +nonce;
            MessageDigest digest = MessageDigest.getInstance("SHA-256");
            byte[] hash = digest.digest(input.getBytes("UTF-8"));

            StringBuffer hexString = new StringBuffer();
            for (int i = 0; i < hash.length; i++) {
                String hex = Integer.toHexString(0xff & hash[i]);
                if(hex.length() == 1) hexString.append('0');
            hexString.append(hex);
            }
            return hexString.toString();
        }
        catch(Exception e) {
            throw new RuntimeException(e);
        }
    }
    public void mineBlock(int difficulty) {
        nonce = 0;
        String target = new String(new char[difficulty]).replace('\0','0');
        while (!currentHash.substring(0, difficulty).equals(target)) {
            nonce++;
            currentHash = calculateHash();
        }
```

```
    }
    public String toString() {
        String s = "Block # : " + index + "\r\n";
        s = s + "PreviousHash : " + previousHash + "\r\n";
        s = s + "Timestamp : " + timestamp + "\r\n";
        s = s + "Data : " + data + "\r\n";
        s = s + "Nonce : " + nonce + "\r\n";
        s = s + "CurrentHash : " +currentHash + "\r\n";
        return s;
    }
}
```

BlockChainMain.java(参见例 10.1B)是主程序,用于创建区块链并将两个区块添加到区块链中。图 10.11 给出了 BlockChainMain.java 的编译、执行和输出结果。

例 10.1B　BlockChainMain.java 文件的内容

```
//Example 10.1B BlockChainMain.java
import java.util.*;

public class BlockChainMain {
    public static ArrayList<Block> blockchain = new ArrayList<Block>();
    public static int difficulty = 5;

    public static void main(String[] args) {
        Block b = new Block(0, null, "My First Block"); //The genesis block
        b.mineBlock(difficulty);
        blockchain.add(b);
        System.out.println(b.toString());

        Block b2 = new Block(1, b.currentHash, "My Second Block");
        b2.mineBlock(difficulty);
        blockchain.add(b2);
        System.out.println(b2.toString());
    }
}
```

图 10.11　BlockChainMain.java 程序的编译、执行和输出结果

可以修改 BlockChainMain.java 程序以验证这些区块,参见例 10.1C。例 10.1C 中的 BlockChainMain2.java 程序使用了一个名为 validateBlock()的新函数,该函数可以根据区块的索引、前一区块的哈希值和当前区块的哈希值来验证区块。图 10.12 中的第一个图给出了 BlockChainMain2.java 程序的编译、执行和输出结果。你可以看到,添加的两个区块均有效。

例 10.1C　BlockChainMain2.java 文件的内容

```java
//Example 10.1C BlockChainMain2.java
import java.util.*;

public class BlockChainMain2 {

    public static ArrayList<Block> blockchain = new ArrayList<Block>();
    public static int difficulty = 5;
    public static void main(String[] args) {
        Block b = new Block(0, null, "My First Block");
        b.mineBlock(difficulty);
        blockchain.add(b);
        System.out.println(b.toString());
        System.out.println("Current Block Valid: " + validateBlock(b,null));
        Block b2 = new Block(1, b.currentHash, "My Second Block");
        b2.mineBlock(difficulty);
        blockchain.add(b2);
        //b2.data="My Third Block";
        System.out.println(b2.toString());
        System.out.println("Current Block Valid: " + validateBlock(b2,b));
    }
    public static boolean validateBlock(Block newBlock, Block previousBlock) {

        if (previousBlock == null){ //The first block
            if (newBlock.index != 0) {
               return false;
            }

            if (newBlock.previousHash != null) {
               return false;
            }

            if (newBlock.currentHash == null ||
                !newBlock.calculateHash().equals(newBlock.currentHash)) {
               return false;
            }

            return true;

        } else{                     //The rest blocks
            if (newBlock != null ) {
              if (previousBlock.index + 1 != newBlock.index) {
                return false;
              }

              if (newBlock.previousHash == null ||
                 !newBlock.previousHash.equals(previousBlock.currentHash)) {
                return false;
              }

              if (newBlock.currentHash == null ||
                 !newBlock.calculateHash().equals(newBlock.currentHash)) {
                return false;
              }
```

```
            return true;
        }
        return false;
    }
}
```

现在，如果取消注释以下代码行，那么效果等同于在创建区块之后手动修改区块：

```
//b2.data="My Third Block";
```

当重新编译并执行程序时，将显示第二个区块是无效的，如图 10.12 的第二个图所示。

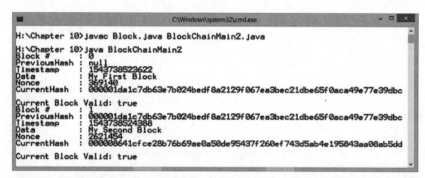

图 10.12　BlockChainMain2.java 程序的编译、执行和输出结果

例 10.1D 中的 BlockChainMain3.java 程序展示了如何验证整个区块链。为此，需要验证区块链中的所有区块。新的 validateChain()函数将循环遍历区块链中的所有区块，并使用之前的 validateBlock()函数验证每个区块。图 10.13 给出了 BlockChainMain3.java 程序的编译、执行和输出结果。

例 10.1D　BlockChainMain3.java 文件的内容

```
//Example 10.1D BlockChainMain3.java
import java.util.*;

public class BlockChainMain3 {
```

```
    public static ArrayList<Block> blockchain = new ArrayList<Block>();
    public static int difficulty = 5;

    public static void main(String[] args) {
        Block b = new Block(0, null, "My First Block");
        b.mineBlock(difficulty);
        blockchain.add(b);
        System.out.println(b.toString());

        Block b2 = new Block(1, b.currentHash, "My Second Block");
        b2.mineBlock(difficulty);
        blockchain.add(b2);
        System.out.println(b2.toString());
        System.out.println("Current Chain Valid: "+validateChain(blockchain));
    }
    public static boolean validateChain(ArrayList<Block> blockchain) {
        if (!validateBlock(blockchain.get(0), null)) {
            return false;
        }

        for (int i = 1; i < blockchain.size(); i++) {
            Block currentBlock = blockchain.get(i);
            Block previousBlock = blockchain.get(i - 1);

            if (!validateBlock(currentBlock, previousBlock)) {
                return false;
            }
        }

        return true;
    }
    public static boolean validateBlock(Block newBlock, Block previousBlock) {
        //The same code as before
        ... ...
    }
}
```

图 10.13　BlockChainMain3.java 程序的编译、执行和输出结果

10.8　Java 区块链交易示例

现在可以开始使用区块链做一些有趣的事情了。例 10.2(分为例 10.2A、例 10.2B、例 10.2C、

例 10.2D 和例 10.2E)展示了如何使用区块链记录交易，其中包含 4 个 Java 程序：Block2.java、Transaction.java、Wallet.java 和 BlockChainMain4.java。Block2.java(参见例 10.2A)与先前的 Block.java 相似，但是不使用 String 数据，而是在每个区块内使用 ArrayList<Transaction>。ArrayList 用于存储多笔交易。

例 10.2A　Block2.java 文件的内容

```java
//Example 10.2A Block2.java
public class Block2 {
    public int index;
    public long timestamp;
    public String currentHash;
    public String previousHash;
    public String data;
    public ArrayList<Transaction> transactions = new
        ArrayList<Transaction>(); //our data will be a simple message.
    public int nonce;

    public Block2(int index, String previousHash, ArrayList<Transaction> transactions) {
        this.index = index;
        this.timestamp = System.currentTimeMillis();
        this.previousHash = previousHash;
        this.transactions = transactions;
        nonce = 0;
        currentHash = calculateHash();
    }

    public String calculateHash(){
        try {
            data="";
            for (int j=0; j<transactions.size();j++){
                Transaction tr = transactions.get(j);
                data = data + tr.sender+tr.recipient+tr.value;
            }
            String input = index + timestamp + previousHash + data + nonce;
            MessageDigest digest = MessageDigest.getInstance("SHA-256");
            byte[] hash = digest.digest(input.getBytes("UTF-8"));

            StringBuffer hexString = new StringBuffer();
            for (int i = 0; i < hash.length; i++) {
                String hex = Integer.toHexString(0xff & hash[i]);
                if(hex.length() == 1) hexString.append('0');
                hexString.append(hex);
            }
            return hexString.toString();
        }   catch(Exception e) {
            throw new RuntimeException(e);
        }
    }

    public void mineBlock(int difficulty) {
        nonce = 0;
        String target = new String(new char[difficulty]).replace('\0','0');
```

```
        while (!currentHash.substring(0, difficulty).equals(target)) {
            nonce++;
            currentHash = calculateHash();
        }
    }

    public String toString() {
        String s = "Block # : " + index + "\r\n";
        s = s + "PreviousHash : " + previousHash + "\r\n";
        s = s + "Timestamp : " + timestamp + "\r\n";
        s = s + "Transactions : " + data + "\r\n";
        s = s + "Nonce : " + nonce + "\r\n";
        s = s + "CurrentHash : " +currentHash + "\r\n";
        return s;
    }
}
```

Transaction.java(参见例 10.2B)仅记录一笔交易,其中包括发送方、接收方和值。

例 10.2B　Transaction.java 文件的内容

```
//Example 10.2B Transaction.java
import java.util.*;

public class Transaction {
    public String sender;
    public String recipient;
    public float value;

    public Transaction(String from, String to, float value) {
        this.sender = from;
        this.recipient = to;
        this.value = value;
    }
}
```

Wallet.java(参见例 10.2C)用于为用户创建数字钱包。可首先使用 generateKeyPair()为用户生成一对公钥/私钥,然后使用 getBalance()获取用户的余额,最后使用 send()以及发送方和接收方的公钥向另一个用户发送一些数字硬币。getBalance()会遍历整个区块链,为用户搜索交易:如果发送交易,就从余额中扣除交易额;如果收到交易,就将交易额添加到余额中。为了简单起见,每个数字钱包都包含 100 枚数字硬币。

例 10.2C　Wallet.java 文件的内容

```
//Example 10.2C Wallet.java
import java.security.*;
import java.util.*;

public class Wallet {

    public String privateKey;
    public String publicKey;
    private float balance=100.0f;
```

```java
        private ArrayList<Block2> blockchain = new ArrayList<Block2>();

    public Wallet(ArrayList<Block2> blockchain) {
        generateKeyPair();
        this.blockchain = blockchain;
    }

    public void generateKeyPair() {
        try {
            KeyPair keyPair;
            String algorithm = "RSA"; //DSA DH etc
            keyPair = KeyPairGenerator.getInstance(algorithm).generateKeyPair();
            privateKey = keyPair.getPrivate().toString();
            publicKey = keyPair.getPublic().toString();

        }catch(Exception e) {
            throw new RuntimeException(e);
        }
    }

    public float getBalance() {
        float total = balance;
         for (int i=0; i<blockchain.size();i++){
            Block2 currentBlock = blockchain.get(i);
            for (int j=0; j<currentBlock.transactions.size();j++){
                Transaction tr = currentBlock.transactions.get(j);
                if (tr.recipient.equals(publicKey)){
                    total += tr.value;
                }
                if (tr.sender.equals(publicKey)){
                    total -= tr.value;
                }
            }
        }
        return total;
    }

    public Transaction send(String recipient,float value ) {
      if(getBalance() < value) {
        System.out.println("!!!Not Enough funds. TransactionDiscarded.");
        return null;
      }

        Transaction newTransaction = new Transaction(publicKey,recipient, value);
        return newTransaction;
    }

}
```

BlockChainMain4.java 是使用 Wallet.java 进行区块链交易的主程序,它仅创建两个数字钱包并显示它们的余额,如图 10.14 所示。

例 10.2D　BlockChainMain4.java 文件的内容

```java
//Example 10.2D BlockChainMain4.java
```

```
import java.util.*;

public class BlockChainMain4 {

    public static ArrayList<Block2> blockchain = newArrayList<Block2>();
    public static ArrayList<Transaction> transactions = new
        ArrayList<Transaction>();
    public static int difficulty = 5;

    public static void main(String[] args) {
        Wallet A = new Wallet(blockchain);
        Wallet B = new Wallet(blockchain);
        System.out.println("Wallet A Balance: " + A.getBalance());
        System.out.println("Wallet B Balance: " + B.getBalance());
    }
}
```

```
H:\Chapter 10>javac Block2.java Wallet.java Transaction.java BlockChainMain4.java

H:\Chapter 10>java BlockChainMain4
Wallet A Balance: 100.0
Wallet B Balance: 100.0
```

图 10.14　BlockChainMain4.java 的编译、执行和输出结果

接下来可以添加一些交易。例 10.2E 添加了两笔交易：A 首先向 B 发送了 10 枚数字硬币，然后又向 B 发送了 20 枚数字硬币。可以使用前面的 validateChain() 来验证整个区块链。图 10.15 给出了程序的编译、执行和输出结果。如你所见，这两笔交易均已通过，整个区块链是有效的。

例 10.2E　BlockChainMain5.java 文件的内容

```
//Example 10.2E BlockChainMain5.java
import java.util.*;

public class BlockChainMain5 {

    public static ArrayList<Block2> blockchain = new ArrayList<Block2>();
    public static ArrayList<Transaction> transactions = new
        ArrayList<Transaction>();
    public static int difficulty = 5;

    public static void main(String[] args) {
        Wallet A = new Wallet(blockchain);
        Wallet B = new Wallet(blockchain);
        System.out.println("Wallet A Balance: " + A.getBalance());
        System.out.println("Wallet B Balance: " + B.getBalance());

        System.out.println("Add two transactions... ");
        Transaction tran1 = A.send(B.publicKey, 10);
        if (tran1!=null){
            transactions.add(tran1);
        }
        Transaction tran2 = A.send(B.publicKey, 20);
        if (tran2!=null){
            transactions.add(tran2);
```

```
      }

      Block2 b = new Block2(0, null, transactions);
      b.mineBlock(difficulty);
      blockchain.add(b);

      System.out.println("Wallet A Balance: " + A.getBalance());
      System.out.println("Wallet B Balance: " + B.getBalance());
      System.out.println("Blockchain Valid : " + validateChain(blockchain));

  }
  public static boolean validateChain(ArrayList<Block2> blockchain) {
     if (!validateBlock(blockchain.get(0), null)) {
       return false;
     }

     for (int i = 1; i < blockchain.size(); i++) {
       Block2 currentBlock = blockchain.get(i);
       Block2 previousBlock = blockchain.get(i - 1);

       if (!validateBlock(currentBlock, previousBlock)) {
         return false;
       }
     }

     return true;
  }
  public static boolean validateBlock(Block2 newBlock, Block2 previousBlock) {
      //The same as before
  }
}
```

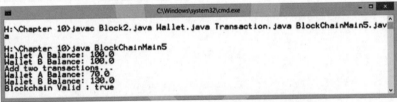

图 10.15　BlockChainMain5.java 的编译、执行和输出结果

如果将第二笔交易改为 200(表示 A 向 B 发送 200 枚数字硬币)，那么由于超出 A 的余额，因此这笔交易将被丢弃，如图 10.16 所示。

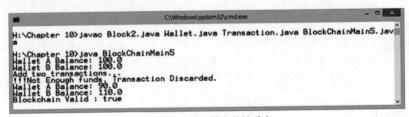

图 10.16　第二笔交易被丢弃

对于 Java 开发人员来说，一些十分流行的区块链库如下。

- BitcoinJ：https://github.com/bitcoinj/bitcoinj。
- Web3j：https://github.com/web3j/web3j。
- EthereumJ：https://github.com/ethereum/ethereumj。
- HyperLedger Fabric：https://github.com/hyperledger/fabric-sdk-java。

10.9　Java BitcoinJ 示例

BitcoinJ 是用于开发 Java Bitcoin 应用的开源库。BitcoinJ 允许维护钱包并发送和接收交易，而不需要本地的 Bitcoin Core 副本。为了使用 BitcoinJ，首先需要从以下链接下载 BitcoinJ JAR 文件 bitcoinj-core-0.14.4-bundled.jar：

https://bitcoinj.github.io/getting-started-java

也可以直接从以下链接下载 BitcoinJ JAR 文件：

https://search.maven.org/remotecontent?filepath=org/bitcoinj/bitcoinjcore/0.14.4/bitcoinj-core-0.14.4-bundled.jar
https://jar-download.com/artifacts/org.bitcoinj

然后，还需要从以下链接下载用于 Java(SLF4J)库的 Simple Logging Facade：

https://www.slf4j.org/download.html

将下载的文件解压缩到文件夹中，命名为 slf4j-simple-1.7.25.jar。如果解压后得到的版本号稍有不同，请不必担心。

例 10.3 中的演示程序由以下链接中的 BitcoinJ 示例 DumpWallet.java 修改而来：

https://github.com/bitcoinj/bitcoinj/blob/master/examples/src/main/java/org/bitcoinj/examples/DumpWallet.java

例 10.3　DumpWallet1.java 演示程序

```java
//Example 10.3 DumpWallet1.java
//Modified from https://github.com/bitcoinj/bitcoinj/blob/master/
//examples/src/main/java/org/bitcoinj/examples/DumpWallet.java
import java.io.File;
import org.bitcoinj.wallet.Wallet;
/**
 * DumpWallet loads a serialized wallet and prints information about
what it contains.
 */
public class DumpWallet1 {
    public static void main(String[] args) throws Exception {
        String walletfile="/path/to/your/walletfile";
        File f=new File(walletfile);
        Wallet wallet = Wallet.loadFromFile(f);
        System.out.println(wallet.toString());
    }
}
```

将 DumpWallet1.java、bitcoinj-core-0.14.4-bundled.jar 和 slf4j-simple-1.7.25.jar 文件放入同一个文件夹。然后，可以通过输入以下命令编译并执行 DumpWallet1.java 程序：

```
javac -classpath ".;bitcoinj-core-0.14.4-bundled.jar;slf4j-simple-
1.7.25.jar" DumpWallet1.java
java -classpath ".;bitcoinj-core-0.14.4-bundled.jar;slf4j-simple-1.7.25.
jar" DumpWallet1
```

这里还需要数字钱包文件才能使程序正常运行。请参阅 10.10 节以了解如何创建数字钱包文件。还有更多的 BitcoinJ 示例，可从如下 GitHub 网站下载完整的源代码：

```
https://github.com/bitcoinj/bitcoinj
```

以下网站提供了有关如何使用 BitcoinJ 库构建 GUI 钱包的简单教程：

```
https://bitcoinj.github.io/simple-gui-wallet
```

testnet

在真正运行程序之前，在模拟环境中对程序进行测试是个好主意。比特币社区提供了名为 testnet 的模拟比特币网络。使用 testnet 可以发送和接收比特币。Testnet 中的比特币没有价值，可从以下网站免费获得：

```
https://testnet-faucet.mempool.co/
http://tpfaucet.appspot.com/
```

有关 BitcoinJ 库的更多信息，请参阅以下资源：

```
https://bitcoinj.github.io/
https://github.com/bitcoinj/bitcoinj
```

10.10 Java Web3j 示例

Web3j 是用来与 Ethereum 客户端集成的轻量级 Java 库。借助 Web3j，可以创建数字钱包、管理钱包、发送以太币以及创建智能合约。使用 Web3j 库的最简单方法是从以下 GitHub 网站下载命令行工具：

```
https://github.com/web3j/web3j/releases/tag/v4.0.1
```

在下载的文件中查找名为 web3j-4.0.1.zip 的压缩文件。关于 Web3j 命令行工具的更多信息，可通过以下链接找到：

```
https://docs.web3j.io/command_line.html
```

将 web3j-4.0.1.zip 文件解压缩到文件夹中，在本例中解压路径为 E:\web3j-4.0.1\。主命令行工具在 E:\web3j-4.0.1\bin\文件夹中，名为 web3j.bat。打开 MS-DOS 终端，转到 E:\文件夹，然后输入以下命令以创建数字钱包，结果如图 10.17 所示。

```
.\web3j-4.0.1\bin\web3j wallet create
```

图 10.17　创建数字钱包

接下来为数字钱包设置密码。完成设置后，将创建如下数字钱包文件：

```
UTC--2018-11-26T13-18-32.250132200Z--ccded263b9310c875d615bf66ba678e121c26362.json
```

接下来显示数字钱包的默认位置，其中 %USERPROFILE% 表示用户文件夹，例如 C:\Users\xiaop。

```
%USERPROFILE%\AppData\Roaming\Ethereum\testnet\keystore
```

可以使用 type 命令(或任何文本编辑器)来显示数字钱包文件的内容，如图 10.18 所示。数字钱包中的地址是唯一的，可以用于发送以太币或创建智能合约。数字钱包中的 crypto 用于指定正在使用的加密算法以及私钥。

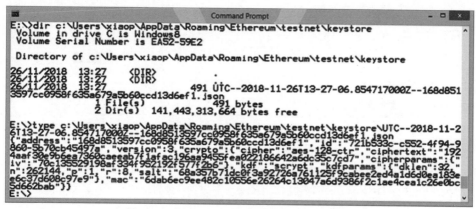

图 10.18　使用 type 命令显示数字钱包文件的内容

数字钱包中的内容如下：

```
{
"address":"168d8513597cc0958f635a679a5b60ccd13d6ef1",
"id":"721b533c-c552-4f94-9860-5b70cb45497a",
"version":3,
"crypto":{"cipher":"aes-128-ctr","ciphertext":
"1924aaf30e9b6ea7360caeeab7f1afac196aa9455fea022186642a6dc35c7cd7",
"cipherparams":{"iv":"70c135529198af334f952192f577f2b6"},
"kdf":"scrypt","kdfparams":{"dklen":32,"n":262144,"p":1,"r":8,"salt":
"68a357b71dc0f3a92726a761125f9cabee2ed4a1d6d0ea183ee6c37d608c97e9"},
"mac":"6dab6ec9ee482c10556e26264c13047a6d9386f2c1ae4cea1c26e0bc5d662bab"}
}
```

可以使用 Etherscan 网站显示交易，URL 如下：

```
https://etherscan.io/address/<your address>
```

图 10.19 展示了数字钱包文件中指定的 Etherscan 网站(https://etherscan.io/address/0x168d8513597cc0958f635a679a5b60ccd13d6ef1)，因为刚刚开始，所以目前还没有交易。

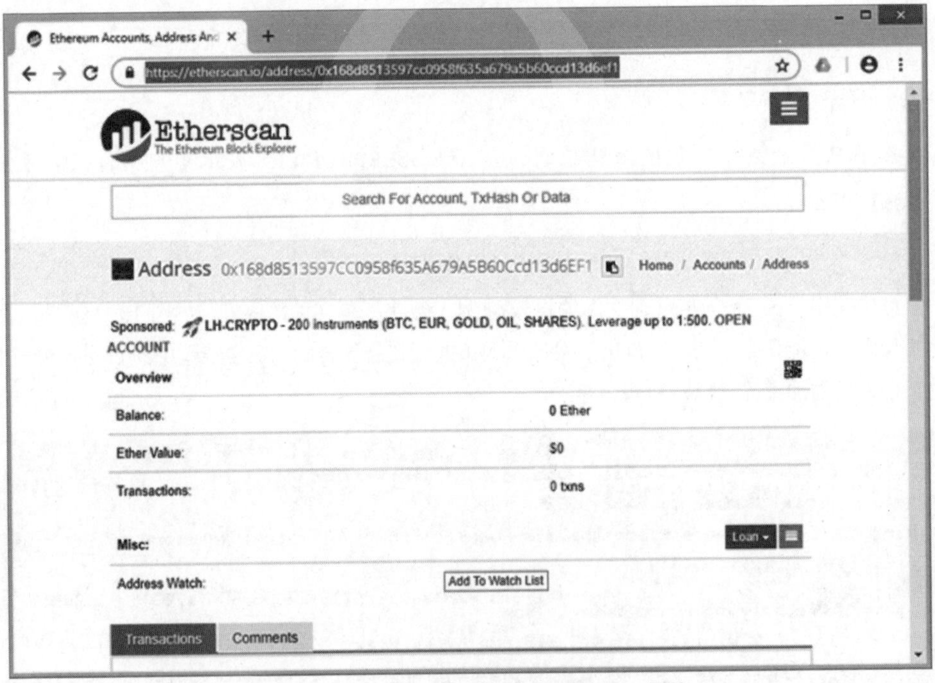

图 10.19　数字钱包文件中指定的 Etherscan 网站

在向任何人发送以太币以前，必须先获得一些以太币。可以通过挖矿或从其他人那里获取以太币。一旦有了以太币，就可以使用以下命令将其发送给他人：

```
.\web3j-4.0.1\bin\web3j wallet send <walletfile> 0x<address>|<ensName>
```

上述命令将要求输入数字钱包的密码、确认接收方地址、指定要发送的以太币数量和以太币单位(例如 ether 或 wei)。

有关智能合约以及 Web3j 示例的详情，请参阅以下资源：

https://docs.web3j.io/smart_contracts.html
https://github.com/web3j/sample-project-gradle
https://github.com/web3j/examples

以下是 Web3j 官方网站：

https://web3j.io/
https://github.com/web3j/web3j
https://docs.web3j.io/

10.11　Java EthereumJ 示例

EthereumJ 是 Ethereum 协议的纯 Java 实现。EthereumJ 库允许使用 Java 与 Ethereum 区块链进行交互。获取和使用 EthereumJ 的最简单方法是使用 Git。有关如何下载、安装和使用 Git 的更多详细信息，可参见附录 C。

在 Windows 中，只需要运行 Git Bash 程序，然后在 Git Bash 命令行中输入以下命令即可下载并运行一个简单的 EthereumJ 入门示例：

```
git clone https://github.com/ether-camp/ethereumj.starter
cd ethereumj.starter
./gradlew run
```

注意，./gradlew run 命令将下载相当多的文件，可能需要几分钟，该命令还将配置和运行本地 REST 服务器。为了检查结果，请在 Git Bash 命令行中输入以下命令，以查看本地区块链的信息：

```
curl -w "\n" -X GET http://localhost:8080/bestBlock
```

使用下面的命令下载或克隆整个 EthereumJ 项目的源代码：

```
git clone https://github.com/ethereum/ethereumj
```

下载完之后，可以转到 ethereumj 子文件夹并运行名为 TestNetSample 的示例程序，部分结果如下：

```
cd ethereumj
./gradlew run -PmainClass=org.ethereum.samples.TestNetSample
Starting a Gradle Daemon, 1 incompatible Daemon could not be reused, use
--status for details
Building version: 1.13.0-SNAPSHOT (from branch develop)
publishing if master || develop current branch: null
[buildinfo] Properties file path was not found! (Relevant only for
builds running on a CI Server)

:ethereumj-core:processResources
This will be printed after the build task even if something else calls
the build task
:ethereumj-core:classes
```

```
:ethereumj-core:run
20:09:17.397 INFO [sample] Starting EthereumJ!
20:09:17.444 INFO [general] Starting EthereumJ...
20:09:26.092 INFO [general] External address identified: 59.72.70.14
20:09:26.170 INFO [discover] Pinging discovery nodes...

20:09:26.339 INFO [general] EthereumJ node started:
enode://23b940843d8adf1fb0ccfe4a781e9e35850be7f3aaf6dd5e3b1f2371c1361
cc2baaac2d8c95882c34b7ae5ff18bf8a504d50e99f22ad4ee60963fa4930d0c9e
8@59.72.70.14:30303
20:09:26.355 INFO [general] DB is empty - adding Genesis
20:09:26.424 INFO [general] Genesis block loaded
20:09:26.439 INFO [ethash] Kept caches: cnt: 1 epochs: 0...0
20:09:26.471 INFO [general] Bind address wasn't set, Punching to
identify it...
20:09:47.519 WARN [general] Can't get bind IP. Fall back to 0.0.0.0:
java.net.ConnectException: Connection timed out: connect
20:09:47.519 INFO [discover] Discovery UDPListener started
20:09:47.688 INFO [net] Listening for incoming connections, port:[30303]
20:09:47.688 INFO [net] NodeId:
[23b940843d8adf1fb0ccfe4a781e9e35850be7f3aaf6dd5e3b1f2371c1361
cc2baaac2d8c95882c34b7ae5ff18bf8a504d50e99f22ad4ee60963fa4930d0c9e8]
20:09:47.741 INFO [discover] Reading Node statistics from DB: 0 nodes.
20:09:48.085 INFO [discover] Received response.
20:09:49.001 INFO [discover] New peers discovered.
20:09:57.570 INFO [net] TCP: Speed in/out 3Kb / 3Kb(sec), packets in/
out 91/150, total in/out: 31Kb / 37Kb
20:09:57.570 INFO [net] UDP: Speed in/out 8Kb / 6Kb(sec), packets in/
out 438/465, total in/out: 81Kb / 63Kb
20:10:07.575 INFO [net] TCP: Speed in/out 3Kb / 4Kb(sec), packets in/
out 98/162, total in/out: 66Kb / 78Kb
20:10:07.575 INFO [net] UDP: Speed in/out 4Kb / 2Kb(sec), packets in/
out 178/192, total in/out: 123Kb / 89Kb
20:10:48.541 INFO [discover] Write Node statistics to DB: 829 nodes.
```

你还可以运行其他示例程序，例如：

```
./gradlew run -PmainClass=org.ethereum.samples.BasicSample
./gradlew run -PmainClass=org.ethereum.samples.FollowAccount
./gradlew run -PmainClass=org.ethereum.samples.PendingStateSample
./gradlew run -PmainClass=org.ethereum.samples.PriceFeedSample
./gradlew run -PmainClass=org.ethereum.samples.PrivateMinerSample
./gradlew run -PmainClass=org.ethereum.samples.TransactionBomb
```

有关 EthereumJ 的更多详细信息，请参阅以下页面：

https://github.com/ethereum/ethereumj

10.12 Java Ethereum 智能合约示例

在本章的最后一个 Java 示例中，你将学习如何创建 Ethereum 智能合约。为此，需要使用 Solidity，这是 Ethereum 智能合约指定的编程语言，可以从下面的网站下载(在本书中，下载的

Solidity 版本为 0.5.7)：

```
https://github.com/ethereum/solidity/releases
```

对于 Windows，只需要查找名为 solidity-windows.zip 的存档文件，下载并解压到本地文件夹中。在本例中，指定的解压路径为 E:\solidity-windows。solidity-windows 文件夹中的 solc.exe 就是编译 Solidity 程序所需的可执行文件。

对于其他操作系统，只需要按照说明进行操作即可。

你还需要 Web3j 库。下载详情请参阅 10.10 节。同样，这里假设 Web3j 库已经下载并解压到名为 E:\web3j-4.0.1\的本地文件夹中。

下面创建名为 E:\contracts\solidity\的文件夹，使用文本编辑器在这个文件夹中创建一个名为 Greeter.sol 的文件，并将例 10.4 所示的程序保存到这个文件中。这是一个简单的智能合约应用，本质上只有一个函数 greet()，该函数只返回名为 greeting 的字符串变量中的值。

例 10.4　Solidity 智能合约程序

```solidity
pragma solidity >0.4.17;
contract mortal {
    address owner;
    function mortal() public { owner = msg.sender; }
    function kill() public { if (msg.sender == owner) selfdestruct(owner); }
}
contract greeter is mortal {
    string greeting;
    // constructor
    function greeter(string _greeting) public {
        greeting = _greeting;
    }
    // getter
    function greet() public constant returns (string memory) {
        return greeting;
    }
}
```

在 Windows 终端转到 E:\contracts\solidity\文件夹，然后输入以下命令，编译 Solidity 智能合约程序：

```
E:\solidity-windows\solc Greeter.sol --bin --abi --optimize -o ../build
```

这将在 E:\contracts\build\文件夹中创建以下两个.bin 文件和两个.abi 文件：

```
E:\contracts\build\Greeter.bin
E:\contracts\build\Greeter.abi
E:\contracts\build\Mortal.bin
E:\contracts\build\Mortal.abi
```

有关 Solidity 编程的更多信息，请访问以下网站：

```
https://solidity.readthedocs.io/
```

然后，转到 E:\contracts\build\ 文件夹并输入以下命令，为 Ethereum 智能合约创建 Java 项目：

```
E:\web3j-4.0.1\bin\web3j solidity generate ./greeter.bin ./greeter.abi -o ../../src/main/java
```

这将在 E:\src\ 文件夹中创建 Java 项目，这个文件夹具有以下结构并包含名为 Greeter.java 的 Java 程序：

```
E:\
    \src
        \main
            \java
                \Greeter.java
```

例 10.5 展示了生成的 Greeter.java 程序。

例 10.5　Greeter.java 程序

```java
import java.math.BigInteger;
import java.util.Arrays;
import java.util.Collections;
import org.web3j.abi.FunctionEncoder;
import org.web3j.abi.TypeReference;
import org.web3j.abi.datatypes.Function;
import org.web3j.abi.datatypes.Type;
import org.web3j.abi.datatypes.Utf8String;
import org.web3j.crypto.Credentials;
import org.web3j.protocol.Web3j;
import org.web3j.protocol.core.RemoteCall;
import org.web3j.protocol.core.methods.response.TransactionReceipt;
import org.web3j.tx.Contract;
import org.web3j.tx.TransactionManager;
import org.web3j.tx.gas.ContractGasProvider;

/**
 * <p>Auto generated code.
 * <p><strong>Do not modify!</strong>
 * <p>Please use the <a href="https://docs.web3j.io/command_line.html">web3j command line tools</a>,
 * or the org.web3j.codegen.SolidityFunctionWrapperGenerator in the
 * <a href="https://github.com/web3j/web3j/tree/master/codegen">
 * codegen module</a> to update.
 *
 * <p>Generated with web3j version 4.0.1.
 */
public class Greeter extends Contract {
    private static final String BINARY = "608060405234801561001057600080fd5b506040516102f03803806102f08339810180604052602081101561003357600080fd5b8101908080516401000000008111156100's4b57600080fd5b820160208101848111156100's5e57600080fd5b8151640100000000811182820187101715610078576000 80fd5b505060008054600160a01b0319163317905580519093506100a49250600191506020840190610ab565b50506101465b828054600181600116156101000203166002900490600052602060002090601f016020900481019282601f10610ec57805160ff1916838001178
```

```
55610119565b828001600101855582156101195791820 15b828111156101195782518255
91602001919060010190610 0fe565b506101259291506102 9565b5090565b6101439190
5b8082111561012557600081 5560010161012f565b90565b 61019b806101556000396000
f3fe6080604052348015610 01057600080fd5b506004361 06100365760003560e01c8063
41c0e1b51461003b578063c fae321714610045575b60008 0fd5b6100436100c2565b005b
61004d6100da565b604080 51602080825283518183015 2835191928392908301918 50190
80838360005b83811015610 0875781810151838201526 0200161006f565b50505050 9050
90810190601f1680156100 b457808203805160018360 20036101000a0319168152602001
91505b5092505050604051 80910390f35b6000546001 600160a01b0316331415610 0d857
33ff5b565b600180546040 80516020601f600260001 9610100878916150 2019095169490
9404938401819004810282 018101909252828152606 0939092909183018 2828015610165
5780601f10610 13a5761010080835404028 352916020019161016556 5b82019190600052
602060000209 05b8154815290600101906 02001808311610148578 29003601f168201915b
5050505050909 050905 6fea165627a7a72305820 855e7ee1fb28333dc0c79 be44c87e61ad0
a03ed6e47185 2ff59c26020f966aa3002 9";

    public static final String FUNC_KILL = "kill";

    public static final String FUNC_GREET = "greet";

    @Deprecated
    protected Greeter(String contractAddress, Web3j web3j, Credentials credentials, BigInteger gasPrice, BigInteger gasLimit) {
        super(BINARY, contractAddress, web3j, credentials, gasPrice,gasLimit);
    }

    protected Greeter(String contractAddress, Web3j web3j, Credentials credentials, ContractGasProvider contractGasProvider) {
        super(BINARY, contractAddress, web3j, credentials,contractGasProvider);
    }

    @Deprecated
    protected Greeter(String contractAddress, Web3j web3j, TransactionManager transactionManager, BigInteger gasPrice, BigInteger gasLimit) {
        super(BINARY, contractAddress, web3j, transactionManager,
            gasPrice, gasLimit);
    }

    protected Greeter(String contractAddress, Web3j web3j, TransactionManager transactionManager, ContractGasProvider contractGasProvider) {
        super(BINARY, contractAddress, web3j, transactionManager,
            contractGasProvider);
    }

    public RemoteCall<TransactionReceipt> kill() {
        final Function function = new Function(
            FUNC_KILL,
            Arrays.<Type>asList(),
            Collections.<TypeReference<?>>emptyList());
        return executeRemoteCallTransaction(function);
    }

    public RemoteCall<String> greet() {
        final Function function = new Function(FUNC_GREET,
```

```java
                Arrays.<Type>asList(),
                Arrays.<TypeReference<?>>asList(new
                        TypeReference<Utf8String>() {}));
        return executeRemoteCallSingleValueReturn(function, String.class);
    }

    @Deprecated
    public static Greeter load(String contractAddress, Web3j web3j,
Credentials credentials, BigInteger gasPrice, BigInteger gasLimit) {
        return new Greeter(contractAddress, web3j, credentials,
            gasPrice, gasLimit);
    }

    @Deprecated
    public static Greeter load(String contractAddress, Web3j web3j,
TransactionManager transactionManager, BigInteger gasPrice, BigInteger
gasLimit) {
        return new Greeter(contractAddress, web3j, transactionManager,
            gasPrice, gasLimit);
    }

    public static Greeter load(String contractAddress, Web3j web3j,
Credentials credentials, ContractGasProvider contractGasProvider) {
        return new Greeter(contractAddress, web3j, credentials,
            contractGasProvider);
    }

    public static Greeter load(String contractAddress, Web3j web3j,
TransactionManager transactionManager, ContractGasProvider
        contractGasProvider) {
        return new Greeter(contractAddress, web3j, transactionManager,
            contractGasProvider);
    }

    public static RemoteCall<Greeter> deploy(Web3j web3j, Credentials
credentials, ContractGasProvider contractGasProvider, String _greeting)
{
        String encodedConstructor = FunctionEncoder.encodeConstructor(
Arrays.<Type>asList(new org.web3j.abi.datatypes.Utf8String(_greeting)));
        return deployRemoteCall(Greeter.class, web3j, credentials,
            contractGasProvider, BINARY, encodedConstructor);
    }

    public static RemoteCall<Greeter> deploy(Web3j web3j,
TransactionManager transactionManager, ContractGasProvider
contractGasProvider, String _greeting) {
        String encodedConstructor = FunctionEncoder.encodeConstructor(
Arrays.<Type>asList(new org.web3j.abi.datatypes.Utf8String(_greeting)));
        return deployRemoteCall(Greeter.class, web3j, transactionManager,
            contractGasProvider, BINARY, encodedConstructor);
    }

    @Deprecated
    public static RemoteCall<Greeter> deploy(Web3j web3j, Credentials
credentials, BigInteger gasPrice, BigInteger gasLimit, String _greeting)
{
```

```
        String encodedConstructor = FunctionEncoder.encodeConstructor(
    Arrays.<Type>asList(new org.web3j.abi.datatypes.Utf8String(_greeting)));
        return deployRemoteCall(Greeter.class, web3j, credentials,
            gasPrice, gasLimit, BINARY, encodedConstructor);
    }

    @Deprecated
    public static RemoteCall<Greeter> deploy(Web3j web3j, TransactionManager
        transactionManager, BigInteger gasPrice, BigInteger gasLimit,
String _greeting) {
        String encodedConstructor = FunctionEncoder.encodeConstructor(
    Arrays.<Type>asList(new org.web3j.abi.datatypes.Utf8String(_greeting)));
        return deployRemoteCall(Greeter.class, web3j, transactionManager,
            gasPrice, gasLimit, BINARY, encodedConstructor);
    }
}
```

接下来使用 Web3j 创建数字钱包，参见 10.10 节。在出现提示时输入密码，并将数字钱包保存到 E:\drive 文件夹中。

```
E:\web3j-4.0.1\bin\web3j wallet create
```

数字钱包如下所示，其中 d517e874a888b58d02dad75c26f2a7ddec14f07b 是数字钱包的 ID。

```
E:\UTC--2019-04-12T03-47-43.931058900Z--d517e874a888b58d02dad75c26f2a7dd
ec14f07b.json
```

接下来访问 Infura 网站(https://infura.io/)并注册账户，创建一个新的项目，然后复制密钥。Infura 是在线平台，提供了多种工具来将应用连接到 Ethereum。有关在 Web3j 中使用 Infura 的更多信息，请访问以下网站：

```
https://docs.web3j.io/infura.html
```

例 10.6 所示的程序可以部署并执行智能合约。在这个程序中，rinkebyKey 是从 Infura 获得的项目 ID 或令牌 ID，而 walletFilePassword 和 walletId 是使用 Web3j 创建的，请使用自己的值更新它们。

例 10.6　Greeter.java 智能合约程序

```java
import java.io.IOException;
import java.util.concurrent.ExecutionException;
import org.web3j.crypto.CipherException;
import org.web3j.crypto.Credentials;
import org.web3j.crypto.WalletUtils;
import org.web3j.protocol.Web3j;
import org.web3j.protocol.http.HttpService;
import org.web3j.tx.Contract;

public class Greeting {
    public static void main(String[] args) throws IOException,
CipherException, ExecutionException, InterruptedException {

        String rinkebyKey = "498d65b077ea40ae9aeb2bbb014947cc";
        String rinkebyUrl = "https://rinkeby.infura.io/" + rinkebyKey;
```

```
        Web3j web3j = Web3j.build(new HttpService(rinkebyUrl));

        String walletFilePassword = "0000000000";
        String walletId = "d517e874a888b58d02dad75c26f2a7ddec14f07b";
        String walletSource = "E:\\UTC--2019-04-12T03-47-43.931058900Z--"
                            + walletId + ".json";

        Credentials credentials = WalletUtils.loadCredentials(
           walletFilePassword, walletSource);

        try {
           //Deploy Smart Contract with a Hello Smart Contract message
           Greeter greeter = Greeter.deploy(web3j, credentials, Contract.
             GAS_PRICE, Contract.GAS_LIMIT, "Hello Smart Contract!!!").send();
           //Display Smart Contract address
           System.out.println(greeter.getContractAddress());
           //Execute Smart Contract's greet() function
           System.out.println(greeter.greet().send());
        }catch(Exception e){
           System.out.println(e.toString());
        }
      }
    }
```

可以通过输入以下命令来编译并执行 Greeting.java 和 Greeter.java 程序：

```
javac -classpath ".;E:\\web3j-4.0.1\\lib\\*" Greeter.java Greeting.java
java -classpath ".;E:\\web3j-4.0.1\\lib\\*" Greeting
```

如果一切正常，那么应该能在屏幕上看到 Hello Smart Contract 消息，否则将显示一条错误消息以说明出错原因。常见的错误是"没有足够的资金支付天然气*价格+价值"。这是因为账户中没有足够的以太币。可以从以下链接获取真正的以太币：

```
https://github.com/ethereum/wiki/wiki/Getting-Ether
https://www.ethereum.org/ether
```

也可从以下链接获取以太币以进行测试：

```
https://faucet.rinkeby.io/
```

10.13 更进一步：选择区块链平台

如果想进一步利用区块链，那么无论应用是什么——加密货币、医疗保健、制造业、供应链或物联网，下一个最重要的步骤将是选择合适的区块链平台，然后开发自己的去中心化应用——DApps。分布式应用不同于传统的集中式应用，例如 Google、Facebook 或 Amazon，集中式应用的内容由中央实体拥有。对于分布式应用，内容归用户所有。以下是一些主流的区块链平台。

Bitcoin(https://bitcoin.org/en/)　这是最早的区块链平台，仅仅用于加密货币比特币。比特币的软件实现称为 Bitcoin Core(https://bitcoincore.org/)，也称为比特币客户端。Bitcoin Core 是用 C++编写的。如今，比特币是最成功的数字货币。

Ethereum(https://www.ethereum.org/)　Ethereum 于 2013 年 11 月由俄裔加拿大程序员 Vitalik Buterin 创建。与 Bitcoin 平台不同，Ethereum 平台可以做的不仅仅是加密货币。Ethereum 是用图灵完备的语言编写的，其中包括七种不同的编程语言。Ethereum 具有智能合约功能，是开源平台，开发人员可以基于区块链技术构建和部署去中心化应用。Ethereum 拥有自己的加密货币以太币(ETH)和编程语言 Solidity，Solidity 是一种用于编写智能合约的编程语言。

Eris(https://monax.io/platform/)　Eris 建立在 Ethereum 区块链基础之上，是另一个免费、开放的区块链平台，用于构建、测试、维护和操作去中心化应用。通过 Eris，可以轻轻松松地实施智能合约。

IBM Blockchain(https://www.ibm.com/blockchain)　IBM Blockchain 是基于开源的 Hyperledger Fabric 的公共云服务，旨在为企业客户构建安全的区块链网络。Hyperledger Fabric 与传统的区块链网络不同，传统的区块链网络不支持对企业至关重要的私人交易或合同保密。Hyperledger Fabric 解决了这个问题，可将私有交易保持为私有的，并将特定的数据只提供给那些需要知道这些数据的人。

NEO(https://neo.org/)　NEO 是非营利性的由社区驱动的区块链平台，旨在通过利用数字化资产、数字身份和智能合约来创建"智能经济"。通过使用分布式网络，Neo 使用了一种有趣的共识机制，从而提高了可伸缩性。

10.14　小结

本章首先介绍了区块链的概念以及如何验证区块链和挖矿，然后介绍了区块链的工作原理以及它们的用途，接下来介绍了一些 Java 区块链示例以及 BitcoinJ、Web3j 和 EthereumJ 开源库的简单用法。最后，当决定在自己的业务中进一步使用区块链时，你需要选择区块链平台。尽管存在一些问题，但区块链是一项令人着迷的技术，并且具有从根本上改变世界的潜力。

10.15　本章复习题

1. 区块链是什么时候发明的？是谁发明了区块链？
2. 用适当的图描述区块内部的内容以及两个相邻区块之间的关系。
3. 如何验证区块链？
4. 什么是区块链共识？
5. 挖矿是什么意思？什么是工作量证明？
6. 区块链可以用来做什么？
7. 区块链技术有什么问题？
8. 什么是比特币和比特币现金？
9. 什么是智能合约？
10. 区块链平台 Bitcoin、Ethereum 和 Hyperledger Fabric 之间有什么区别？

第 11 章

面向大数据应用的Java编程

"数据就是新的石油。"
——Clive Humby

11.1 什么是大数据
11.2 大数据的来源
11.3 大数据的三大 V
11.4 大数据分析带来的好处
11.5 什么是 Hadoop
11.6 Hadoop 的关键组件
11.7 在树莓派集群上实现 Hadoop
11.8 Java Hadoop 示例
11.9 小结
11.10 本章复习题

11.1 什么是大数据

数据在我们的生活中至关重要，我们一直在使用数据。例如，保存在文件中的实验数据、保存在数据库中的学生记录或员工记录、保存在电子表格中的销售数据以及常见的 Word 文件、Excel 文件、PowerPoint 文件、声音文件和电影文件，这些传统的数据文件可以使用标准的个人计算机进行存储、分析和显示。传统的数据文件的大小从千字节(KB，2^{10} 字节)、兆字节(MB，2^{20} 字节)到吉字节(GB，2^{30} 字节)不等，有时甚至达到太字节(TB，

2^{40} 字节)。但是，随着 Internet 的迅速发展和移动用户的增加，可以使用拍字节(PB，2^{50} 字节)甚至艾字节(EB，2^{60} 字节)来测量数据。人们把这些数据存储称为大数据。大数据太大、太复杂，无法使用传统的计算机软硬件进行存储和分析。

根据维基百科提供的数据，自 20 世纪 80 年代以来，世界人均数据量大约每 40 个月就翻一番。截至 2012 年，每天大约生成 2.5 EB 的数据。根据国际数据公司(IDC)的一份报告，2013—2020 年，全球数据量以指数形式增长，从 4.4 ZB(泽字节，2^{70} 字节)增加至 44 ZB，到 2025 年将会有 163 ZB 的数据，如图 11.1 所示。

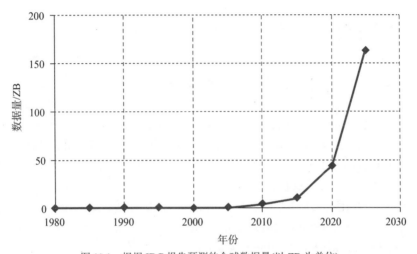

图 11.1 根据 IDC 报告预测的全球数据量(以 ZB 为单位)

Java 作为一种现代的高级编程语言，非常适合大数据应用。Java 是目前最流行的大数据框架 Hadoop 采用的语言。许多 Hadoop 关键模块(例如 MapReduce)都在 Java 虚拟机(JVM)上运行。Java 可以在许多设备和平台上运行。Java 建模并实现了堆栈数据结构，从而可以快速重建统计数据。Java 可以自动执行垃圾收集和内存分配。Java 具有丰富的联网功能。Java 具有很强的安全性，是一种安全的编程语言。

11.2 大数据的来源

如今，如下几个地方正在生成大数据。

社交媒体网站 Facebook、Twitter、Snapchat、WhatsApp、YouTube、谷歌、雅虎等网站每天都会从它们位于世界各地的数十亿用户中产生大量数据。

电子商务网站 亚马逊、eBay、阿里巴巴等在线购物网站也会产生大量数据，可以对这些数据进行分析，了解消费者的购物习惯并进行销售预测。

电信公司 Verizon、AT&T、中国移动、Nippon、EE、沃达丰和 Telefonica 等电信巨头也将通过存储通信记录和客户信息生成大量数据。

股票市场 世界各地的股票市场通过存储每日交易也在产生大量数据。

物联网(IOT) 随着数十亿台设备连接到物联网,物联网每天也会从启用传感器的设备产生大量数据。

医院 从病历中产生大量的数据。

银行 从客户的交易中产生大量的数据。

气象站 从卫星图像生成大量数据以进行天气预报。

政府部门 政府部门掌管着公民的大量个人信息。

11.3 大数据的三个 V

大数据的三个典型特征是容量(Volume)、速度(Velocity)和多样性(Variety),它们被称为"大数据的三个 V"。

容量 大数据容量巨大,通常范围从数十太字节到数百拍字节。例如,Facebook 有 20 亿用户,微信有 10 亿用户,YouTube 有 10 亿用户,WhatsApp 有 10 亿用户,Instagram 有 10 亿用户,阿里巴巴有 6 亿用户,Twitter 有 3 亿用户,Snapchat 有 1.8 亿用户。这些用户每天都会产生数十亿张图片、帖子、视频、推文等。

速度 大数据的产生速度也是惊人的。例如,Facebook 每 60 秒就有 317 000 次状态更新、400 个新用户、147 000 次照片上传以及 54 000 个共享链接。大数据也在以越来越快的速度增长。根据之前提到的 IDC 报告,估计数据量将每两年翻一番。

多样性 大数据可以包含许多可用的数据类型,包括结构化数据类型,如文本、图片和视频;半结构化数据类型,如 XML 数据;以及非结构化数据类型,如手写文本、绘图、语音记录和测量数据。所有这些形式的数据都需要进行额外的预处理以得出含义并支持元数据。

11.4 大数据分析带来的好处

大数据确实是现代世界最重要的新兴数字技术之一。大数据分析可以带来很多好处。例如,通过分析保存在 Facebook 等社交网络中的信息,可以更好地理解用户对产品广告以及社会、经济问题的反应。通过分析电子商务网站上的信息,企业和零售商可以根据消费者的喜好和产品感知更好地开展生产活动。通过分析数百万患者的既往病史,医生可以更轻松、更早地进行诊断,医院也可以提供更好、更快的服务。

11.5 什么是 Hadoop

Hadoop 是由 Apache 软件基金会开发的一种开源的大数据框架。Hadoop 允许用户使用简单的编程模型在跨计算机集群的分布式环境中存储和处理大型数据集。Hadoop 旨在从一台计算机扩展到数千台计算机，每台计算机都有自己的本地计算和存储。Hadoop 集群由一个主节点和多个从节点组成。Hadoop 可以处理各种形式的结构化和非结构化数据，这使大数据分析变得更加灵活。Hadoop 使用同名的分布式文件系统，从而在集群中的各个节点之间提供快速的数据访问。Hadoop 还具有容错功能，因此即使单个节点发生故障，应用也可以继续运行。凭借这些优势，Hadoop 已成为大数据分析的关键数据管理平台，例如预测分析、数据挖掘和机器学习应用。Hadoop 主要是用 Java 编写的，一些本机代码是用 C 编写的，命令行实用工具是用 shell 脚本编写的。

迄今为止，金融公司已使用 Hadoop 构建了一些用于评估风险、构建投资模型和创建交易算法的应用。零售商则使用 Hadoop 分析结构化和非结构化数据，以更好地了解和服务客户。电信公司使用基于 Hadoop 的分析技术对其基础架构进行预测性维护，以支持面向客户的运营方式。通过分析客户行为和账单，他们可以为现有客户提供新的服务。Hadoop 已在许多其他领域得到广泛应用，并且是大数据中事实上的标准。

11.6 Hadoop 的关键组件

Hadoop 由如下几个关键组件组成：HDFS(Hadoop 分布式文件系统)、MapReduce、Hadoop Common 和 YARN(Yet Another Resource Negotiator)。

11.6.1 HDFS

HDFS 用来提供数据存储，它被设计为运行在低成本、高容错性的硬件上。在 HDFS 中，文件被拆分为许多块，这些块被复制到 DataNode 中。默认情况下，每个块的大小为 64 MB，并被复制到集群的三个 DataNode 中。HDFS 使用 TCP/IP 套接字进行通信。集群中的客户端使用远程过程调用(Remote Procedure Call，RPC)相互通信。

HDFS 提供如下 5 个服务：

- NameNode(主服务)
- 辅助 NameNode(主服务)
- JobTracker(主服务)
- DataNode(从属服务)
- TaskTracker(从属服务)

NameNode 是主节点，DataNode 是对应的从节点。Hadoop 集群有一个 NameNode 和一个 DataNode 集群。NameNode 可以跟踪文件并管理文件系统。NameNode 包含详细信息，包括块的总数、特定块的位置(用于确定数据的哪些部分存储在哪个节点中)、副本被存储在何处，等等。

DataNode 将数据存储为块，以供客户端读写。DataNode 是从属守护程序。DataNode 每三秒将一次心跳消息发送到 NameNode，以表明它们是活跃的。如果 NameNode 在两分钟内未收到来自 DataNode 的心跳，就假定 DataNode 已死。

Secondary NameNode 也称为检查点节点，是 NameNode 的辅助节点，用于通过从 NameNode 提取相关文件(例如 edit 和 fsimage)来定期更新文件系统的元数据并合并它们。

JobTracker 接收来自客户端的 MapReduce 执行请求。JobTracker 与 NameNode 通信以了解数据的位置。NameNode 会将元数据提供给 JobTracker。

TaskTracker 从 JobTracker 获取任务和代码。TaskTracker 会将代码应用于文件，这一过程又称为映射(mapping)。

11.6.2 MapReduce

MapReduce 用于数据处理。MapReduce 是用 Java 编写的软件框架，可用于创建处理大量数据的应用。与 HDFS 相似，MapReduce 也被构建为具有容错功能，并且能够在大型集群环境中工作。MapReduce 将输入数据拆分为较小的任务(映射任务的过程)，这些任务可以在并行进程中执行。然后，减少映射任务的输出，并将结果保存到 HDFS 中。例如，假设要进行字数统计，也就是计算每个单词在一组文档中使用的次数，MapReduce 会将这项工作分为两个阶段：映射阶段和简化阶段。映射阶段对每个文档中的单词进行计数，简化阶段则将每个文档中的计数聚合为跨越整个集合的单词计数。

MapReduce 是 Hadoop 的本地批处理引擎，涉及的基本步骤如下：

(1) 从 HDFS 读取数据。
(2) 将数据分成小块，并在节点之间分布。
(3) 在每个节点上应用计算。
(4) 将中间结果保存到 HDFS 中。
(5) 重新分配中间结果并按照键进行分组。
(6) 通过汇总和组合每个节点的结果来减少每个键的值。
(7) 将最终结果保存到 HDFS 中。

11.6.3 Hadoop Common

Hadoop Common 是一组共享的 Java 工具和库，其他 Hadoop 模块需要使用这些工具和库。这些 Java 库能够提供文件系统和操作系统级别的抽象，并包含启动 Hadoop 所需的 Java 文件和脚本。

11.6.4 Hadoop YARN

Hadoop YARN 是用于作业调度和集群资源管理的框架。YARN 使得在 Hadoop 集群中运行更加多样化的工作负载成为可能。

11.6.5 Hadoop 集群概述

图 11.2 给出了多节点 Hadoop 集群的示意图。典型的小型 Hadoop 集群包括一个主节点和多个从节点。主节点由 JobTracker、TaskTracker、NameNode 和 DataNode 组成，从节点由 DataNode 和 TaskTracker 组成。

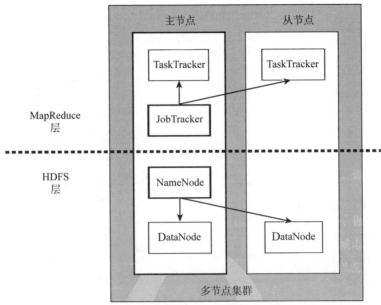

图 11.2 多节点 Hadoop 集群的示意图

11.7 在树莓派集群上实现 Hadoop

Hadoop 只能在类似 Linux 的操作系统中运行，并且需要 Java Runtime Environment (JRE) 1.6 或更高版本。Hadoop 还要求在集群的节点之间为启动和关闭脚本设置 SSH。图 11.3 展示了如何在双节点的树莓派集群中设置 Hadoop，设置详情参见表 11.1。树莓派提供了极好的廉价解决方案，用于构建自己的 Hadoop 数据集群以供学习和实践。即使这个系统处理的数据量不是非常大，也仍然能演示集群的结构。

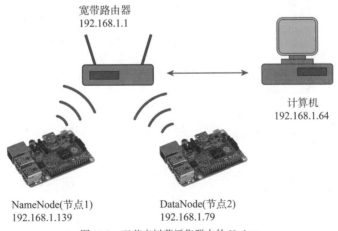

图 11.3 双节点树莓派集群中的 Hadoop

表 11.1 双点节树莓派集群中的 Hadoop 设置详情

名称	IP 地址	Hadoop 服务
节点 1	192.168.1.139	NameNode
		辅助 NameNode
		JobTracker
		DataNode
		TaskTracker
节点 2	192.168.1.79	DataNode
		TaskTracker

11.7.1 树莓派的安装和配置

树莓派的安装和配置请参阅第 7 章的 7.6 节。

一旦启动并运行树莓派，就请打开树莓派的终端窗口。剩余过程将通过终端窗口来完成。

在终端窗口中输入 sudo nano /etc/hostname，并更改为以下内容：

```
node1
```

这会将树莓派的主机名更改为 node1。在这里，nano 是 Linux 文本编辑器，sudo 则指示系统以超级用户身份修改文件。你还可以使用其他 Linux 文本编辑器，例如 vi、vim、pico、emacs 或 sublime。

在终端窗口中输入 sudo nano /etc/hosts，附加以下内容，从而设置 node1 的 IP 地址。

```
192.168.1.139          node1
```

Java 应该预装了 Raspbian。可以通过在树莓派的终端窗口中输入 java –version 以再次进行检查。

```
java -version

java version "1.8.0_65"
Java(TM) SE Runtime Environment (build 1.8.0_65-b17)
Java HotSpot(TM) Client VM (build 25.65-b01, mixed mode)
```

现在，可通过输入以下命令来重新启动树莓派：

```
sudo reboot
```

11.7.2 Hadoop 的安装和配置

可按照以下步骤在集群上安装和配置 Hadoop：

(1) 准备 Hadoop 用户账号和组。

(2) 配置 SSH。

(3) 下载并安装 Hadoop。

(4) 配置环境变量。

(5) 配置 Hadoop。

(6) 启动和停止 Hadoop 服务。

(7) 测试 Hadoop。

(8) 在 Web 浏览器中使用 Hadoop。

准备 Hadoop 用户账号和组

为了安装 Hadoop，需要准备 Hadoop 用户账号和组。在树莓派的终端窗口中，输入以下命令：首先创建一个名为 hadoop 的 Hadoop 组；然后将一个名为 hduser 的用户添加到这个 Hadoop 组中，并使 hduser 成为超级用户或管理员；最后提示输入 hduser 用户的密码和其他信息。

```
$sudo addgroup hadoop
$sudo adduser --ingroup hadoop hduser
$sudo adduser hduser sudo
```

配置 SSH

输入以下命令，创建一对 SSH RSA 密钥，密钥的密码为空，这样 Hadoop 节点之间就可以相互通信，而不需要提示输入密码。

```
$su hduser
$mkdir ~/.ssh
$ssh-keygen -t rsa -P ""
$cat ~/.ssh/id_rsa.pub > ~/.ssh/authorized_keys
$exit
```

输入以下命令，验证 hduser 用户可以登录 SSH：

```
$su hduser
$ssh localhost
$exit
```

下载并安装 Hadoop

输入以下命令，下载并安装 Hadoop 2.9.2：

```
$cd ~/
$wget http://mirror.vorboss.net/apache/hadoop/common/hadoop-2.9.2/hadoop-2.9.2.tar.gz
$sudo mkdir /opt
$sudo tar -xvzf hadoop-2.9.2.tar.gz -C /opt/
$cd /opt
$sudo mv hadoop-2.9.2 hadoop
$sudo chown -R hduser:hadoop hadoop
```

配置环境变量

输入 sudo nano /etc/bash.bashrc，在 .bashrc 文件的末尾添加以下代码行：

```
export JAVA_HOME=$(readlink -f /usr/bin/java | sed "s:bin/java::")
export HADOOP_HOME=/opt/hadoop
```

```
export HADOOP_INSTALL=$HADOOP_HOME
export YARN_HOME=$HADOOP_HOME
export PATH=$PATH:$HADOOP_INSTALL/bin
```

然后输入 source ~/.bashrc 以应用更改。现在切换到 hduser 用户,验证 Hadoop 可执行文件在/opt/Hadoop/bin 文件夹之外是可访问的。

```
$su hduser
$hadoop version

hduser@node1 /home/hduser $ hadoop version
Hadoop 2.9.2
Subversion https://git-wip-us.apache.org/repos/asf/hadoop.git -r
826afbeae31ca687bc2f8471dc841b66ed2c6704
Compiled by ajisaka on 2018-11-13T12:42Z
Compiled with protoc 2.5.0
From source with checksum 3a9939967262218aa556c684d107985
This command was run using /opt/hadoop/share/hadoop/common/
hadoop-common-2.9.2.jar
```

输入 sudo nano /opt/hadoop/etc/hadoop/hadoop-env.sh,然后附加以下代码行:

```
export JAVA_HOME=$(readlink -f /usr/bin/java | sed "s:bin/java::")
export HADOOP_HEAPSIZE=250
```

配置 Hadoop

现在转到/opt/hadoop/etc/hadoop/目录,并使用文本编辑器以超级用户身份编辑以下配置文件:

core-site.xml
```
  <configuration>
    <property>
      <name>fs.default.name</name>
      <value>hdfs://node1:9000</value>
    </property>
  </configuration>
```

mapred-site.xml
```
  <configuration>
    <property>
      <name>mapreduce.framework.name</name>
      <value>yarn</value>
    </property>
  </configuration>
```

hdfs-site.xml
```
  <configuration>
    <property>
      <name>dfs.replication</name>
      <value>1</value>
    </property>
    <property>
      <name>dfs.namenode.name.dir</name>
      <value>file:/opt/hadoop/hadoop_data/hdfs/namenode</value>
    </property>
```

```xml
    <property>
      <name>dfs.datanode.name.dir</name>
      <value>file:/opt/hadoop/hadoop_data/hdfs/datanode</value>
    </property>
</configuration>
```

yarn-site.xml
```xml
<configuration>
    <property>
      <name>yarn.nodemanager.aux-services</name>
      <value>mapreduce_shuffle</value>
    </property>
    <property>
      <name>yarn.nodemanager.aux-services.mapreduce.shuffle.class</name>
      <value>org.apache.hadoop.mapred.ShuffleHandler</value>
    </property>
</configuration>
```

输入以下命令，创建和格式化 HDFS 文件系统：

```
$sudo mkdir -p /opt/hadoop/hadoop_data/hdfs/namenode
$sudo mkdir -p /opt/hadoop/hadoop_data/hdfs/datanode
$sudo chown hduser:hadoop /opt/hadoop/hadoop_data/hdfs -R
$sudo chmod 750 /opt/hadoop/hadoop_data/hdfs

$cd $HADOOP_INSTALL
$hdfs namenode -format
```

启动和停止 Hadoop 服务

输入以下命令，切换到 hduser 用户并启动 Hadoop 服务。jps 命令将显示所有正在运行的服务：

```
$su hduser
$cd $HADOOP_HOME/sbin
$./start-dfs.sh
$./start-yarn.sh
$jps
2082 NameNode
2578 ResourceManager
2724 Jps
2344 SecondaryNameNode
2683 NodeManager
2189 DataNode

$./stop-dfs.sh
$./stop-yarn.sh
```

测试 Hadoop

输入以下命令，运行 Hadoop 提供的一个名为 pi 的示例，并计算 π 的值：

```
$cd $HADOOP_INSTALL/bin
$./hadoop jar /opt/hadoop/share/hadoop/mapreduce/hadoop-mapreduce-examples-2.9.2.jar pi 16 1000
```

下面运行 Hadoop 提供的另一个名为 wordCount 的示例。

输入以下命令，将 LICENSE.txt 文件从/opt/hadoop/文件夹中的本地文件系统复制到 Hadoop

的分布式文件系统中，作为根目录中的 license.txt。

```
$hdfs dfs -copyFromLocal /opt/hadoop/LICENSE.txt /license.txt
```

输入以下命令，在根目录中列出 Hadoop 分布式文件系统的内容。

```
$hdfs dfs -ls /
```

输入以下命令，运行 wordCount 示例并将结果保存到名为 license-out.txt 的文件夹中。

```
$cd /opt/hadoop/bin
$./hadoop jar /opt/hadoop/share/hadoop/mapreduce/hadoop-mapreduce-examples-2.9.2.jar wordcount /license.txt /license-out.txt
```

图 11.4 给出了在 license.txt 文件上运行 wordCount 示例后的部分输出结果。

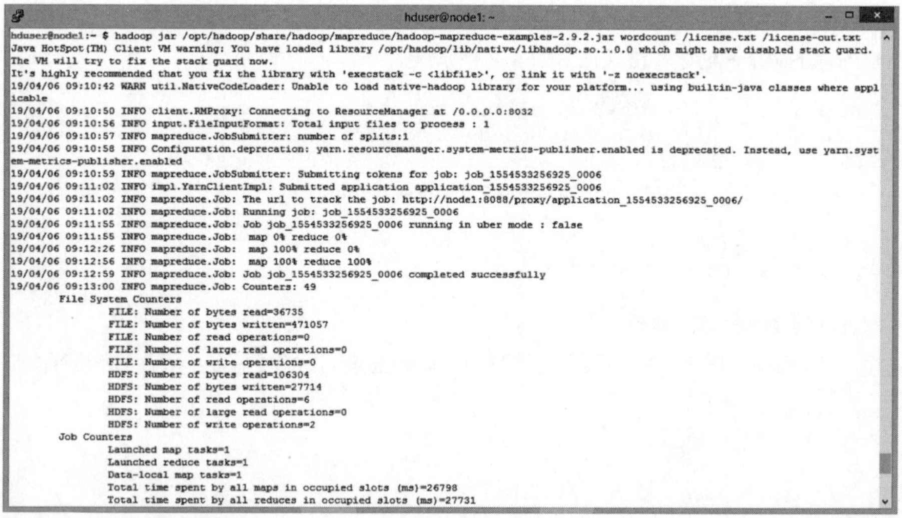

```
hduser@node1: ~
hduser@node1:~ $ hadoop jar /opt/hadoop/share/hadoop/mapreduce/hadoop-mapreduce-examples-2.9.2.jar wordcount /license.txt /license-out.txt
Java HotSpot(TM) Client VM warning: You have loaded library /opt/hadoop/lib/native/libhadoop.so.1.0.0 which might have disabled stack guard.
The VM will try to fix the stack guard now.
It's highly recommended that you fix the library with 'execstack -c <libfile>', or link it with '-z noexecstack'.
19/04/06 09:10:42 WARN util.NativeCodeLoader: Unable to load native-hadoop library for your platform... using builtin-java classes where appl
icable
19/04/06 09:10:50 INFO client.RMProxy: Connecting to ResourceManager at /0.0.0.0:8032
19/04/06 09:10:56 INFO input.FileInputFormat: Total input files to process : 1
19/04/06 09:10:57 INFO mapreduce.JobSubmitter: number of splits:1
19/04/06 09:10:58 INFO Configuration.deprecation: yarn.resourcemanager.system-metrics-publisher.enabled is deprecated. Instead, use yarn.syst
em-metrics-publisher.enabled
19/04/06 09:10:59 INFO mapreduce.JobSubmitter: Submitting tokens for job: job_1554533256925_0006
19/04/06 09:11:02 INFO impl.YarnClientImpl: Submitted application application_1554533256925_0006
19/04/06 09:11:02 INFO mapreduce.Job: The url to track the job: http://node1:8088/proxy/application_1554533256925_0006/
19/04/06 09:11:02 INFO mapreduce.Job: Running job: job_1554533256925_0006
19/04/06 09:11:55 INFO mapreduce.Job: Job job_1554533256925_0006 running in uber mode : false
19/04/06 09:11:55 INFO mapreduce.Job:  map 0% reduce 0%
19/04/06 09:12:26 INFO mapreduce.Job:  map 100% reduce 0%
19/04/06 09:12:56 INFO mapreduce.Job:  map 100% reduce 100%
19/04/06 09:12:59 INFO mapreduce.Job: Job job_1554533256925_0006 completed successfully
19/04/06 09:13:00 INFO mapreduce.Job: Counters: 49
        File System Counters
                FILE: Number of bytes read=36735
                FILE: Number of bytes written=471057
                FILE: Number of read operations=0
                FILE: Number of large read operations=0
                FILE: Number of write operations=0
                HDFS: Number of bytes read=106304
                HDFS: Number of bytes written=27714
                HDFS: Number of read operations=6
                HDFS: Number of large read operations=0
                HDFS: Number of write operations=2
        Job Counters
                Launched map tasks=1
                Launched reduce tasks=1
                Data-local map tasks=1
                Total time spent by all maps in occupied slots (ms)=26798
                Total time spent by all reduces in occupied slots (ms)=27731
```

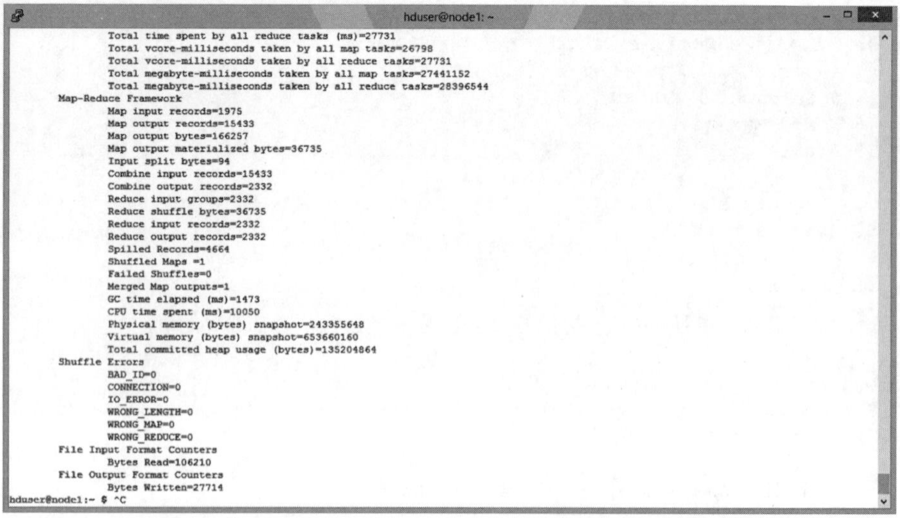

```
hduser@node1: ~
                Total time spent by all reduce tasks (ms)=27731
                Total vcore-milliseconds taken by all map tasks=26798
                Total vcore-milliseconds taken by all reduce tasks=27731
                Total megabyte-milliseconds taken by all map tasks=27441152
                Total megabyte-milliseconds taken by all reduce tasks=28396544
        Map-Reduce Framework
                Map input records=1975
                Map output records=15433
                Map output bytes=166257
                Map output materialized bytes=36735
                Input split bytes=94
                Combine input records=15433
                Combine output records=2332
                Reduce input groups=2332
                Reduce shuffle bytes=36735
                Reduce input records=2332
                Reduce output records=2332
                Spilled Records=4664
                Shuffled Maps =1
                Failed Shuffles=0
                Merged Map outputs=1
                GC time elapsed (ms)=1473
                CPU time spent (ms)=10050
                Physical memory (bytes) snapshot=243355648
                Virtual memory (bytes) snapshot=653660160
                Total committed heap usage (bytes)=135204864
        Shuffle Errors
                BAD_ID=0
                CONNECTION=0
                IO_ERROR=0
                WRONG_LENGTH=0
                WRONG_MAP=0
                WRONG_REDUCE=0
        File Input Format Counters
                Bytes Read=106210
        File Output Format Counters
                Bytes Written=27714
hduser@node1:~ $ ^C
```

图 11.4　在文件 license.txt 上运行 wordCount 示例后的部分输出

输入以下命令，将 license-out.txt 文件从 Hadoop 分布式文件系统复制到本地文件系统，然后使用 nano 打开 license-out.txt/part-r-00000 文件。

```
$hdfs dfs -copyToLocal /license-out.txt ~/
$nano ~/license-out.txt/part-r-00000
```

图 11.5 给出了执行上述命令后的输出结果。

图 11.5　将 license-out.txt 文件复制到本地文件系统并使用 nano 打开 license-out.txt/part-r-00000 文件

如果不想打开 license-out.txt/part-r-00000 文件，那么可以使用以下命令仅显示内容。

```
$cat ~/license-out.txt/part-r-00000
```

以下命令显示了如何删除其中包含非空文件的目录：

```
$hdfs dfs -rm -r /license-out.txt
```

在 Web 浏览器中使用 Hadoop

可以使用 Web 浏览器查看 Hadoop 及其应用集群，如图 11.6 所示。

为了将节点 node2 添加到 Hadoop 群集中，请添加另一个树莓派，最好使用完全相同的树莓派。然后，可以使用 Win32 Disk Imager 将 node1 的树莓派 SD 卡烧录到 node2，如图 11.7 所示。

```
https://sourceforge.net/projects/win32diskimager/files/latest/download
```

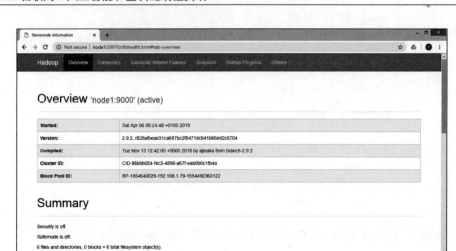

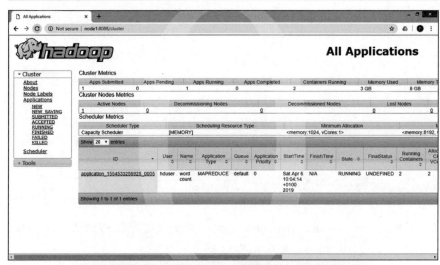

图 11.6　带有 Web 浏览器的 Hadoop 软件的 Web 视图和应用集群

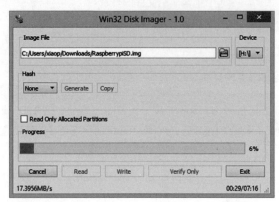

图 11.7　使用 Win32 Disk Imager 烧录 node1 的树莓派 SD 卡

node2 启动并运行后，请记住编辑/etc/hosts 文件以包括 node2 的 IP 地址，并相应地将 IP 地址更改为自己的 IP 地址。你还需要为 node2 重新配置 SSH。

```
$sudo nano /etc/hosts
192.168.1.139 node1
192.168.1.76 node2
```

最后，输入以下命令，删除 HDFS 存储并在 node1 上添加权限：

```
$sudo rm -rf /opt/hadoop/hadoop_data
$sudo mkdir -p /opt/hadoop/hadoop_data/hdfs/namenode
$sudo mkdir -p /opt/hadoop/hadoop_data/hdfs/datanode
$sudo chown hduser:hadoop /opt/hadoop/hadoop_data/hdfs -R
$sudo chmod 750 /opt/hadoop/hadoop_data/hdfs
```

在 node2 上输入类似的命令：

```
$sudo rm -rf /opt/hadoop/hadoop_data
$sudo mkdir -p /opt/hadoop/hadoop_data/hdfs/datanode
$sudo chown hduser:hadoop /opt/hadoop/hadoop_data/hdfs -R
$sudo chmod 750 /opt/hadoop/hadoop_data/hdfs
```

有关 Hadoop 的更多信息，请参阅以下资源：

```
http://hadoop.apache.org/docs/stable/
https://hadoop.apache.org/docs/stable/hadoop-project-dist/hadoop-common/SingleCluster.html
https://hadoop.apache.org/docs/stable/hadoop-project-dist/hadoop-common/ClusterSetup.html
https://www.javatpoint.com/hadoop-tutorial
https://howtodoinjava.com/hadoop/hadoop-big-data-tutorial/
https://www.guru99.com/bigdata-tutorials.html
```

11.8 Java Hadoop 示例

你也可以为 Hadoop 编写 Java 程序。例 11.1 所示的 Java 程序用来进行单词计数，其中包含三个类：MyMapper 类代表 Hadoop MapReduce 的 Map 阶段，该阶段实现了 map() 方法以计算每个文档中单词的数量。MyReducer 类代表 Hadoop MapReduce 的 Reduce 阶段，该阶段实现了 reducer() 方法，从而将每个文档中的单词计数聚合为所有文档的单词计数。主类 MyWordCounter 将调用 MyMapper 和 MyReducer 类并显示最终结果。

例 11.1　Java Hadoop 单词计数程序

```java
import java.util.*;
import java.io.IOException;
import org.apache.hadoop.conf.*;
import org.apache.hadoop.io.*;
import org.apache.hadoop.mapreduce.*;
import org.apache.hadoop.fs.Path;
```

```java
import org.apache.hadoop.mapreduce.lib.input.FileInputFormat;
import org.apache.hadoop.mapreduce.lib.output.FileOutputFormat;
import org.apache.hadoop.util.GenericOptionsParser;

//The MyMapper class - MapReduce's Map Phase
class MyMapper extends Mapper<Object, Text, Text, IntWritable>{

    private final static IntWritable one = new IntWritable(1);
    private Text word = new Text();

    public void map(Object key, Text value, Context context) throws IOException, InterruptedException {
      StringTokenizer itr = new StringTokenizer(value.toString());
      while (itr.hasMoreTokens()) {
         word.set(itr.nextToken());
         context.write(word, one);
      }
    }
}
//The MyReducer class -- MapReduce's Reduce Phase
class MyReducer extends Reducer<Text,IntWritable,Text,IntWritable> {
    private IntWritable result = new IntWritable();
    public void reduce(Text key, Iterable<IntWritable> values, Context context) throws IOException, InterruptedException {
        int sum = 0;
        for (IntWritable val : values) {
          sum += val.get();
        }
        result.set(sum);
        context.write(key, result);
      }
    }

  //The main class
  public class MyWordCounter {

  public static void main(String[] args) throws Exception {
    Configuration conf = new Configuration();
    String[] otherArgs = new GenericOptionsParser(conf, args).getRemainingArgs();
    if (otherArgs.length != 2) {
      System.err.println("Usage: MyWordCounter <in> <out>");
      System.exit(2);
    }

    //For the latest Hadoop versions 3.x.y
    Job job = new Job.getInstance(conf, "word count");
    //For Hadoop version 2.9.2 or earlier
    //Job job = new Job(conf, "word count");

    job.setJarByClass(MyWordCounter.class);
    job.setMapperClass(MyMapper.class);
    job.setCombinerClass(MyReducer.class);
    job.setReducerClass(MyReducer.class);
    job.setOutputKeyClass(Text.class);
    job.setOutputValueClass(IntWritable.class);
    FileInputFormat.addInputPath(job, new Path(otherArgs[0]));
```

```
        FileOutputFormat.setOutputPath(job, new Path(otherArgs[1]));
        System.exit(job.waitForCompletion(true) ? 0 : 1);
    }
}
```

为了编译以前的 Java Hadoop 程序,需要将 Hadoop 类路径包含到编译类路径中。在树莓派的终端窗口中输入以下命令,显示 Hadoop 类路径的内容:

```
$hadoop classpath
/opt/hadoop/etc/hadoop:/opt/hadoop/share/hadoop/common/lib/*:/opt/
hadoop/share/hadoop/common/*:/opt/hadoop/share/hadoop/hdfs:/opt/hadoop/
share/hadoop/hdfs/lib/*:/opt/hadoop/share/hadoop/hdfs/*:/opt/hadoop/
share/hadoop/yarn:/opt/hadoop/share/hadoop/yarn/lib/*:/opt/hadoop/
share/hadoop/yarn/*:/opt/hadoop/share/hadoop/mapreduce/lib/*:/opt/
hadoop/share/hadoop/mapreduce/*:/usr/lib/jvm/jdk-8-oracle-arm32-vfphflt/
jre//lib/tools.jar:/opt/hadoop/contrib/capacity-scheduler/*.jar
```

输入以下命令,编译 Java 程序,其中包含了 Hadoop 类路径:

```
$javac -cp $(hadoop classpath) MyWordCounter.java
```

输入以下命令,列出所有已生成的类文件:

```
$ls -la My*.class
-rw-r--r-- 1 hduser hadoop 1682 Apr 10 04:45  MyMapper.class
-rw-r--r-- 1 hduser hadoop 1690 Apr 10 04:45  MyReducer.class
-rw-r--r-- 1 hduser hadoop 1728 Apr 10 04:45  MyWordCounter.class
-rw-r--r-- 1 hduser hadoop 1734 Apr 10 04:41  'MyWordCounter$MyMapper.class'
-rw-r--r-- 1 hduser hadoop 1743 Apr 10 04:41  'MyWordCounter$MyReducer.class'
```

然后使用以下命令将所有类文件包含到名为 Test.jar 的 JAR 文件中:

```
$jar -cvf Test.jar My*.class
```

现在,输入以下命令,在/license.txt 文件上运行 Test.jar 文件,并将结果保存到名为/license-out.txt 的文件中,与前面的操作完全相同。

```
$hadoop jar Test.jar MyWordCounter /license.txt /license-out.txt
```

同样,输入以下命令,将 license-out.txt 文件从 Hadoop 分布式文件系统复制到本地文件系统,并使用 nano 打开 license-out.txt/part-r-00000 文件:

```
$hdfs dfs -copyToLocal /license-out.txt ~/
$nano ~/license-out.txt/part-r-00000
```

最后,输入以下命令,删除/license-out.txt 路径:

```
$hdfs dfs -rm -r /license-out.txt
```

例 11.2 是例 11.1 的另一个版本,用来计算文档中的字符数。它们的主要区别在于 MyMapper 类计算的是字符而不是单词的数量。MyReducer 类和主类非常相似。与前面的例 11.1 不同的是,MyMapper 和 MyReducer 类都是从 MapReduceBase 类扩展而来的,MapReduceBase 类在 Hadoop

的早期版本中很常见。

例 11.2　Java Hadoop 字符计数程序

```java
import java.io.IOException;
import java.util.Iterator;
import org.apache.hadoop.io.IntWritable;
import org.apache.hadoop.io.Text;
import org.apache.hadoop.mapred.MapReduceBase;
import org.apache.hadoop.mapred.OutputCollector;
import org.apache.hadoop.mapred.Mapper;
import org.apache.hadoop.mapred.Reducer;
import org.apache.hadoop.mapred.Reporter;
import org.apache.hadoop.io.LongWritable;
import org.apache.hadoop.fs.Path;
import org.apache.hadoop.mapred.FileInputFormat;
import org.apache.hadoop.mapred.FileOutputFormat;
import org.apache.hadoop.mapred.JobClient;
import org.apache.hadoop.mapred.JobConf;
import org.apache.hadoop.mapred.TextInputFormat;
import org.apache.hadoop.mapred.TextOutputFormat;

//The MyMapper class - MapReduce's Map Phase
class MyMapper extends MapReduceBase implements
Mapper<LongWritable,Text,Text, IntWritable>{
    public void map(LongWritable key, Text value,OutputCollector<Text,
IntWritable> output, Reporter reporter) throws IOException{
        String line = value.toString();
        String tokenizer[] = line.split("");
        for(String SingleChar : tokenizer)
        {
            Text charKey = new Text(SingleChar);
            IntWritable One = new IntWritable(1);
            output.collect(charKey, One);
        }
    }
}
//The MyReducer class - MapReduce's Reduce Phase
class MyReducer extends MapReduceBase implements
Reducer<Text,IntWritable, Text,IntWritable> {
    public void reduce(Text key, Iterator<IntWritable> values, OutputCol
lector<Text,IntWritable> output, Reporter reporter) throws IOException {
        int sum=0;
        while (values.hasNext()) {
            sum+=values.next().get();
        }
        output.collect(key,new IntWritable(sum));
    }
}
public class MyCharCounter {
    public static void main(String[] args) throws IOException{
        JobConf conf = new JobConf(MyCharCounter.class);
        conf.setJobName("Character Count");
        conf.setOutputKeyClass(Text.class);
        conf.setOutputValueClass(IntWritable.class);
```

```java
            conf.setMapperClass(MyMapper.class);
            conf.setCombinerClass(MyReducer.class);
            conf.setReducerClass(MyReducer.class);
            conf.setInputFormat(TextInputFormat.class);
            conf.setOutputFormat(TextOutputFormat.class);
            FileInputFormat.setInputPaths(conf,new Path(args[0]));
            FileOutputFormat.setOutputPath(conf,new Path(args[1]));
            JobClient.runJob(conf);
    }
}
```

下面来看另一个 Java Hadoop 程序。假设一个名为 grades.txt 的文本文件包含四门不同学科的学生成绩，如下所示。在每一行中，数据都用制表符进行分隔。第一项是学生的名字，第二项至第五项是学生取得的各科分数。

```
Tony        56    76    83    42
William     33    91    82    73
Alan        76    39    65    89
Tom         51    68    77    52
John        88    54    94    98
```

例 11.3 所示的 Java Hadoop 总成绩程序用来从 grades.txt 文件读取学生成绩并计算每个学生的总成绩。在这里，MyMapper 类计算的是字符而不是单词。MyReducer 类和主类非常相似。

例 11.3　Java Haddoop 总成绩程序

```java
import java.util.*;
import java.io.IOException;
import org.apache.hadoop.fs.Path;
import org.apache.hadoop.conf.*;
import org.apache.hadoop.io.*;
import org.apache.hadoop.mapred.*;
import org.apache.hadoop.util.*;

//Mapper class
class MyMapper extends MapReduceBase implements
    Mapper<LongWritable, Text, Text, IntWritable>
    {
        //Map function
        public void map(LongWritable key, Text value,
        OutputCollector<Text, IntWritable> output,
        Reporter reporter) throws IOException {
            String line = value.toString(); //read each line
            String lasttoken = null;
            //separate items in each line
            StringTokenizer s = new StringTokenizer(line,"\t");
            String name = s.nextToken(); //get the student name

            while(s.hasMoreTokens()) {
                lasttoken = s.nextToken(); //get all the subject marks
            }
            int mark = Integer.parseInt(lasttoken);
            output.collect(new Text(name), new IntWritable(mark));
        }
```

```
        }
    //Reducer class
    class MyReduce extends MapReduceBase implements Reducer< Text,
IntWritable, Text, IntWritable > {

        //Reduce function
        public void reduce( Text key, Iterator <IntWritable> values,
        OutputCollector<Text, IntWritable> output, Reporter reporter)
throws IOException {
            int sum=0;
            while (values.hasNext()) {
                sum+=values.next().get(); //add all the subject marks
                                          //for each student
            }
            //return the total marks of each student
            output.collect(key, new IntWritable(sum));
        }
    }

//The main class
public class MyMarker {


    //Main function
    public static void main(String args[])throws Exception {
        JobConf conf = new JobConf(MyMarker.class);

        conf.setJobName("max_eletricityunits");
        conf.setOutputKeyClass(Text.class);
        conf.setOutputValueClass(IntWritable.class);
        conf.setMapperClass(MyMapper.class);
        conf.setCombinerClass(MyReduce.class);
        conf.setReducerClass(MyReduce.class);
        conf.setInputFormat(TextInputFormat.class);
        conf.setOutputFormat(TextOutputFormat.class);
        FileInputFormat.setInputPaths(conf, new Path(args[0]));
        FileOutputFormat.setOutputPath(conf, new Path(args[1]));

        JobClient.runJob(conf);
    }
}
```

输入以下命令,编译并执行以上 Java 程序:

```
$javac -cp $(hadoop classpath) MyMarker.java
$jar -cvf Test.jar My*.class
$hdfs dfs -copyFromLocal ~/marks.txt /marks.txt
$hadoop jar Test.jar MyMarker /marks.txt /marks-out.txt
$hdfs dfs -copyToLocal /marks-out.txt ~/
$cat ~/marks-out.txt/part-r-00000
```

最后,输入以下命令,删除/marks-out.txt 目录:

```
$hdfs dfs -rm -r /marks-out.txt
```

11.9 小结

本章介绍了大数据的概念、来源、三个典型特征以及优势,还介绍了大数据开源软件 Hadoop,并展示了如何在树莓派上下载、设置和使用 Hadoop。

11.10 本章复习题

1. 什么是大数据?
2. 解释术语千字节、兆字节、吉字节、太字节、拍字节、艾字节和泽字节。
3. 大数据的来源有哪些?
4. 大数据的三个 V 是什么?
5. Hadoop 是什么?
6. Hadoop 的关键组件是什么?
7. 什么是 HDFS、MapReduce、Hadoop Common 和 Hadoop YARN?
8. Hadoop 提供的五项服务是什么?
9. 如何为 Hadoop 设置 SSH?
10. 如何启动和停止 Hadoop?

附录 A

Java 文档和归档工具以及在线资源

本附录展示如何使用 Java 提供的工具来自我记录和归档代码，并且列出用于学习 Java 的最重要的一些在线资源。

A.1 Javadoc 教程

文档在许多科学和工程学科中都很重要，因此维护文档变成了专业人士的一种习惯。对于软件而言，情况略有不同，因为软件可能会不断变化。在创建软件之后创建文档，然后在每次更新软件时进行修改，这样做效率低下且烦琐。为了解决此问题，Java JDK 提供了名为 javadoc 的工具，从而允许用户使用预定义的注释格式从 Java 源代码生成 HTML 格式的 Java 代码文档。

回想一下，Java 使用//创建单行注释，使用/*和*/创建多行注释。

```
//This is a single line comment

/*
This is a multiple
line comment
*/
```

调用 javadoc 命令后，就可以使用/*和*/创建 Java 文档。在下面的示例中，位于代码开头的注释用来创建 Java 程序的文档。@author 标记指定了程序的作者，@version 标记指定了版本，@since 标记指定了程序的开发日期，@see 标记指定了程序的 URL。

```
/*
 * The HelloWorld program implements an application that
 * simply displays "Hello World!" to the standard output.
```

```
 *
 * @author Dr Perry Xiao
 * @version 1.0
 * @since 2018-12-08
 * @see <a href="http://www.yourcompany.com/yourApp/">Hello World</a>
 */
public class HelloWorld {
    public static void main(String[] args) {
        System.out.println("Hello World!");
    }
}
```

如果程序中有方法，也可以使用 javadoc 为方法创建文档，如下所示。@param 标记指定了方法的输入参数，@return 标记指定了方法的输出。

```
/*
 * This method is used to print title and name on screen
 *
 * @param title This is the title
 * @param name This is the name
 */
  public void printName (String title, String name)
  {
      System.out.println("Hello " + title + " " +name);
  }
```

可以将上述两段代码放在一起，从而创建一个完整的 Java 程序。以下是带有 javadoc 注释的一个完整的程序：

```
/*
 * The HelloWorld program implements an application that
 * simply displays "Hello World!" to the standard output.
 *
 * @author Dr Perry Xiao
 * @version 1.0
 * @since 2018-12-08
 * @see <a href="http://www.yourcompany.com/yourApp/">Hello World</a>
 */
public class HelloWorld {
    /*
     * This method is used to print title and name on screen
     *
     * @param title This is the title
     * @param name This is the name
     */
    public void printName (String title, String name)
    {
        System.out.println("Hello " + title + " " +name);
    }
    /*
     * This is the main method of the program
     *
     * @param args This is the command line parameter,
                  where args[0] is title, and args[1] is name
     */
```

```
    public static void main(String[] args) {
        HelloWorld hw = new HelloWorld();
        hw.printName(args[0],args[1]);
    }
}
```

为了创建上述程序的 Java 文档，只需要在 Windows 终端窗口中输入以下命令：

```
javadoc HelloWorld.java
```

图 A.1 给出了执行以上命令后的结果：创建了 Java 文档以及相应的 HTML 文件和 resources 子目录。Java 文档采用的是 HTML 格式，名为 index.html。

图 A.2 展示了浏览器中的 index.html 文件，其中分为如下几个部分：标题和类信息、构造函数摘要和方法摘要、构造函数详细信息和方法详细信息以及页脚。

图 A.1 使用 javadoc 命令创建的 Java 文档以及相应的 HTML 文件和 resources 子目录

附录 A　Java 文档和归档工具以及在线资源　347

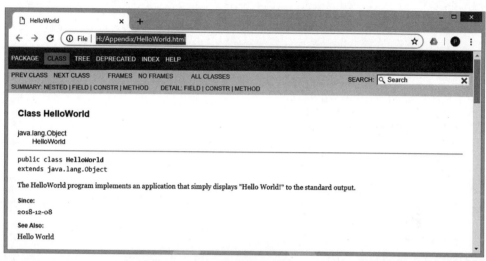

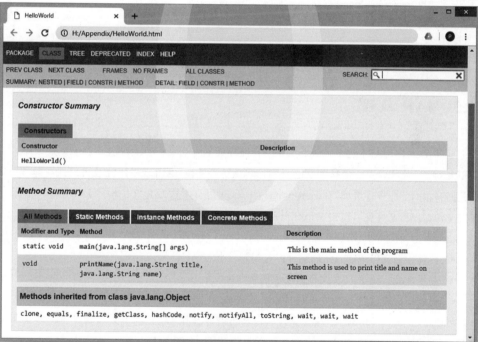

图 A.2　Web 浏览器中的 index.html 文件：自上而下分别是标题和类信息、构造函数摘要和方法摘要、构造函数详细信息和方法详细信息以及页脚

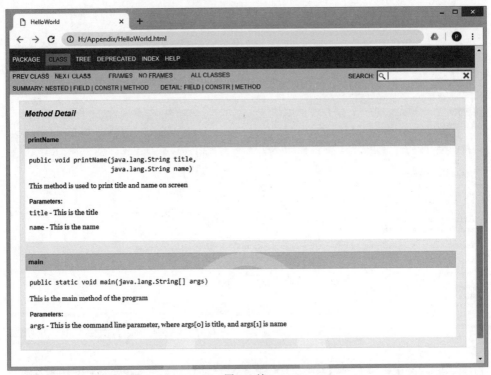

图 A.2(续)

A.2 JAR 教程

Java Archive (JAR) 格式允许将多个文件压缩并捆绑到单个 JAR 文件中。JAR 是一种独立于平台的文件格式,基于流行的 ZIP 算法,可模仿 UNIX TAR(磁带存档)文件格式。jar 和 tar 工具拥有相同的命令行选项。Java Runtime (JRE) 可以直接从 JAR 文件运行 Java 程序,而无须解压文件。JAR 文件也可以由 WinZIP 或 WinRAR 程序直接打开。

在项目中使用 JAR 文件有很多好处。

- **能够压缩文件**:通过压缩文件,可得到较小的项目。
- **易于部署**:单个 JAR 文件更易于部署和分发。
- **认证**:可通过对 JAR 文件进行数字签名来为用户提供认证。

可以使用 jar 工具从多个 Java 文件或目录创建 JAR 文件。例如,以下命令将从两个 Java 类 HelloWorld.class 和 HelloWorld2.class 创建名为 hello.jar 的 JAR 文件:

```
jar cvf hello.jar HelloWorld.class HelloWorld2.class
```

jar 工具的命令行选项如表 A.1 所示(注意,JAR 文件中显示的-前缀是可选的)。

表 A.1　jar 工具的命令行选项

选项	说明
-c	创建新的归档文件
-v	使用标准输出生成详细输出
-f	指定文件名，比如 hello.jar

要想了解关于 jar 工具的命令行选项的更多信息，只需要在命令行中单独输入 jar 即可：

```
jar
```

以下命令将从所有 Java 类(*.class)创建名为 hello.jar 的 JAR 文件：

```
jar cvf hello.jar *.class
```

以下命令将从所有 Java 类(*.class)和子目录 images 创建名为 hello.jar 的 JAR 文件：

```
jar cvf hello.jar *.class images
```

以下命令将从子目录 ProjectA 中的所有 Java 类(*.class)创建名为 hello.jar 的 JAR 文件：

```
jar cvf hello.jar ProjectA/*.class
```

以下命令将从名为 DIR1、DIR2 和 DIR3 的三个子目录创建名为 hello.jar 的 JAR 文件。与前面仅添加类文件不同，以下命令会将所有文件添加到 JAR 文件中：

```
jar cvf hello.jar DIR1 DIR2 DIR3
```

jar 工具最重要的用途之一是创建可执行的 JAR 文件，这样就可以通过双击可执行的 JAR 文件来运行 Java 程序。

为了创建可执行的 JAR 文件，首先需要创建清单文件，清单文件指定了要运行的 Java 主类。例如，以下清单文件指定 Java 主类为 HelloWorld.class、类路径为 hello.jar：

```
Main-Class: HelloWorld
Class-Path: hello.jar
```

将以上内容保存在名为 hello.mf 的文本文件中。以下命令会将清单文件和所有 Java(*.class) 文件组合在一起，创建名为 hello.jar 的可执行的 JAR 文件：

```
jar cmf hello.mf hello.jar *.class
```

其中，-m 选项的含义为：JAR 文件将包含来自指定的清单文件的清单信息。

为了运行 JAR 文件，只需要在文件资源管理器中双击 JAR 文件，或在 Windows 终端窗口中输入以下命令即可：

```
java -jar hello.jar
```

A.3 一些十分重要的 Java 在线资源

Java 官方网站：

https://www.java.com/
https://www.oracle.com/java/

Java SE JDK 下载网站：

https://www.oracle.com/technetwork/java/javase/downloads/index.html

Java SE 文档网站：

https://www.oracle.com/technetwork/java/javase/documentation/api-jsp-136079.html

Java JDK 10 API 网站：

https://docs.oracle.com/javase/10/docs/api/overview-summary.html

Javadoc 网站：

https://www.oracle.com/technetwork/java/javase/tech/index-137868.html
https://www.tutorialspoint.com/java/java_documentation.htm

命令行 jar 工具 jar.exe：

http://docs.oracle.com/javase/7/docs/technotes/tools/windows/jar.html

Java 参考书籍：

https://whatpixel.com/best-java-books/
https://dzone.com/articles/10-all-time-great-books-for-java-programrs-best
https://www.journaldev.com/6162/5-best-core-java-books-for-beginners
https://www.javacodegeeks.com/2011/10/top-10-java-books-you-dont-want-to-miss.html

附录 B

Apache Maven教程

在开发 Java 软件时，有许多任务需要完成，比如：下载依赖项、将 JAR 文件放在类路径中、将源代码编译为二进制字节码、运行测试、将字节码打包为可部署的 JAR 文件、将 JAR 文件部署到远程存储库服务器，等等。Apache Maven 是基于项目对象模型(Project Object Model，POM)的软件项目管理和理解工具，它可以自动化所有这些任务。Maven 主要用来管理基于 Java 的项目，但它也可以在使用其他编程语言(例如 C#和 Ruby)的项目中使用。作为 Maven 的替代品，Ant 和 Gradle 也是主流的软件项目管理工具。

有关 Ant 的更多信息，请参见 http://ant.apache.org/。

有关 Gradle 的更多信息，请参见 https://gradle.org/guides/#getting-started。

有关 Ant、Maven 和 Gradle 之间的对比情况，请参见 https://www.baeldung.com/ant-maven-gradle。

B.1 下载 Maven

可以从以下网址下载 Apache Maven 的压缩文件 apache-maven-3.6.0-bin.zip：

```
https://maven.apache.org/download.cgi
```

下载后，只需要将文件解压到目录 C:\apache-maven-3.6.0\。解压后，即可在 bin 子目录中找到 Maven 程序 mvn.jar。要想测试 Maven，只需要运行以下命令即可：

```
mvn -version
```

不过，你首先需要将 C:\apache-maven-3.6.0\bin\添加到环境变量 PATH 中，也可使用完整的路径运行 Maven 程序。无论采用哪种方式，结果都应显示 Maven 的版本、Java 的版本以及有关操作系统的信息，如图 B.1 所示。

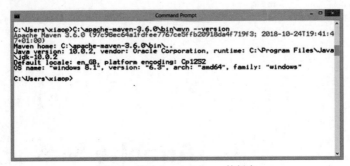

图 B.1　显示 Apache Maven 的版本

B.2　创建 Maven 项目

使用以下命令创建名为 my-app 的 Maven 项目，组 ID 为 com.mycompany.app，如图 B.2 所示。

```
mvn archetype:generate -DgroupId=com.mycompany.app -DartifactId=my-app
-DarchetypeArtifactId=maven-archetype-quickstart -DinteractiveMode=false
```

成功创建 Maven 项目后，将自动创建名为 my-app 的项目目录，图 B.3 展示了 my-app 项目目录的结构，其中主要包括 pom.xml、App.java 和 AppTest.java 文件。

图 B.2　创建 Maven 项目

图 B.2(续)

```
my-app
|--pom.xml
`-- target
`-- src
    |-- main
    |    `-- java
    |         `-- com
    |              `-- mycompany
    |                   `-- app
    |                        `-- App.java
    `-- test
         `-- java
              `-- com
                   `-- mycompany
                        `-- app
                             `-- AppTest.java
```

图 B.3　my-app 项目目录的结构

pom.xml 文件是 Maven 项目的核心，除了描述项目的详细信息并管理依赖项之外，还配置了用于构建软件的插件，如下所示：

```
<project xmlns="http://maven.apache.org/POM/4.0.0"
         xmlns:xsi="http://www.w3.org/2001/XMLSchema-instance"
         xsi:schemaLocation="http://maven.apache.org/POM/4.0.
                             http://maven.apache.org/maven-v4_0_0.xsd">
  <modelVersion>4.0.0</modelVersion>
  <groupId>com.mycompany.app</groupId>
```

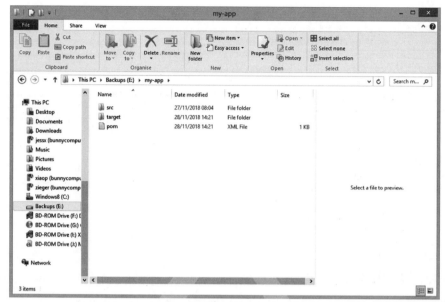

```xml
<artifactId>my-app</artifactId>
<packaging>jar</packaging>
<version>1.0-SNAPSHOT</version>
<name>my-app</name>
<url>http://maven.apache.org</url>
<properties>
  <maven.compiler.source>1.8</maven.compiler.source>
  <maven.compiler.target>1.8</maven.compiler.target>
</properties>

<dependencies>
  <dependency>
    <groupId>junit</groupId>
    <artifactId>junit</artifactId>
    <version>3.8.1</version>
    <scope>test</scope>
  </dependency>
</dependencies>
</project>
```

App.java 文件是 Maven 项目的主 Java 程序,用于指定将要执行的操作。在本例中,要执行的操作仅仅是在屏幕上显示 Hello World!。AppTest.java 用于调用 App.java 程序。

以下展示了 App.java 文件的代码:

```java
package com.mycompany.app;

/**
 * Hello world!
 *
 */
public class App
{
    public static void main( String[] args )
    {
        System.out.println( "Hello World!" );
    }
}
```

以下展示了 AppTest.java 文件的代码:

```java
package com.mycompany.app;

import junit.framework.Test;
import junit.framework.TestCase;
import junit.framework.TestSuite;

/**
 * Unit test for simple App.
 */
public class AppTest
    extends TestCase
{
    /**
     * Create the test case
     *
```

```
     * @param testName name of the test case
     */
    public AppTest( String testName )
    {
        super( testName );
    }

    /**
     * @return the suite of tests being tested
     */
    public static Test suite()
    {
        return new TestSuite( AppTest.class );
    }
    /**
     * Rigourous Test :-)
     */
    public void testApp()
    {
        assertTrue( true );
    }
}
```

B.3 编译和构建 Maven 项目

输入以下命令,编译和构建 Maven 项目,如图 B.4 所示。这将编译 Maven 项目并将二进制字节码打包到/my-app/target/子目录下名为 my-app-1.0-SNAPSHOT.jar 的可部署 JAR 文件中。

```
mvn package
```

B.4 运行 Maven 项目

输入以下命令,运行 Maven 项目,将显示 Hello World!,如图 B.5 所示。

```
java -cp target/my-app-1.0-SNAPSHOT.jar com.mycompany.app.App
```

有关 Maven 的更多信息,请参阅以下资源:

```
https://maven.apache.org/guides/getting-started/maven-in-five-minutes.html
https://www.javatpoint.com/maven-tutorial
http://tutorials.jenkov.com/maven/maven-tutorial.html
https://www.guru99.com/maven-tutorial.html
https://examples.javacodegeeks.com/enterprise-java/maven/create-java-project
-with-maven-example/
```

```
E:\my-app>cd ..

E:\>cd my-app

E:\my-app>c:\apache-maven-3.6.0\bin\mvn package
[INFO] Scanning for projects...
[INFO]
[INFO] ----------------------< com.mycompany.app:my-app >----------------------
[INFO] Building my-app 1.0-SNAPSHOT
[INFO] --------------------------------[ jar ]---------------------------------
[INFO]
[INFO] --- maven-resources-plugin:2.6:resources (default-resources) @ my-app ---
[WARNING] Using platform encoding (Cp1252 actually) to copy filtered resources,
i.e. build is platform dependent!
[INFO] skip non existing resourceDirectory E:\my-app\src\main\resources
[INFO]
[INFO] --- maven-compiler-plugin:3.1:compile (default-compile) @ my-app ---
[INFO] Changes detected - recompiling the module!
[WARNING] File encoding has not been set, using platform encoding Cp1252, i.e. b
uild is platform dependent!
[INFO] Compiling 1 source file to E:\my-app\target\classes
[INFO]
[INFO] --- maven-resources-plugin:2.6:testResources (default-testResources) @ my
-app ---
[WARNING] Using platform encoding (Cp1252 actually) to copy filtered resources,
i.e. build is platform dependent!
[INFO] skip non existing resourceDirectory E:\my-app\src\test\resources
[INFO]
[INFO] --- maven-compiler-plugin:3.1:testCompile (default-testCompile) @ my-app
---
[INFO] Changes detected - recompiling the module!
[WARNING] File encoding has not been set, using platform encoding Cp1252, i.e. b
uild is platform dependent!
[INFO] Compiling 1 source file to E:\my-app\target\test-classes
[INFO]
[INFO] --- maven-surefire-plugin:2.12.4:test (default-test) @ my-app ---
[INFO] Surefire report directory: E:\my-app\target\surefire-reports
```

```
[INFO]
[INFO] --- maven-surefire-plugin:2.12.4:test (default-test) @ my-app ---
[INFO] Surefire report directory: E:\my-app\target\surefire-reports

-------------------------------------------------------
 T E S T S
-------------------------------------------------------
Running com.mycompany.app.AppTest
Tests run: 1, Failures: 0, Errors: 0, Skipped: 0, Time elapsed: 0.022 sec

Results :

Tests run: 1, Failures: 0, Errors: 0, Skipped: 0

[INFO]
[INFO] --- maven-jar-plugin:2.4:jar (default-jar) @ my-app ---
[INFO] Building jar: E:\my-app\target\my-app-1.0-SNAPSHOT.jar
[INFO] ------------------------------------------------------------------------
[INFO] BUILD SUCCESS
[INFO] ------------------------------------------------------------------------
[INFO] Total time:  3.746 s
[INFO] Finished at: 2019-04-04T22:38:43+01:00
[INFO] ------------------------------------------------------------------------

E:\my-app>
```

图 B.4　编译和构建 Maven 项目

```
E:\my-app>java -cp target/my-app-1.0-SNAPSHOT.jar com.mycompany.app.App
Hello World!

E:\my-app>_
```

图 B.5　运行 Maven 项目

附录 C

Git和GitHub教程

Git 是使用最为广泛的分布式版本控制系统之一。为了使用 Git，首先需要下载并将 Git 安装到计算机上。对于 Windows 操作系统，可以从以下网站下载 Git 安装程序(在编写本书时，安装文件名为 Git-2.20.1-64-bit.exe)：

https://gitforwindows.org/

对于其他操作系统，请访问以下网站以获取详细信息：

https://www.atlassian.com/git/tutorials/install-git

双击安装文件 Git-2.20.1-64-bit.exe 以安装 Git，并在安装过程中接受所有默认设置。安装成功后，运行名为 Git GUI 的程序，如图 C.1(a)所示。选择 Create New Repository，将弹出一个新的窗口，如图 C.1(b)所示。选择项目所在的目录，在这里选择的是 E:\MyProject，然后单击 Create 按钮，将出现 Git GUI 程序的主界面，如图 C.1(c)所示。从 Repository 菜单中选择 Git Bash，Git Bash 是终端程序。为了使用 Git，可以使用标准命令(例如 pwd)来显示当前项目目录，使用 dir 或 ls 命令可以列出其中的内容，如图 C.1(d)所示。

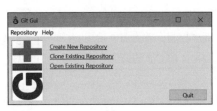

(a)

图 C.1　Git GUI 程序和 Git Bash 终端程序

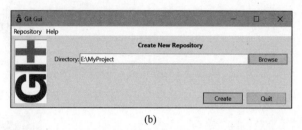

(b)

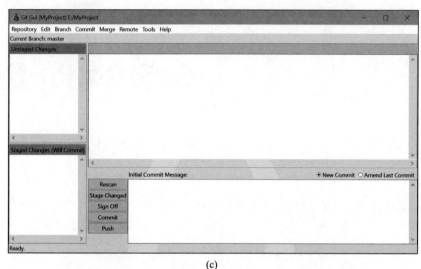

(c)

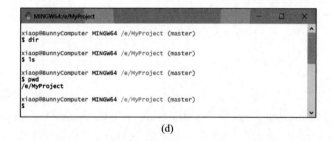

(d)

图 C.1(续)

有关 Git 的更多详细信息，请参阅如下免费的在线书籍 *Pro Git*：

https://git-scm.com/book/en/v2

下面开始使用 Git 进行版本控制。首先需要使用 git config --global 命令在 Git Bash 终端程序中配置姓名和电子邮件地址，如图 C.2 的第一个图所示。还可以使用 git config --list 命令列出所有配置。

然后输入 git init 以初始化版本控制。git init 命令将生成名为.git 的隐藏目录，其中包含有关项目的版本控制的所有详细信息。还可以输入 git status 命令以获取项目的版本控制状态，如图C.2的第二个图所示。每个Git项目都是一个存储库。

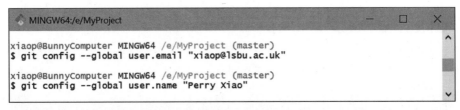

图C.2　用于在Git Bash终端程序中配置姓名和电子邮件的git config --global以及git init和git status命令

现在，可以将Java程序添加到项目中。下面以第2章中例2.1所示的HelloWorld.java程序为例。使用Windows资源管理器将HelloWorld.java程序添加到E:\MyProject项目目录中。在项目中提供README文件来解释项目的含义是很有用的。可以使用文本编辑器在项目目录中创建名为README.md(不是README.md.txt)的文本文件，也可以在Git Bash终端程序中输入以下命令：

```
echo "# MyProject" >> README.md
```

要将文件添加到Git版本控制系统中，请在Git Bash终端程序中输入以下命令：

```
git add .
```

这会将当前目录中的所有文件添加到Git版本控制系统中。请参考图C.3中的第一个图。由于git add命令不会添加子目录，因此如果想要添加所有内容，包括文件和子目录，请输入命令

```
git add -all
```

或

```
git add -A
```

还可以使用git rm <文件名>或git reset <文件名>命令从版本控制系统中删除任何文件。当对项目满意时，可以提交项目，更改将会保存在Git中，如图C.3中的第二个图所示。可以随意提交多次，还可以使用git log命令跟踪所有更改，以及使用git reset --soft HEAD~1命令还原提交。

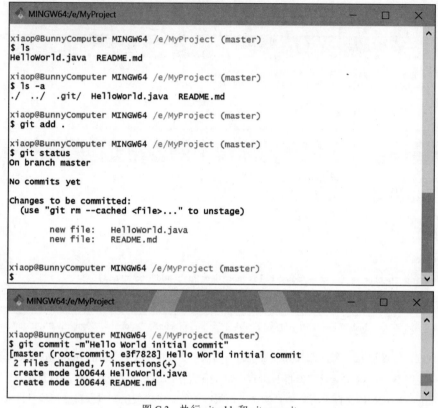

图 C.3　执行 git add . 和 git commit

为了与他人共享项目，需要使用远程版本控制服务器。最佳选择是 GitHub，GitHub 是目前最流行的基于 Web 的远程版本控制服务器之一。为了使用 GitHub，只需要访问网站 https://github.com/ 并在 GitHub 上创建账户，参见图 C.4(a)。注册并登录后，单击 Start a project 按钮并为项目创建新的存储库，参见图 C.4(b)。为项目命名，并使用所有默认设置来创建存储库，参见图 C.4(c)。之后将出现快速设置页面，从中可以找到新创建的存储库的 HTTPS 链接地址，如下所示：

```
https://github.com/PerryXiao2015/MyProject.git
```

如果想要使用 SSH 链接，请输入下面的命令：

```
git@github.com :PerryXiao2015/MyProject.git
```

注意：SSH 是一种通信协议，可通过身份验证和加密提供安全的远程登录。SSH 旨在替代传统的不安全远程登录协议，例如 Telnet 和 rlogin。HTTPS 是网络协议 HTTP 的安全版本。有关安全性的更多详细信息，参见第 9 章。

可以使用以下命令将 GitHub 链接设置为本地的 Git 源，并将本地存储库上传或推送到 GitHub 的远程存储库：

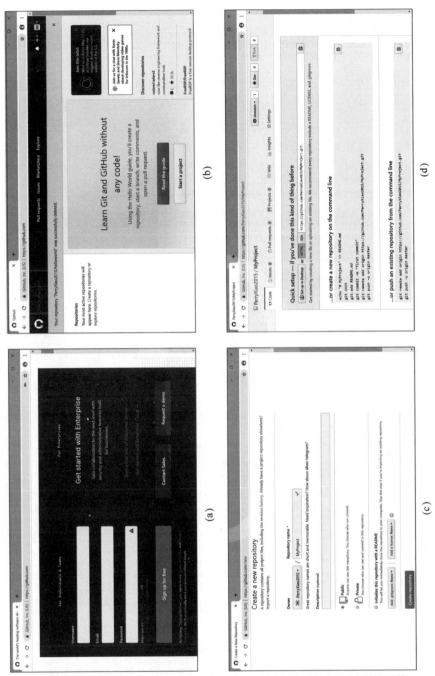

图 C.4 在 GitHub 网站上注册账户、启动项目、创建新的存储库并进行快速设置

```
git remote add origin https://github.com/PerryXiao2015/MyProject.git
git push -u origin master
```

或

```
git remote add origin git@github.com:PerryXiao2015/MyProject.git
git push -u origin master
```

在将本地存储库推送到 GitHub 远程存储库之前，需要通过生成 对公钥/私钥来确保存储库的内容已加密。图 C.5 中的第一个图展示了如何生成 RSA 密钥对。RSA 是一种公钥加密算法，已被广泛应用于安全数据传输。你还可以使用其他不同的算法来生成公钥/私钥。生成的密钥对存储在.ssh 隐藏目录中。默认情况下，私钥存储在名为 id_rsa 的文件中，而公钥存储在名为 id_rsa.pub 的文件中。图 C.5 中的第二个图展示了如何显示.ssh 隐藏目录和公钥。公钥的内容应以 ssh-rsa 开头。

图 C.5　用于生成 RSA 公钥/私钥对的 Git 命令，以及用于显示.ssh 隐藏目录，从而为 Git 项目创建 Hub 远程存储库并将项目推送至远程存储库的 Git 命令

首先复制公钥，然后进入 GitHub 网站，从项目设置中选择 Deploy Keys，添加一个新的密钥。对这个新的密钥进行命名，将公钥粘贴到 Key 文本框中，然后单击 Add key 按钮，如图 C.6 所示。

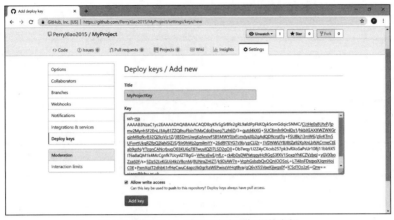

图 C.6　在 GitHub 项目站点中添加新的密钥

图 C.7(a)展示了如何添加 GitHub 项目站点以及如何将本地 Git 存储库推送到远程 GitHub 存储库。图 C.7(b)展示了 GitHub 登录界面。输入用户名和密码，然后 GitHub 项目站点将使用你的文件进行更新，如图 C.7(c)所示。

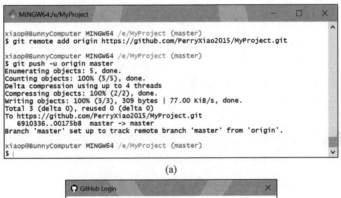

(a)

(b)

图 C.7　使用 Git 命令添加 GitHub 站点、GitHub 登录界面以及更新后的 GitHub 项目站点

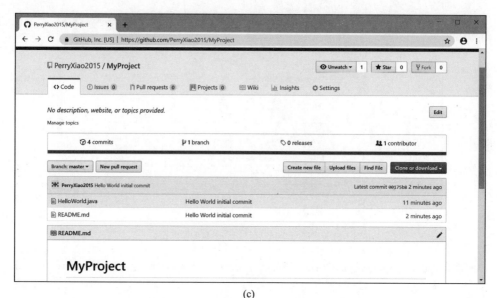

(c)

图 C.7(续)

现在修改 HelloWorld.java 程序，使其类似于第 2 章的例 2.2，这里使用了 args[0]。如果执行 git status 命令，那么将以红色标记表示 HelloWorld.java 程序已被修改(参见图 C.8)。此处的红色标记表示尚未上传更改以进行提交。可以使用 git add . 或 git add -A 命令将更改的文件添加到本地存储库。如果再次执行 git status 命令，那么将以绿色标记表示 HelloWorld.java 程序已被修改，这里的绿色标记表示已准备好提交更改。

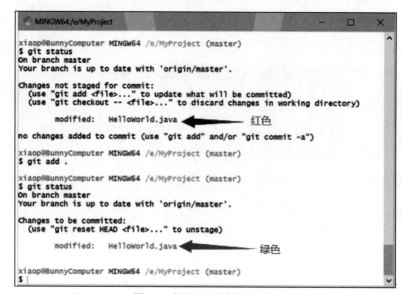

图 C.8　使用 Git 命令更新更改

可以再次提交程序并将其推送到 GitHub 主存储库。在图 C.9 中可以看到，GitHub 站点上的文件已更新。GitHub 网站还显示了已提交的次数，以及每次提交的详细信息和对应的文件。每一次提交都代表项目代码的不同版本。

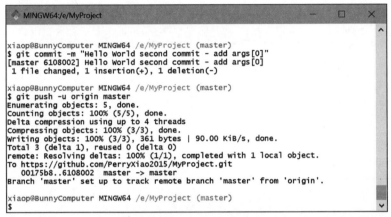

图 C.9　使用 Git 命令提交更改，将本地存储库推送到 GitHub 远程存储库，并将结果推送到更新的 GitHub 项目站点中

还可以使用 git log --online 命令在 Git Bash 窗口中显示提交历史，如图 C.10 中的第一个图所示。每一次提交都有唯一的数字，以十六进制格式表示。如果发生灾难性事件，那么始终可以使用 git checkout <unique number>命令恢复到程序的先前版本，如图 C.10 中的第二个图所示。现在，本地存储库文件将还原为以前的版本。

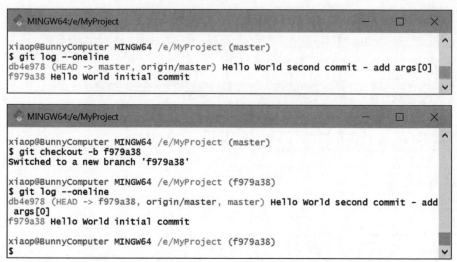

图 C.10　使用 Git 命令显示通过在线日志文件提交的历史记录并给出程序的不同版本

C.1　Git 术语

Repository(或 Repo)　存储项目的所有文件和子目录。

Stage　将文件和子目录添加到存储库中。

Commit　将更改保存到存储库中。

Branch　分支。

Master　所有存储库的主要分支。

Origin　存储库主版本的常规名称。

Head　存储库的最新提交。

Clone　制作存储库的副本。

Fork　创建存储库的副本。

Push　使用本地存储库更新远程存储库。

Pull　从远程存储库获取和下载内容,并立即更新本地存储库以匹配这些内容。

CheckOut　切换到不同版本的存储库。

C.2　备忘单

表 C.1 所示的备忘单总结了 Git 命令。

表 C.1　Git 命令备忘单

命令类别	命令	作用
配置	git config --global user.name "your_NAME"	在 Git 中设置姓名
	git config --global user.email "YourEmail@wherever.com"	在 Git 中设置电子邮件地址
	git config --list	列出 Git 配置
启动存储库	git init	初始化 Git
	git status	获取 Git 的状态
暂存文件	git add \<filename\> \<filename\> …	将文件添加到本地存储库
	git add .	将当前目录(不包括子目录)中的所有文件添加到本地存储库
	git add –all	将当前目录和子目录中的所有文件添加到本地存储库
	git add –A	将当前目录和子目录中的所有文件添加到本地存储库
	git rm \<filename\>	从本地存储库中删除文件
	git reset \<filename\>	从本地存储库中删除文件
生成 SSH 公钥/私钥对	ssh-keygen -t rsa -b 4096 -C "your_email@example.com"	生成 RSA 密钥对
	ls –la ~/.ssh	列出 .ssh 隐藏目录中的内容
	cat ~/.ssh/id_rsa.pub	显示 RSA 公钥的内容
提交到存储库	git commit --m "your message here"	将更改的内容保存到本地存储库
	git commit --amend –m "your amend message here"	修改提交
	git reset --soft HEAD~1	恢复提交
	git log	显示提交的历史记录
从 GitHub 推送和提取	git remote add origin \<your GitHut link\>	将GitHub链接设置为本地的Git源
	git push –u origin master	将本地存储库上传到GitHub远程存储库
	git clone \<project name\>	获取现有 Git 存储库的副本
	git pull	从远程存储库获取和下载内容，并立即更新本地存储库以匹配这些内容
	git log --oneline	显示远程存储库的提交历史
	git remote rm origin	删除远程的源
分支	git branch	创建分支
	git branch \<branch name\>	使用给定名称创建分支
	git checkout \<branch name\>	将当前分支设置为主项目
	git merge \<branch name\>	将变更从一个分支合并到另一个分支